Companion Animals and Us

Over the past 30 years there has been a tremendous growth in interest in the multidisciplinary field of human–companion animal interactions and relationships. This is not surprising considering that pets are kept in at least half of the households in Western societies. What, then, is so special about the relationship people have with their pets? Are we very different from our ancestors in the ways we feel about animals? What does pet-keeping tell us about ourselves and our relationships with people? Can pets be good for our health? Do they help promote empathy for other humans? These questions and more are explored in this book. *Companion Animals and Us* brings together some of the newest research from a wide variety of disciplines, including anthropology, history, psychology, sociology, human and veterinary medicine. This book will make fascinating reading for anyone interested in understanding more about the human–pet relationship.

ANTHONY L. PODBERSCEK is a Postdoctoral Research Associate in the Department of Clinical Veterinary Medicine at the University of Cambridge. He is Editor-in-Chief of the journal *Anthrozoös*, a council member of the International Society for Anthrozoology, and an editorial board member of the *Journal of Applied Animal Welfare Science*.

ELIZABETH S. PAUL is a Research Fellow in the Department of Clinical Veterinary Science at the University of Bristol. She is on the editorial board for the journals *Anthrozoös* and *Society and Animals*.

JAMES A. SERPELL is the Marie Moore Associate Professor of Humane Ethics and Animal Welfare at the University of Pennsylvania. He is the Secretary of the International Society for Anthrozoology, and has authored, edited or co-edited several other books, including *Companion Animals in Society* (1988), *Animals and Human Society* (1994), *The Domestic Dog* (1995), and *In the Company of Animals* (1996).

Companion Animals and Us

Exploring the relationships between people and pets

Edited by

ANTHONY L. PODBERSCEK
ELIZABETH S. PAUL
JAMES A. SERPELL

CAMBRIDGE
UNIVERSITY PRESS

PUBLISHED BY THE PRESS SYNDICATE OF THE UNIVERSITY OF CAMBRIDGE
The Pitt Building, Trumpington Street, Cambridge, United Kingdom

CAMBRIDGE UNIVERSITY PRESS
The Edinburgh Building, Cambridge, CB2 2RU, UK
40 West 20th Street, New York, NY 10011–4211, USA
10 Stamford Road, Oakleigh, VIC 3166, Australia
Ruiz de Alarcón 13, 28014 Madrid, Spain
Dock House, The Waterfront, Cape Town 8001, South Africa

http://www.cambridge.org

First published 2000
Reprinted 2001

Printed in the United Kingdom at the University Press, Cambridge

Typeset in Swift 9/13 *System* QuarkXPress® [SE]

A catalogue record for this book is available from the British Library

Library of Congress Cataloguing in Publication data

Companion animals and us: exploring the relationships between people
and pets / edited by Anthony L. Podberscek, Elizabeth S. Paul, and
James A. Serpell.
p. cm.
Based on papers from a conference held at Downing College,
Cambridge in 1996 as well as invited ones.
Includes index.
ISBN 0 521 63113 0 (hb.)
1. Pets – Social aspects Congresses. 2. Pet owners Congresses.
3. Human–animal relationships Congresses. I. Podberscek, Anthony
L. (Anthony Louis), 1963– . II. Paul, Elizabeth S. (Elizabeth
Shaun), 1964– . III. Serpell, James, 1952– .
SF411.5.C648 2000 99-32199 CIP
636.088′7–dc21

ISBN 0 521 63113 0 hardback

To Aline H. Kidd PhD (1922–1999)

Contents

Part II
The nature of the relationship

Part III
Pets, families and interactions

Part IV
Welfare and ethics

Contributors

ARNOLD ARLUKE, *Department of Sociology/Anthropology, Northeastern University, Boston, MA 02115, USA*

PIERS BEIRNE, *Department of Criminology, University of Southern Maine, 96 Falmouth Street, PO Box 9300, Portland, ME 04104–9300, USA*

LILIANE BODSON, *Histoire des connaissances zoologiques, Université de Liège, Place du 20–Août 72, B-4000 Liège, Belgium*

SHEILA BONAS, *School of Health and Social Sciences, Coventry University, Priory Street, Coventry CV1 5FB, UK*

GLYN M. COLLIS, *Department of Psychology, University of Warwick, Coventry CV4 7AL, UK*

NORINE DRESSER, *Department of English and American Studies, California State University, 3093 St George Street, Los Angeles, CA 90027, USA*

TIMOTHY J. EDDY, *Department of Psychology, PO Box 43131, University of Southwestern Louisiana, Lafayette, LA 70504–3131, USA*

MARIE-JOSÉ ENDERS-SLEGERS, *Faculty of Social Sciences, Department of Clinical and Health Psychology, University of Utrecht, PO Box 80140, Utrecht, The Netherlands*

PHILIPPE ERIKSON, *Départment d'Ethnologie, Université de Paris X–Nanterre, 200 Avenue de la République, 92001 Nanterre Cedex, France*

ERIKA FRIEDMANN, *Department of Health & Nutrition Sciences, Brooklyn College, City University of New York, Brooklyn, New York, NY 11210, USA*

SAMUEL D. GOSLING, *Department of Psychology, Mezes Hall 330, University of Texas, Austin, TX 78712, USA*

RACHAEL M. HARKER, *Centre for Primary Health Care Studies, Postgraduate Medical School, University of Warwick, Coventry CV4 7AL, UK*

ERICK L. LAURENT, *Gifu Keizai University, Kitakatach 5–50, Ogaki-shi 503, Gifu-ken, Japan*

JUNE MCNICHOLAS, *Department of Psychology, University of Warwick, Coventry CV4 7AL, UK*

SOPHIA MENACHE, *Department of History, University of Haifa, Haifa 31905, Israel*

ELIZABETH S. PAUL, *Department of Clinical Veterinary Science, University of Bristol, Langford House, Langford, Bristol BS18 7DU, UK*

ANTHONY L. PODBERSCEK, *Department of Clinical Veterinary Medicine, University of Cambridge, Madingley Road, Cambridge CB3 0ES, UK*

JAMES A. SERPELL, *Department of Clinical Studies, School of Veterinary Medicine, University of Pennsylvania, 3900 Delancey Street, Philadelphia, PA 19104–6010, USA*

JOANNA SWABE, *Amsterdam School for Social Research, University of Amsterdam, Oude Hoogstraat 24, 1012 CE Amsterdam, The Netherlands*

SUE A. THOMAS, *School of Nursing, Georgetown University, 3700 Reservoir Road, N.W., Washington DC 20007, USA*

DENNIS C. TURNER, *I.E.T., Vorderi Siten 30, PO Box 32, CH-8816 Hirzel, Switzerland*

ANTHONY L. PODBERSCEK, ELIZABETH S. PAUL AND JAMES A. SERPELL

1

Introduction

To me you are still nothing more than a little boy who is just like a hundred thousand other little boys. And I have no need of you. And you, on your part, have no need of me. To you, I am nothing more than a fox like a hundred thousand other foxes. But if you tame me, then we shall need each other. To me, you will be unique in all the world. To you, I shall be unique in all the world . . .

(*The Little Prince* by Antoine de Saint-Exupéry, 1945: 64).

Not so long ago, the idea of studying social relationships between humans and other animals would have been regarded as tantamount to heresy. In Europe, until the early modern period, animals were viewed as irrational beings placed on earth solely for the economic benefit of mankind, and most scholars would have insisted that affectionate relationships between people and animals were not only distasteful but depraved. Happily, those days are now gone. Attitudes to animals have changed, and, during the past three decades, the subject of relations between people and other animals has become a respectable area of research. The field of 'anthrozoology', as it is often called, now crosses a wide variety of academic disciplines, including anthropology, art and literature, education, ethology, history, psychology, sociology, philosophy, and human and veterinary medicine. In 1991, the International Society for Anthrozoology (ISAZ) was formed in Cambridge, England, its stated aim being to promote the study of all aspects of human–animal relationships by encouraging and publishing research, holding meetings, and disseminating information. To facilitate this process, there are now two academic journals dedicated to publishing original research in the field: *Anthrozoös* (published since 1987), and *Society & Animals* (published since 1993). In addition, ISAZ publishes a biennial Newsletter containing review articles and book reviews.

But why study relationships between people and animals in the first place? What purpose does it serve? The key to answering these questions lies in the unique ability of anthrozoology to create theoretical and conceptual bridges that not only link together widely separated disciplines but also span the gulf between the world of humans and the life of the rest of the planet. As the fox intimates in his speech to the little Prince, it is through the medium of social relationships that we find our true connection with others, irrespective of whether those others are human or non-human. Poised as we are on the brink of environmental catastrophe, the importance of establishing or reinforcing this sense of connection and identity with other lives can hardly be over-emphasized. In the past, medicine, psychology, sociology, anthropology and the humanities have all been guilty of studying humans in isolation, as if our species somehow evolved in the absence of interactions with anyone or anything except other humans. The existence of relationships with beings outside this strictly 'human' domain was either denied or dismissed as aberrant. And yet humans have been dependent on animals as sources of food, raw materials, companionship, and religious and artistic inspiration since the Palaeolithic Period, and animals have continued to mould the shape of human culture and psychology ever since. We are who we are as much because of our relationships with non-human animals as because of the human ones, and we do ourselves a great disservice – and probably great harm – by denying or ignoring this.

Anthrozoology is still a young science, and the primary goal of this book is to introduce readers to the richness of this emerging, interdisciplinary field by bringing together a collection of diverse and eclectic research papers and reviews representing the broad theme of human–pet interactions and relationships. While this compilation is designed to be illustrative rather than exhaustive, we hope that the breadth of topics covered, and the results achieved so far, will not only help stimulate discussion and debate, but also encourage the field to move ahead with new and ground-breaking research.

For convenience, the book has been divided into four parts. Part I addresses fundamental questions about the origins of the human–companion animal relationship, and highlights some of the cultural differences and similarities that exist in how these relationships are perceived. In Chapter 2, a novel interpretation of why Amazonian Indians tame animals and keep them as pets is presented. Chapter 3 explores the importance of pets in the lives of the ancient Greeks and Romans through an examination of pet epitaphs. Chapter 4 moves to the Middle Ages and examines the positive impact of aristocratic hunting on the nobility's attitudes and attachments to dogs. In Chapter 5 the importance of insects in the lives of Japanese children is examined from both historical and contemporary perspectives. Chapter 6 looks at the reasons why many people impose human celebrations and festivals upon their pets. Finally, Chapter 7 looks at conceptions of animality in various cultures, and discusses the potential for pets to help people psychologically by reconnecting them with the natural world via the 'animal within'.

Part II deals with other aspects of our relationships with pets. Chapter 8 takes a critical look at the evidence for the potential health benefits that pet owners are thought to derive from their animals, including the possible mechanisms responsible. In Chapter 9 conceptual issues to do with human and pet personalities are discussed. In addition, literature is reviewed on whether human personality can influence pet personality and whether the personalities of pet owners are significantly different from those of non-owners. Chapter 10 considers the emotive question of whether or not love of pets is associated with love of people.

Part III focuses on the role of the pet in contemporary Western families. Chapter 11 asks whether people sometimes adopt or purchase pets because of a lack of one or more social provisions in their relationships with humans. Chapter 12 examines the role of the pet in family networks: are human–pet relationships similar to human–human relationships in terms of the social provisions they provide? And this theme is further explored in Chapter 13 in relation to pets and the elderly. All of these chapters use Weiss' theory of social provisions as their theoretical framework. The human–cat relationship forms the focus of Chapter 14, in which owner assessments of cat behaviour, together with independent observations of the interactions between owners and different breeds of cats, are reported.

Part IV examines some important welfare and ethical issues concerning our relations with companion animals. Chapter 15 reports on an understudied aspect of animal abuse: secondary victimization in pet owners; while Chapter 16 explores the ethical dilemmas that veterinarians face on a day-to-day basis when dealing with pets and their owners. Finally, Chapter 17 re-evaluates the topic of bestiality, and proposes a new definition of this ancient and long-tabooed practice.

The idea for this book was conceived at a conference organized by the first editor at Downing College, Cambridge, in 1996 for the International Society for Anthrozoology. Although the conference covered many kinds of animal–human relationships, the wealth of material addressing different facets of the human–pet relationship was particularly striking, and we decided that this would be an opportune time to re-examine the field critically. Thanks are due to Pauline and David Appleby for their invaluable help with conference organization, and to the Royal Society for the Prevention of Cruelty to Animals (RSPCA), Waltham, Pedigree Petfoods, Universities Federation for Animal Welfare (UFAW) and the World Society for the Protection of Animals (WSPA) for providing funding for this meeting. This book contains chapters based on talks presented at the conference as well as some invited ones which we considered complementary to the book's overall theme.

All of these contributions were subjected to peer review prior to publication. We thank Alexa Albert, Ron Anderson, Warwick Anderson, Frank Ascione, Alan Beck, Marc Bekoff, Penny Bernstein, John Bradshaw, Juliet Clutton-Brock, Mary Ann Elston, Nienke Endenburg, Bruce Fogle, Lynette Hart, Hal Herzog, Adelma Hills, Robert Hubrecht, Tim Ingold, Elizabeth Lawrence, Richard Lobban, June McNicholas, Jill Nicholson, Harriet Ritvo, Irene Rochlitz, Andrew Rowan, Clinton Sanders, Boria Sax, Ken Shapiro, Aki Takumi and Cindy Wilson for their careful reviews and helpful, constructive comments on one or more manuscripts. We are grateful to the contributors to this book for their enthusiasm and quality of work. Thanks also to Tracey Sanderson at CUP for her support and patience.

Contact for International Society for Anthrozoology
Dr Debbie Wells, Membership Secretary & Treasurer
School of Psychology
The Queen's University of Belfast
Belfast BT7 1NN
Northern Ireland
UK
e-mail: d.wells@qub.ac.uk
Web page: http://www.soton.ac.uk/~azi/isaz1.htm

Part I
History and culture

PHILIPPE ERIKSON

2

The social significance of pet-keeping among Amazonian Indians

INTRODUCTION

The passionate relationship native lowland South Americans maintain with a wide array of pets has long been a favourite topic of chroniclers and scholars, such as Im Thurn (1882) and Guppy (1958), to name but two. Yet, its social significance and contrasted stance with regard to hunting have until now attracted surprisingly little attention.

Most studies devoted to native hunting in the neotropics emphasize the importance of predation in the symbolic universe of Amazonian peoples. Yet, despite its valorization as a key metaphor for social life, hunting, as a unilateral appropriation of wildlife, seems to clash with the great emphasis Amazonian ideology generally places on reciprocity. Along with its positive aspects, hunting therefore also engenders a kind of conceptual discomfort. A number of institutions (linked to shamanism, hunting ethic and hunting rites, prohibitions and so forth) tend to reduce the logical consequences of this imbalance. But these might not fully suffice to give 'a good conscience' to the hunter and his society (Hugh-Jones, 1996; Erikson, 1997). This chapter suggests that Amerindians solve the problem with their household animals, which I propose to consider as the semantic counterpoint of prey animals.

Considering pets and prey animals as two complementary facets of human–animal relations in Amazonia allows one to understand hunting as more than a simple means of obtaining protein. It also offers an alternative to the widely accepted interpretation of Amazonian pet-keeping as a kind of proto-domestication. Before turning to hunting itself, let us briefly examine the connections between wildlife and household animals, and the place the latter occupy in Amerindian ideology.

Fig. 2.1. Matis couple in a hammock with their pet sloth.

PETS AND PREY

Tamed and hunted animals usually belong to the same species. In most cases, prospective pets are brought back to the camp or village by the hunter who has just killed their mother. The little animal is then often given to the hunter's wife, who premasticates food for the fledglings, or breast-feeds the mammals. Men, women, prey, and household animals thus appear to be in a complementary distribution. But one could also consider that the relation between wildlife and pets has normative as well as factual groundings. For the Kalapalo (Basso, 1977: 102): 'birds, monkeys and turtles are the only wildlife kept as pets. Other animals are occasionally captured and briefly held in the village . . . but such animals are not referred to as *itologu* [pets], except in jest'. Interestingly enough, these animals, along with fish, happen to comprise the only flesh judged edible by the Kalapalo. Other animals are neither hunted nor 'tamed' in the Kalapalo sense of the term, being denied the status of *itologu*. Numerous other instances could be cited in which the favourite prey, however uncommon, also provides the favourite pet – e.g. sloth for the Matses (Fig. 2.1), capybara for the Txicão, coati for the Aché.

If there is indeed a continuity between those animals that are hunted and those that are tamed, they appear as mirror-images of each other rather than as analogues. In fact, the captive animal is so differentiated from its wild counterpart that it sometimes bears a name which is no longer that of its original species. Among the Wayãpi, the *yele* (*Touit pur-*

purata) and the *tapi'ilaanga* (*Piontes melanocephala*) parakeets, once tame, are respectively called *kala* and *paila paila*, the *kule* (*Amazona farinosa*) parrot becomes *palakut*, and the capuchin monkey *kai'i* (*Cebus apella*) becomes *maka* (Grenand, 1980).

According to Reichel-Dolmatoff (1978: 252), for the Desana: 'animals can be classified into *vai-mera bara*, edible animals, *vai-mera nyera*, bad, that is inedible animals, and *vai-mera ehora*, "fed animals" or "pets".' Given that the first two categories cover the entire spectrum of animal species, the simple existence of the third bears witness to its importance (and to the importance of the criteria of its definition). Once tamed, an animal changes status to the extent of nearly ceasing to belong to its original species.

Ethnography provides numerous cases in which pets appear as symmetrical to prey animals. Among the Maquiritare, for example, the feathers of birds killed for food are thrown away (in order to guarantee the reproduction of wildlife), while those belonging to tamed birds are carefully stored (Wilbert, 1972: 143). The Matis display little (if any) sensitivity when it comes to finishing off wounded prey animals, but very strictly prohibit even the slightest mistreatment of household animals, however boisterous their behaviour (Erikson, 1988). Furthermore, in dream symbolism, tamed animals often appear as being to women what wildlife is to men (Perrin, 1976: 146). Finally, that household animals and prey animals are complementary opposites, like the sexes with which they are associated, can be made explicit, as in the transmission of local rights among the Kaiapo: 'just as only men receive the *o mry* [rights to receive specific parts of certain animals in food distribution], only women receive the rights to raise certain animals (*o krit*) such as wild pigs, parrots, coati and ocelots' (Verswyver, 1983: 312).

It is therefore tempting to imagine that raising the young of animals killed in hunting is part of the logic of re-establishing a 'natural' balance, a way of cancelling (or at least compensating for) the destructive effects of hunting by their symbolic opposite. In taking care of young animals, and therefore keeping them alive, women would be playing the reverse of the destructive role of their male companions. Turning to the ideology of hunting, we may now seek the origin of this need to counterbalance its effects.

THE HUNTER'S PROBLEM

In Amazonian societies, one finds a widespread belief in the existence of 'masters of animals' (Rodrigues Barbosa, 1890; Zerries, 1954),

Fig. 2.2. Matis hunter carrying a pet spider monkey (*Ateles pamiscus*).

who seem to relate to wild species just as humans do with their household pets (Ahlbrinck, 1956: 123; Clastres, 1972: 39; Weiss, 1974: 256; Menget, 1988: 68; (Fig. 2.2)). Grenand (1982: 208) states that the Wayãpi consider that the real food of humans – i.e. meat – necessarily comes from the forest (i.e. the domain of spirits (*ayã*) who see animals as their *ima*, their 'domestic animals'). The consequence of this is clearly articulated in another text by Grenand (1980: 44): 'hunting is a risky business since a man most often kills an animal which does not belong to him, wherefore the perpetual fear of the spirits' retaliation'.

The idea that the relations between wildlife and spirits are of the same nature as those between humans and tamed animals is an important point, to which I will return. For the time being, let us simply point out that obtaining meat, apart from a strenuous chase, also implies a fearsome interaction with the 'masters of animals'. Since entering into competition with them is obviously out of the question, Amerindian hunters are left with three types of strategy to avoid the wrath of wronged spirits. They can pretend that they are eminently generous, they can negotiate with them, or they can contract an alliance with them. The following review of these strategies will show why none of them is totally satisfying.

ATTEMPTS AT RESOLUTION

Resolution through the 'gift'

The first option amounts to denying the problem by maintaining that humans benefit freely from the spirits' assistance. This is the solution, for example, which Dumont (1972: 20) points out in the case of the Panaré. Analysing the opposition between savannah and forest, he writes that if the former is 'the place where one neither gives nor receives' (since there is commerce with the Creoles), the latter 'is perceived as exactly the opposite since one helps oneself to it at will. The forest shelters the masters of plants and wild animals, and they are basically generous since they give to the Panaré even though the Panaré have nothing to offer in return, as long as they appropriate the "children of the forest" with moderation'.

The Matsiguenga seem to have adopted a similar viewpoint since, according to Renard Casevitz (pers. commun.), the Matsiguenga hunter never refuses prey sent (i.e. offered) to him by the spirits. He simply *has* to kill it. Nonetheless, this sounds more like a double bind (a paradoxical injunction) than a solution to the hunter's dilemma: whether or not you kill, you expose yourself to the spirits' wrath. Besides, in spite of their 'reassuring' talk, neither the Panaré nor the Matsiguenga are exempt from a certain fear of the spirits. In the evening, when a Panaré has not yet returned from the forest, 'there is a high degree of nervous tension within the group, whereas this attitude does not prevail at all if they know the hunter has stayed in the savannah' (Dumont, 1972: 19). The Matsiguenga too, even though they may have accepted the gift of the spirits, remain prudent, since the hunter should not bring back or even touch the game he has slaughtered: 'the convoy is always *innocent* of the death of the victims' (Renard-Casevitz, 1972: 245, my emphasis).

It is likely that the absence of a counter-gift is what produces the hunters' fear. If reciprocity had its rights, the spirits would have to eat humans (a theme which is amply covered in mythology); one would literally have to pay with one's own person (Siskind, 1973: 154 *et seq.*). It is to avoid such a dilemma that certain ethnic groups try to re-establish the balance through the influence of shamanism.

Resolution through negotiation

The most flagrant (and famous) example of this second solution is that of the Desana (Tucano). Their shamans gather with the master of the animals (*wai-maxsë*) in order to exchange human souls for meat for their group (Reichel-Dolmatoff, 1968: 107, 160), or in order to copulate with the 'peccaries', which the Tapirapé shamans also do (Wagley, 1977: 194). Ideally speaking, the situation is thus balanced and in harmony. Nonetheless, despite the shamans' negotiations, the Desana fear the vengeance of the *wai-maxsë* (Reichel-Dolmatoff, 1968: 146) since in addition to animal souls, his maloca (communal long-house) harbours diseases which he does not hesitate to blow onto humans if he considers himself wronged by a hunter (*ibid.*: 107). For my purpose, it is therefore particularly important to point out that for the Desana, taming has a prophylactic function (*ibid.*: 217–18):

> In many of the malocas, animals of the forest are kept, either small mammals or birds, which the indigenous people bring up with much care . . . These animals are thought to possess protective qualities in the sense that they attract contagious diseases to themselves . . . They are rodents: the paca, the agouti and the prea . . ., monkeys . . ., the loriot . . ., the agami . . ., hummingbirds . . ., parrots . . ., the tinamou.

Here, we find the general idea that I am proposing for the whole cultural area, transposed to the field of disease. The same idea seems to prevail among the Piaroa, for whom 'the giving of disease [can be seen] as a metaphor for a state of non-reciprocity' (Overing Kaplan, n.d.: 11), meanwhile claiming their pet parrots sing diseases away in compensation for the food they are given (Monod, 1972: 414).

For the Yagua, obtaining meat similarly implies a shamanic counterpart. This is why they depict cynegetic activity as a form of 'trade, exchange, barter' (*tari-*) rather than as 'labour' (*waria*). Whereas the Desana shaman deals directly with the master of the animals, his Yagua equivalent only deals with the spirits who 'do not themselves engender the different species of animals, being merely intermediaries between the masters of the animals and men' (Chaumeil & Chaumeil, 1977: 16). But in the last resort, killing wildlife indeed makes one contract a worrisome debt, albeit an indirect one since it is mediated through intermediary spirits.

Despite the ritual during which they are remunerated with an offering of cassava-beer and banana-gruel, the Yagua remain indebted to the masters of the animals, who are fearsome: 'some of them are depicted by the Yagua as man-eaters. As owners of the animals, they are the only ones who are really responsible for bad hunting and dearth' (Chaumeil & Chaumeil, 1977: 16). The underlying alternative could be formulated thus: eat (meat) and risk being eaten yourself, or starve. I am certainly far from demonstrating that the Yagua use their tamed animals to re-establish firmer connections of security with the latter's old masters. But it is interesting to note that they justify their consumption of domestic animals (pigs, chickens) by the fact that they do not belong to these masters (Chaumeil, personal communication). Thus, one may suppose that they abstain from eating other edible household animals due, among other reasons, to their connections with supernatural entities.[1]

Resolution through alliance

The option of relying upon spirit generosity or shamanic counter-payment to ensure a safe provision of meat is altogether rare. Predation seems much more frequently envisioned as a kind of alliance, as shown by the numerous rituals in which hosts 'invite' members of their opposite exogamic group who then impersonate animal spirits (Jacopin, 1972: 127). Throughout the Amazon Basin, exchange, alliance and predation are most often seen as interconnected (Bidou, 1972: 91; Lizot, 1977: 67).

For the Achuar, for instance, Descola (1994a) writes that contrary to war, hunting:

> is based on a gentleman's agreement and implies a seduction of affine animals which, whatever its outcome, at least accords them the merit of a social existence . . . In hunting, alliance is kept up through the kind of implicit contract between the *amana* [spirits] and the spirits who are protectors of the animals.

Landaburu (1979: 18), to take another example, recalls the primacy of this model among the Andoke:

> The felling of *chagra* trees, war, cannibalism, sexuality, the acquisition of a woman, all this is viewed as the capture of prey in return for the establishment of an alliance. This alliance, whether it is actually expressed through the obligations of a husband towards the group of his wife, or through the show of counterfeit presents left for the masters of the animals during a hunt . . ., is the effort to resolve the clash of opposites . . .

'Counterfeit presents' might not be enough to resolve antagonisms, and from this arises the tension accompanying this conception of hunting. Unlike alliances between human beings, where reciprocity is the rule, this is an alliance without a counterpart, where trickery reigns. Except for death, the relationship does not lead anywhere, neither to counter-benefits nor filiations: 'as a snared exchange, the relationship of alliance fully assumes the character of a tragic bet' (Descola, 1994a). One could say that this lethal and unilateral alliance is 'against nature'. And the reversal of values would no doubt appear complete if we add that in this perverse alliance (which hunting amounts to) it is the 'sons-in-law' (predators, takers) and not the 'parents-in-law' (masters of game, givers) who receive meat; in total contradiction to the Amazonian norm.

ADEQUATE SOLUTIONS TO THE HUNTER'S DILEMMA

Hunting therefore appears as a type of paradoxical alliance. It is indeed fundamental (in a double sense, since it is essential in the definition of marriage between humans, and indispensable during festivals between allies), but yet basically dishonest. This no doubt explains why the characteristic pride of the good hunter is tainted by a feeling of guilt, or to state it in slightly less psychological terms, conceptual uneasiness. My hypothesis is that there are two ways to remedy this predicament.

The first consists of scrupulously respecting the hunting ethic (often under pain of seeing one's own self transformed into prey). Before the hunt, one is supposed to abstain from all sexual relations (which tends to show that the hunter's double allegiance is indeed perceived as a logical difficulty). In the course of hunting, one owes some consideration towards the prey animal which, among other things, should be 'properly' slaughtered (Ruddle, 1970: 41). Lastly, on his return from the forest, a hunter remains modest in his triumph, despite a legitimate pride which does not fail to be expressed in other circumstances. When coming back laden with meat, it is advisable to appear silent and reserved, in short, to adopt a respectful attitude similar to that maintained in the presence of one's parents-in-law. This behaviour of course conforms to the indigenous morality where the display of inequalities is unwelcome. But it is also a means of assuring later success by tactfully respecting the susceptibility of the spirits and the animals' masters who, in accepting a 'curtailed' alliance, make hunting possible.

The second way to palliate the absence of reciprocity, in my view, resides in taming; if hunting poisons the relations between humans and the masters of game, household animals act as a kind of antidote. The status of household animals and how they are treated are evidence of this; they are coddled, fed, and 'considered true sons of those who bring them up' (Monod-Becquelin, 1975: 102). Thus, consanguinity takes over from affinity, violence gives way to affection, and people feed animals instead of the reverse. Feeding some members of a species certainly makes eating others appear much more legitimate (Fig. 2.3).

The metaphor of filiation also appears as an obvious extension of alliance. Evidently, it involves complementarity rather than compensa-

Fig. 2.3. Matis child with a pet sipi (*Saguinius mystax*) on his head, in front of a heap of smoked monkey meat.

tion, but the result is the same: interaction with the animal kingdom becomes less one-sided, and thereby less dangerous. Even if the quantity of tamed animals is numerically far from equal to that of slaughtered wildlife, this is irrelevant since we are dealing with ideal – not numerical – relations. 'It is the exchange situation itself that is crucial to the establishment and the maintenance of an alliance, and there need be no balance in numbers' writes Overing Kaplan (n.d.: 15) of Piaroa marriage alliances. This principle seems even more valid in the present case.

One may ask how a relationship conceived as a mode of filiation (pets as surrogate children) could affect a relationship conceived as one of affinity (prey and their spirit masters as in-laws). My answer would be that similar phenomena are found within the social structure. Indeed, if affinity is often considered in Amazonia as a necessary evil (for social reproduction), it is often 'controlled' by the Amerindians who seek somehow to transform their relationships of affinity into ones of consanguinity, especially through teknonymy and the forms of address (Menget, 1977). The most perfect example, or at least the one described in the greatest depth, is that of the Piaroa, whose communities have paradoxically been defined as 'alliance-structured kinship groups' (Overing Kaplan, 1975). By insisting, as they do, on the metaphor of filiation applied to household animals, Amazonians are simply extending to the animal kingdom this same principle of regressive (or ascending) consanguinization of affines.

In short, this metaphor is altogether equivalent to the Piaroa teknonymy: in becoming the co-parents of their household animals, the Amerindians add a grain of consanguinity to their affinity.

This is especially evident in the context in which antagonism between humanity and animality could be interpreted as a battle for 'fertility' (Goldman, 1981: 149; Langdon, 1981: 60–61), an essential notion in this lowland cultural area. Familiarization then appears as a form of ideal conciliation: it works in the sense of the fertility of animals (their young are raised), without running counter to the fertility of humans (since they gain additional 'children'). All told, taming and hunting could therefore be considered as two complementary aspects of a unique phenomenon: the assimilation of animals by human society.

INDIRECT CONFIRMATIONS

It may seem insufficiently convincing to rely solely on a structural analysis to deduce that pets serve as an intellectual counterweight to prey. Yet, this hypothesis is not implausible if one takes into account the semantic sphere to which Amazonian taming belongs. We shall therefore see, on the one hand, that the idea that taming allows for communication with the supernatural is well attested in other contexts, and, on the other, that the idea of a regulation of conflict through adoption and captivity is also widespread among Amazonian peoples.

The taming of the 'spirits'

The notion of *control* is one of the essential components of 'taming', especially when applied to the 'supernatural'. Thus, according to Grenand (1983: 371), the little particles contained in the shaman's rattle are 'domesticated spirits' . More generally, one could say the Wayãpi consider that 'the alliance between shaman and *ayã* (spirits) is ideal when the latter are domesticated' (Grenand & Grenand, 1981–2: 562). Similarly, Ahlbrinck (1956: 131) says that among the Caribs 'in the figurative sense the dance rattle is called a household pet (*ekë*)'.

Among the Bara, the sacred instruments (ordinarily called *he*) are referred to as *minia* (bird, tamed animal) during ceremonies (Hugh-Jones, 1979: 140–2). Similarly, the Txicão say that the bamboo horn played during initiation rituals is the *egu* (household animal) of the instrumentalist (Menget, 1977: 155). Although essential, these instruments are also dangerous and must be completely 'mastered'. Referring to their *jakui* flutes, the Xinguano say 'they carry very far, in other words all the way to the spirits, and people often cry while playing them' (Menget, 1981).

Another example of communication with the supernatural through an intermediary mascot is given by Henry (1964: 73), who notes that among the Xokleng [then called Kaingang]:

> one of the ways in which mortals may establish permanent and friendly relations with the spirits is to adopt one of their children . . . Even if a spirit is adopted in its animal form and never confers power of any kind, it is loved like a pet, and its death is avenged like that of a human being.

A note (unfortunately, rather vague) by Kensinger (1981: 170 n.7) alludes to something similar among the Cashinahua:

> in addition [to using proper hunting etiquette] men attempt to establish cordial relationships with spirit beings encountered during hallucinogenic experiences and dreams. The spirit familiars and pets gained in this way protect and assist the hunter.

What Henry and Kensinger propose only for shamans and some animals seems to me to exist also in an underlying way in all cases of taming. Summarizing the work of Mark Münzel on Kamayura mythology, Schaden (1978: 37) also insists upon the possibility of neutralizing danger – especially that of being devoured – by socializing it:

> The character of *mama'é* spirits who are aggressive and dangerous in the forest reverses when they are in the village, where they behave in a tender, peaceful, and even benign manner. The monstrous *anhang-ù* of the forest can become, in the heart of the village, *anhang-ù katù*, and thus instead of eating people, he gives them something to eat [in favouring the growth of manioc].

All the aforementioned examples neatly suggest that the idea that taming can serve as mediation between humans and spirits is far from being foreign to Amazonian cultures.

The captive as a mediator

The fact that adoption could serve to minimize the consequences of antagonism emerges clearly from the analysis of the role and status of Amazonian captives. Captives are indeed knowingly and systematically integrated into the society that has captured them. For the Txicão, Menget (1977: 146) goes as far as speaking of the 'intellectual and moral necessity of substituting captives for the dead', and suggests that captives make for essential bridges on the road leading from violence to alliance (Menget, n.d.).

The same holds true in the north-west of Amazonia: 'war, among the Yukuna, was so ritualized that although a means of destruction, it especially represented a particular mode of social relations' (Jacopin 1977: 14). Concerning the Makuna, Århem (1981: 162) similarly shows how marriage through abduction (considered, moreover, within the idiom of hunting) regularly spilt over into negotiation, and the 'regularization' of alliance between formerly hostile groups. Such ideas are widespread throughout the Amazonian cultural domain.

Yagua ethnography also shows that the use of captivity and the ensuing alliance are clearly perceived as a mechanism for regulating conflicts. Indeed, when they fought against potential allies (Jebero, Ticuna, Cocama, Omagua), the Yagua considered that it involved a *tari*, that is to say a reciprocal relationship, solidified by the abduction of enemy women. But when confronting ethnic groups with whom peace was out of the question (Mayoruna, Witoto), capturing women or children was also out of the question (Chaumeil, 1985: 148 *et seq.*). One could therefore say that captivity is more than an epiphenomenon of war: it is truly the recognition (or the first step in the establishment) of an alliance.

The same certainly holds true for taking prisoners among animals, especially if one considers that in many Amazonian societies, pets are often designated by the same term as captives and adopted children (Abreu, 1941: 575; Murphy, 1958: 1028; Menget, 1977: 155; Seeger, 1981: 225; Ortiz-Gomez, 1983: 97). Indeed, much as the Yagua distinguish between capturable and non-capturable enemies, the Waorani set peccaries aside from other wildlife. The former are killed with spears and bloodshed, considered as enemies, and are never tamed. The latter are 'softly' hunted with the blow-gun, and willingly adopted as pets (Rival, 1996: 156). In Amazonia, one only tames (or raises as captives) members of those species (or social groups) with which one engages in 'civilized' (albeit lethal) interaction. Taking prisoners is part of making interkilling more humane, and taking especially good care of them guarantees minimal retaliation.

MYTHOLOGICAL AND ESCHATOLOGICAL CONFIRMATIONS

Numerous myths concerning the relations between humans and animals tell of cases in which captivity also retro-acts on antagonism and limits its harmful consequences. A Cashinahua story, for instance, describes the fate of a hunter stuck in the trees and saved from being slaughtered by birds only because their leader happened to be his former pet (D'ans, 1978: 270–9). In a similar vein, the Xinguano believe that in the afterworld sky, the 'shadows of the dead' are in constant battle against birds (Carneiro, 1977; Villas Boas & Villas Boas, 1970: 111–18). One always finds, however, some of them to be volatile enough to take sides with humans, in recognition of the good deeds received on earth: 'after death [the household animals] take care of their former master' (Monod-Becquelin, 1975: 102, n. 73).

Speaking of the relation between the Yawalapiti and their tamed animals, Viveiros de Castro (1978: 19) points out the same thing: 'certain myths tell how deceased birds assist travellers to the sky, in exchange for the care received on earth. This relationship therefore continues after death since the sky belongs to the birds as well as to the souls'. Let us add that this persistence in relationship is accomplished with a reversal of roles: the former benefactor becomes a dependant, just as in the beliefs of many ethnic groups the former hunter himself risks becoming game after death.

The idea of gratitude from pets towards their former masters (deceased) is also elaborated more systematically in the beliefs relating to the psychopompic animals (conductors of souls in the land of the dead). Indeed, besides the earthly satisfactions which these bring, taking care of animals has repercussions on future life as they may in some way pay men back. In return for the good deeds received on Earth, they help them reach the 'sky'. In the Upper-Xingu societies we saw that birds were involved, probably because birds represent the paradigmatic pet (Basso, 1977: 102; Viveiros de Castro, 1978: 18).

It could be objected that even if they were not tamed, birds would still be the ideal vehicles for reaching the sky: they fly! But an explanation founded on the post-mortem continuation of the relation established through taming seems far more convincing than a 'Frazerian' one. This is shown by the Apinayé statement that *urubu* vultures help humans reach their final destination in the sky *because they have been fed* (Da Matta, 1970: 84), and by the numerous instances in which passage into the beyond depends on a non-winged animal. For the Cashibo or the Marubo, for example (Girard, 1963: 242; Melatti & Montagner Melatti, 1975: 100), access to the ultimate abode depends upon other, often-tamed animals: monkeys (the exact species is not specified).

The Aché (Guayaki), according to Clastres (1972: 187), ascribe the role of psychopomp to the coati (*Nasua nasua*) rather than to birds. This example is particularly interesting since coatis also happen to be the Aché's main *bita* (household pets, *ibid.*: 45, 50), thus confirming the previously established correlation between pets and psychopomps. Further evidence can be found in the striking contrast between the Aché data and those concerning another Tupi-speaking group, the Arawete (Viveiros de Castro, 1992). The latter see coatis as harmful necrophages rather than helpful psychopomps and therefore make fires around their graves to keep them away. Interestingly enough, this reversal of roles goes hand in hand with a modification of their status as pets. Rather than having them as favourite prey[2] and favourite pets, like the Aché, the Arawete regard coatis as familiars of the cannibal *Ayi* spirits. It is therefore tempting to imagine that coatis take sides with whichever of the two communities (humans and spirits) controls them through taming. It is true that the Aché justify their belief by invoking (*ibid.*: 224) the speed with which coatis climb vertically along tree trunks. But this is probably only a rationalization. On a deeper level, this belief illustrates the semantic importance of taming relationships, and not the ethological feature that 'explains' it.

It would be rather futile to search for causal relations between practices and beliefs. One obviously cannot say that Amerindians engage in taming in order to assure the passage of the dead into the beyond, or to soften the masters of the animals. However, independently of all explanations of the functional type, there is nonetheless a link between pets and psychopompic animals, on the one hand, and between pets and prey on the other. Taming ensures minimal security in an otherwise harsh universe.

CONCLUSION: PETS, PROTEIN OR PROTECTION?

In view of all that has been said so far, it is hardly surprising that the majority of Amerindians think, as do the Barafiri (Yanoama), 'that the idea of killing and eating . . . a pet, or of eating its eggs, is tantamount to savagery' (Smole, 1976: 185) or, as the Bororo, that 'eating dogs is ridiculous, eating [pet] macaws is scandalous, outrageous' (Crocker, 1977: 183). Even using non-flesh by-products of pets (such as their teeth) can be prohibited, as I have observed among the Matis (Erikson, 1988).

Nevertheless, the hypothesis that taming is a means of constituting a mobile food reserve has often been postulated, most often by superficial observers (Maufrais, 1951; Waisbard, 1969: 207), but sometimes even by contemporary and competent anthropologists. Lyon (1974: 92), for example, in reviewing some examples mentioned in the ethnography, writes: 'in the light of these cases, it is tempting to think of pet-keeping as functioning partially as a form of meat storage among groups who do not have domesticated animals'. Such a position seems contestable, first because even truly domesticated animals are seldom eaten in Amazonia,[3] and second because, as far as our knowledge extends, the groups cited by Lyon (namely the Campa of Peru and Yupa of Colombia) are quite atypical precisely because they seem willingly to eat tamed animals in non-ceremonial contexts: they are the exception rather than the rule.

For lack of data, it is difficult to have an exact idea of the Campa case, but it is worth noting that Denevan (1974: 105) clearly indicates that although Campa pets 'usually end up being eaten during times of hunger', he also mentions the unusual protein shortage in the diet of the Gran Paronal Campa (Denevan, 1974: 104, 107). Furthermore, this unusual edibility of pets has never been mentioned by earlier observers, which suggests it is probably a recent development.

The Yupa example described by Bolinder (1958), Ruddle (1971) and Wilbert (1974: 42) also obliges us to recognize that some Amazonian groups actually do eat tamed animals on a more or less regular basis. That said, the case once again involves a population living under very difficult conditions (Cariage, 1980: 13), where shortage of food is such that 'anthropophagy . . . appears to have evolved into endocannibalism, a last resort actively practiced only when starvation threatens the group's existence' (Ruddle, 1971: 281). It is clear that this form of anthropophagy is an occasional, even desperate practice, and that it has no ritual or even institutional character. The description of the meal which follows the slaying of an old Yupa woman shows this perfectly: 'the ensuing feast is a sad and sentimental occasion, without rejoicing, attended by all members of the starving group, no matter what was their kin relationship to the woman' (Ruddle, 1971: 282).

The consumption of household animals by the Yupa and the Campa is probably just as exceptional and, from their point of view, abnormal. A thoroughgoing cultural materialist approach, in which everything that happens to provide protein is deemed specifically adapted to this end, might deduce the reverse, namely that if the Yupa have grandmothers, it must be because in times of scarcity they furnish a welcome supplement to the diet. But obviously, the mere fact that an institution (in this case taming) can contribute, in exceptional cases, some sort of protein is certainly not enough to set it up as an adaptive technique of production. If one takes the native point of view into account, it appears that if pets have a role to play in providing meat, this is paradoxically because they are kept alive. Pets help future hunters acquire basic ethological knowledge (Laughlin, 1968: 309), and above all, provide an elegant solution to the ethical dilemma faced by people whose livelihood depends on the perilous enterprise of killing animals.

One could therefore say that taming is in some way a sort of sacrifice, rather unusual of course since it consists of *not* killing the animal raised for that purpose: sparing it as a token of good will rather than killing it in offering. Eating pets would therefore appear as a severe infraction of the norm, which one could not possibly interpret as the ultimate purpose of taming. All things considered, it is only possible to see household animals as potential meat by making a total abstraction of the native viewpoint, unless one is to postulate that this is a case of an institution which exists only to deny itself. But if, like Amerindians, one considers 'the symbolic conditions of the preparation for hunting . . . as indispensable requisites for carrying it out' (Descola, 1983: 84), one must then conclude that far from being a first step towards domestication for food, the presence of pets among Amazonian peoples is the expression of their allegiance to a way of life firmly centred on hunting.

Admittedly, as argued, for instance, by Serpell (1996: 72), 'prehistoric pet-keeping may well have paved the way toward animal husbandry and livestock farming', at least in some parts of the world. Yet, in places like Amazonia, it could be argued that, far from being a step towards proto-domestication, traditional modes of relating to pets are a major obstacle to the introduction of true domesticates (Queixalos, 1993; Erikson, 1998), as well as being an obstacle to attempts at domesticating native species (Descola, 1994b).

NOTES

1. Queixalos (1993: 97 & n.38), in a similar vein, reports that the Sikuani systematically mistreat their dogs but never even slightly harm their pets because the latter, unlike dogs, come from the forest and are therefore potentially capable of supernatural means of retaliation. Penteado Coelho (1995: 272) also depicts the striking contrast between the incredibly gentle treatment traditional pets receive among the Waura, and the extreme brutality imposed upon cats and chickens.
2. Hill & Hawkes (1983: 149) point out that 'coatis seem to be much more common in the Aché diet than has been reported for many other South American groups'. Coati are among the most intelligent animals of Amazonia (Filloux, 1956: 104; Brand, 1964: 193), and are easily and frequently tamed (Mason, 1899: 73).
3. Amazonian Indians are generally averse to eating not only pets but also domestic animals such as pigs or chickens (cf. Chapman, 1961; Perrin, 1976; Kracke, 1981; Urban, 1981; Henley, 1982: 47).

REFERENCES

Abreu, J.C. de. (1941). *Rã-txa Huni Kuin. A Lingua dos Caxinauás do rio Ubuaçu, Affluente do Muru*. Prefeitura de Tarauacá, Rio de Janiro: Typographia Leuzinger.
Ahlbrinck, W. (1956). *Encyclopédie Caribe*. Paris: Institut Géographique National.

Århem, K. (1981). *Makuna Social Organization. A Study in Descent, Alliance and the Formation of Corporate Groups in the North-Western Amazon.* Stockholm: Acta Universalis Upsaliensis.

Basso, E. (1977). The Kalapalo dietary system. In *Carib Speaking Indians, Culture, Society and Language*, ed. E. Basso, pp. 98–105. Tucson, AZ: Anthropological Papers of the University of Arizona no. 28.

Bidou, P. (1972). Représentations de l'espace dans la mythologie Tatuyo (Indiens Tucano). *Journal de la Société des Américanistes*, **LXI**, 45–108.

Bolinder, G. (1958). *We Dared the Andes*. New York: Abelard-Schuman.

Brand, D. (1964). The Coati or Pisote (Nasua narica) in the archaeology and ethnology of meso-america. In *Actes du XXXVth Congrès des Américanistes*, pp. 193–202. Paris.

Cariage, P. (1980). Guerre et guerre entre familles chez les Yuko. *Bulletin de la Société Suisse des Américanistes*, **44**, 13–26.

Carneiro, R. (1977). The afterworld of the Kuikuru Indians. In *Colloquia in Anthropology*, ed. R.K. Wetherington, pp. 3–15.

Chapman, A. (1961). Mythologie et éthique chez les Jicaques. *L'Homme*, **I**, 95–101.

Chaumeil, B. & Chaumeil, J-P. (1977). El rol de los instrumentos de musica sagrados en la produccion alimenticia de los Yagua del Nor-Este Peruano. *Amazonia Peruana*, **I**, 101–22.

Chaumeil, J-P. (1985). Echange d'énergie: guerre, identité et reproduction sociale chez les Yagua de l'amazonie péruvienne. *Journal de la Société des Américanistes*, **LXXI**, 143–57.

Clastres, P. (1972). *Chronique des Indiens Guayaki*. Paris: Plon.

Crocker, C. (1977). My brother the parrot. In *The Social Use of Metaphor*, ed. J.D. Sapir & J.C. Crocker, pp. 164–92. Philadelphia: University of Pennsylvania Press.

D'ans, A-M. (1978). *Le Dit des Vrais Hommmes. Mythes, Contes, Légendes et Traditions des Indiens Cashinahua*. Paris: Union Générale d'Edition.

Da Matta, R. (1970). Les présages Apinayé. In *Echanges et Communications. Mélanges Offerts à Cl. Lévi-Strauss à l'occasion de son 60ème Anniversaire*, ed. J. Pouillon & P. Maranda, pp. 77–99. La Haye: Mouton.

Denevan, W. (1974). Campa subsistence in the Gran Pajonal, Eastern Peru. In *Native South America. Ethnology of the Least Known Continent*, ed. P. Lyon, pp. 92–110. Boston: Little, Brown & Company.

Descola, P. (1983). Le jardin de colibri. Procès de travail et catégorisations sexuelles chez les Achuar de l'equateur. *L'Homme*, **XXIII**, 61–89.

Descola, P. (1994a). *In the Society of Nature: a Native Ecology in Amazonia*. Cambridge: Cambridge University Press.

Descola, P. (1994b). Pourquoi les Indiens d'Amazonie n'ont-ils pas domestiqué le pécari? Généalogie des objets et anthropologie de l'objectivation. In *De la Préhistoire aux Missiles Balistiques. L'intelligence Sociale des Techniques*, ed. B. Latour & P. Lemonnier, pp. 329–44. Paris: La Découverte.

Dumont, J-P. (1972). Espaces et déplacements dans l'habitat Panaré. *Journal de la Société des Américanistes*, **LXI**, 17–30.

Erikson, P. (1988). Apprivoisement et habitat chez les Amerindiens Matis (Langue Pano, Amazonas, Brésil). *Anthropozoologica*, **9**, 25–35.

Erikson, P. (1997). De l'acclimatation des concepts et des animaux ou les tribulations d'idées américanistes en Europe. *Terrain*, **28**, 119–24.

Erikson, P. (1998). Du pécari au manioc ou du riz sans porc? Reflexions sur l'introduction de la riziculture et de l'élevage chez les Chacobo (Amazonie bolivienne), In *Résistance et Changements des Comportements Alimentaires*, ed. M. Garrigues-Cresswell & M. A. Martin, pp. 363–78. Special issue of *Techniques & Culture*, no. 31–2.

Filloux, J.C. (1956). *Psychologie des Animaux*. Paris: Presses Universitaires de France.

Girard, R. (1963). *Les Indiens de l'Amazonie Péruvienne*. Paris: Payot.
Goldman, I. (1981). Cubeo dietary rules. In *Food Taboos in Lowland South America*, ed. K. Kensinger & W. Kracke, pp. 144–56. Bennington, VT: Bennington College.
Grenand, P. (1980). *Introduction à l'Étude de l'Univers Wayãpi. Ethnoécologie des Indiens du Haut Oyapock, Guyane Française*. Paris: SELAF.
Grenand, P. (1982). *Ainsi parlaient nos Ancêtres . . . Essai d'ethnohistoire Wayãpi*. Paris: Orstom.
Grenand, F. (1983). *Et l'homme Devint Jaguar. L'univers Imaginaire et Quotidien des Indiens Wayãpi de Guyane*. Paris: L'Harmattan.
Grenand, P. & Grenand, F. (1981–2). La médecine traditionnelle des Wayãpi (Amerindiens de Guyane). *Cahiers Orstom, Série Sciences Humaines*, **XVIII**, 561–7.
Guppy, N. (1958). *Wai-Wai. Through the Forests North of the Amazon*. London: John Murray.
Henley, P. (1982). *The Panare. Tradition and Change on the Amazonian Frontier*. New Haven & London: Yale Univerity Press.
Henry, J. (1964). *Jungle People. A Kaingang Tribe of the Highlands of Brazil*. New York: Random House, Vintage Books.
Hill, K. & Hawkes, K. (1983). Neotropical hunting among the Aché. In *Adaptive Responses in Amazonia*, ed. R. Hames & W. Vickers, pp. 139–88. New York: Academic Press.
Hugh-Jones, S. (1979). *Male Initiation and Cosmology Among the Barasana Indians of the Vaupès Area of Columbia*. Cambridge: Cambridge University Press.
Hugh-Jones, S. (1996). De bonnes raisons ou de la mauvaise conscience? ou: de l'ambivalence de certains Amazoniens envers la consommation de viande. *Terrain*, **26**, 123–48.
Im Thurn, E.F. (1882). Tame animals among the Red Men of (South) America. *Timehri, Journal of the Royal Agricultural and Commercial Society of British Guiana*, **I**, 25–43.
Jacopin, P-Y. (1972). Habitat et territoire Yukuna. *Journal de la Société des Américanistes*, **LXI**, 109–40.
Jacopin, P.-Y. (1977). La parole et la différence, ou l'entrée des Blancs dans la mythologie des Indiens Yukuna. *Bulletin de la Société Suisse des Américanistes*, **41**, 5–19.
Kensinger, K. (1981). Food taboos as markers of age categories in Cashinahua. In *Food Taboos in Lowland South America*, ed. K. Kensinger & W. Kracke, pp. 151–71. Bennington, VT: Bennington College.
Kracke, W. (1981). Don't let the piranha bite your liver: a psychoanalytic approach to Kagwahiv (tupi) food taboos. In *Food Taboos in Lowland South America*, ed. K. Kensinger & W. Kracke, pp. 91–142. Bennington, VT: Bennington College.
Landaburu, J. (1979). *La Langue des Andoke, Amazonie Colombienne*. Paris: SELAF.
Langdon, T. (1981). Food taboos and the balance of oppositions among the Barasana and the Taiwano. In *Food Taboos in Lowland South America*, ed. K. Kensinger & W. Kracke, pp. 56–75. Bennington, VT: Bennington College.
Laughlin, W. (1968). Hunting: an integrative biobehavior system and its evolutionary importance. In *Man the Hunter*, ed. R.B. Lee & I. deVore, Chicago: Aldine.
Lizot, J. (1977). Descendance et affinité chez les Yanomami: antinomie et complémentarité. In *Actes du XLIIème Congrès International des Américanistes*, pp. 55–70. Paris: Société des Américanistes.
Lyon, P. (1974). *Native South America. Ethnology of the Least Known Continent*. Boston: Little, Brown & Company.
Mason, T. (1899). Aboriginal American zoötechney. *American Anthropologist*, **1**, 54.
Maufrais, R. (1951). *Aventures au Mato Grosso*. Paris: Julliard.
Melatti, J.C. & Montagner Melatti, D. (1975). *Relatorio sobre os Indios Marubo*. Brasilia: Trabalhos de ciencias socias, série antropologia social 13.

Menget, P. (1977). Au nom des autres. Classification des relations sociales chez les Txicão du Haut-Xingu (Brésil). Unpublished dissertation of the École des Hautes Études en Sciences Sociales, Paris.
Menget, P. (1981). *Brésil, Musiques du Haut-Xingu.* Paris: Musidisc-Europe, Ocora 558517.
Menget, P. (1988). Note sur l'adoption chez les Txicão du Brésil central. *Anthropologie et Sociétés,* **XII**, 63–72.
Menget, P. (n.d.). Alliance and Violence in the Upper-Xingu. Manuscript, University of Paris X-Nanterre.
Monod, J. (1972). *Un Riche Cannibale.* Paris: Union Générale d'Edition.
Monod-Becquelin, A. (1975). *La Pratique Linguistique des Indiens Trumaï (Haut-Xingu, Mato Grosso, Brésil).* Paris: SELAF.
Murphy, R.F. (1958). Mundurucu religion. *University of California Publications in American Archaeology and Ethnology,* **49**, 1–154.
Ortiz-Gomez, F. (1983). Organisation sociale et mythologie des Indiens Cuiba et Sikuani (Guahibos, llanos de Colombie). Unpublished dissertation of the Ecole des Hautes Etudes en Sciences Sociales, Paris.
Overing Kaplan, J. (1975). *The Piaroa. A People of the Orinoco Basin. A Study in Kinship and Marriage.* Oxford: Clarendon Press.
Overing Kaplan, J. (n.d.). Masters of Land and Masters of Water: Cosmology and Social Structure Among the Piaroa. Manuscript, London School of Economics.
Penteado Coelho (1995). Figuras zoomorfas na arte Waura: anotações para o estudo de uma estética indígena. *Revista do museu de arqueologia e etnologia,* **5**, 267–81.
Perrin, M. (1976). *Le Chemin des Indiens Morts. Mythes et Symboles Guajiro.* Paris: Payot.
Queixalos, F. (1993). Les mythes et les mots de l'identité Sikuani. In *Mémoire de la Tradition,* ed. A. Monod-Becquelin & A. Molinié, pp. 71–106. Nanterre: Société d'Ethnologie.
Reichel-Dolmatoff, G. (1968). *Desana. Le Symbolisme Universel des Indiens Tukano du Vaupès.* Paris: Gallimard.
Reichel-Dolmatoff, G. (1978). Desana animal categories, food restrictions and the concept of color energies. *Journal of Latin American Lore,* **4**, 243–91.
Renard-Casevitz, F-M. (1972). Les Matsiguenga. *Journal de la Société des Américanistes,* **LXI**, 215–53.
Rival, L. (1996). Blowpipes and spears. The social significance of Huaorani technological choices. In *Nature and Society, Anthropological Perspectives,* ed. P. Descola and G. Pálsson, pp. 145–64. London & New York: Routledge.
Rodrigues Barbosa, J. (1890). Poranduba Amazonense. *Anais da Biblioteca Nacional de Rio de Janeiro,* **14**, 48–85.
Ruddle, K. (1970). The hunting technology of the Maraca Indians. *Anthropológica,* **25**, 21–63.
Ruddle, K. (1971). The Yukpa autosubsistence system: a study of shifting cultivation and ancillary activities in Colombia and Venezuela. Unpublished PhD thesis, Ann Arbor, University Microfilms.
Schaden, E. (1978). O indio e a sua visão do mundo. *Revista de Antropologia,* **21**, 33–44.
Seeger, A. (1981). *Nature and Society in Central Brazil. The Suya Indians of Mato Grosso.* Cambridge, MA: Harvard University Press.
Serpell, J. (1996). *In the Company of Animals: a Study of Human–Animal Relationships.* Cambridge: Cambridge University Press.
Siskind, J. (1973). *To Hunt in the Morning.* New York: Oxford University Press.
Smole, W.J. (1976). *The Yanoama Indians. A Cultural Geography.* Austin, TX, & London: Texas University Press.
Urban, G. (1981). The semiotics of tabooed food: Shokleng (gê). In *Food Taboos in*

Lowland South America, ed. K. Kensinger & W. Kracke, pp. 76–90. Bennington, VT: Bennington College.

Verswyver, G. (1983). Cycles in Kaiapo naming practices. *Communication and Cognition*, **16**, 301–23.

Villas Boas, O. & Villas Boas, C. (1970). *Xingu. Os Indios, seus mitos*. Rio de Janeiro: Zahar Editores.

Viveiros de Castro, E. (1978). Alguns aspectos do pensamento Yawalapiti (alto Xingu): classificações e transformações, Boletim do Museu Nacional. *Antropologia*, **26**, 1–41.

Viveiros de Castro, E. (1992). *From the Enemy's Point of View, Humanity and Divinity in an Amazonian Society*. Chicago and London: The University of Chicago Press.

Wagley, C. (1977). *Welcome of Tears. The Tapirapé Indians of Central Brazil*. New York: Oxford University Press.

Waisbard, S. (1969). *Chez les Chasseurs de Têtes (Aguarunas)*. Paris: Société Continentale d'Editions Modernes Illustrées.

Weiss, G. (1974). Campa cosmology. In *Native South America. Ethnology of the Least Known Continent*, ed. P. Lyon, pp. 251–66. Boston: Little, Brown & Company.

Wilbert, J. (1972). *Survivors of El Dorado (4 Indian Cultures of South America)*. New York & London: Praeger Publishers.

Wilbert, J. (1974). *Yupa Folktales*. Latin American Anthropological Studies, 24. Los Angeles: University of California.

Zerries, O. (1954). *Wild und Bushgeister in Südamerika. Eine Untersuchung Jägerzitlicher Phänomene im Kulturbild südamerikanischer Indianer.* Studien zur Kulturkunde, 2, XI. Wiesbaden: Franz Steiner Verlag.

LILIANE BODSON

3

Motivations for pet-keeping in Ancient Greece and Rome: a preliminary survey

INTRODUCTION

Pet-keeping was a widespread and well-accepted phen[illegible] classical antiquity,[1] raising disapproval only when pets [illegible] or were thought to supplant children in human affectio[illegible]ess of the owners' self-respect and consideration for their own species.[4] Besides this, ancient Greeks and Romans of all ages enjoyed animal companions of many different species, from insects to mammals. However, for all the pets found so far in ancient literature and art, owners' explicit statements of their reasons for preferring one species to another and, more fundamentally, for wanting or needing a pet are still lacking, even supposing that they had once been recorded and preserved from destruction. An insight into the reasons for ancient people's interest in pets is provided by pet epitaphs. Some animal companions were offered burials intended not for ritual or apotropaic purposes,[5] for example, but for their own sake. The tombstone or sarcophagus erected at the grave site was carved with a funerary text which not only mentioned the pet's name, but also listed its merits and expressed the mourner's grief.

This chapter seeks to retrieve the classical pet owner's motivations from the content of animal epitaphs. First, animal burial in classical antiquity will be described briefly. After an overview of the animals' qualities and merits and the owners' sorrow, some points relevant to the conceptual background of ancient pet-keeping will be delineated. These points will serve as an introduction to the discussion of the intentional wording of the epitaphs as they disclose grounds for keeping companion animals in ancient Greece and Rome.

ANIMAL BURIAL IN CLASSICAL ANTIQUITY

The principle of funerals for dead animals (pets, domestic or even wild animals) was neither officially forbidden nor morally condemned in ancient Greece and Rome, although not everyone showed sympathy and respect towards the burial place of lower beings. One dog epitaph (Roman Empire) makes this plea: 'Do not laugh, I beg you, you passing by, because it is a mere dog's grave'.[6] Apart from this, explicit criticisms were levelled only at people 'plunged into shameful and intolerable grief'[7] and deemed too ostentatious and extravagant by the standards of pet funerals in Greece or Rome (Georgoudi, 1984: 41). It is worth noting that human funerals, too, were governed by regulations which were intended to impose restrictions on both emotional displays and funerary expenses.

The ancient Greek and Roman burial of animals closely followed the pattern of human burials in two main respects. First, with regard to the grave site, and in contrast to the situation observed, for example, in nineteenth to twentieth-century France, where a 'line between human and canine was carefully drawn' (Kete, 1994: 33),[8] companion animals were not buried in 'pet cemeteries' but, just as with human beings, their graves could be seen everywhere and paid tribute to by passersby: intramurally, close to the market place, as with the fourth century BC Athenian dog buried behind the Stoa of Attalos (Thompson, 1951; Immerwahr, 1973: nos. 63–64); extramurally, along the roadside, as with Cimon's mares facing his own grave,[9] or in human cemeteries.[10] Second, tombstones or sarcophagi of suitable size marked out the burial place (see Fig. 3.1).[11] They were sometimes carved with a portrait of the animal and were engraved with an epitaph identifying the bodies lying therein. Deposits of funerary offerings similar to those intended for human graves were also made, such as, in the Athenian dog grave, for example, a miniature pitcher and lamp, in addition to the purposely and well-chosen offering of a large beef bone. The animals' afterworld was alluded to in the same terms as that of humans: 'Inexorable Hades',[12] 'the house of Acheron',[13] 'the silent paths of night',[14] and 'the dark road of Orcus'.[15] As Rhodope justly proclaimed in the epitaph carved on the sarcophagus designed for her pet dog 'Crown' (after 212 AD), ancient animal companions were buried 'like human beings'.[16]

Whether long or short, epitaphs were written in prose or in verse (funerary epigrams) (Herrlinger, 1930: 1–120)[17] and were intended theoretically to be cut into tombstones (inscriptional epigrams). The fact is that many are not preserved on their original supports, but on papyri and parchments. As with other classes of epigrams, funerary epigrams were often written by renowned poets and were therefore not regarded as mere

HELENAE ALVMNAE
ANIMAE
INCOMPARABILI ET
BENE MERENTI

Fig. 3.1. Grave stele for Helena, about 150–200 AD. Marble, H: 61 cm; W: 31.5 cm. Unknown artist, Italy (Rome). Translation by Vermeule & Neuerburg (1973: 38): 'To Helena, foster child, soul without comparison and deserving of praise.' Printed with permission of the J. Paul Getty Museum, Malibu, California.

pieces of information, but were also appreciated as literature. They were collected by ancient editors of anthologies who, being poets themselves, used to enlarge their collections with so-called fictional epitaphs of their own or of other writers. Literary imitations and actual eulogies complement each other and thus will both be alluded to.

While the earliest record of an animal funeral, namely the cenotaph erected to the memory of Xanthippos' dog, goes back to the early fifth century BC,[18] the earliest extant grave of a dog buried for its own sake dates back to the fourth century BC.[19] As for the epitaphs, they highlight dogs more than any other species and belong mostly to the Hellenistic and Greco-Roman periods (third/second centuries BC to fourth/fifth centuries AD). This may be due to an accidental loss of evidence commonly affecting ancient Greek and Roman material, or it may reflect a deeper change in the human–animal relationship than suspected by Lazenby (1949: 299). Such change occurring by Alexander's time would be responsible, among other things, for an increase in the number of pets, and therefore in their burials. Be that as it may, inscriptional epitaphs written in Greek or Latin have been identified around the entire Mediterranean area: mainland Greece and the Aegean islands, Asia minor, Egypt, Sicily, Rome and Southern Gaul. Both the varied geographical locations of the graves, and the chronology of the direct (epitaphs) and indirect (accounts of animal funerals) information covering roughly ten centuries indicate the long-standing practice of pet burial in the classical civilizations.

Pet-keepers of both sexes and all ages mourned their companions. Their social status, although often unknown, ranged all the way from Athenian aristocrats,[20] Alexander the Great[21] or the Roman emperor Hadrian[22] to the lowest cobbler[23] or slave sailor.[24] In all likelihood, pet-keeping was not a matter of social position or wealth in antiquity.

The animals recorded in epitaphs include insects (cicadas, crickets, etc.), birds (swallows' nestlings, nightingales, parrots, partridges, cocks), and mammals (dolphins, hares, young pigs, goats, dogs and horses). Not all of these animals were pets and, conversely, not all pets (e.g. cheetahs, monkeys, grass snakes, fish) were commemorated in epitaphs.[25] The swallows' nestlings 'robbed by a many-coiled serpent',[26] or dolphins 'thrown on the land by the dark sea-water'[27] were wild animals whose fate inspired reverence[28] due to their cultural status in ancient Greece and Rome. Domestic animals such as the cock, which was his owner's 'alarm-clock',[29] a decoy partridge,[30] hounds,[31] watchdogs,[32] war and race horses,[33] or the goat stolen by a wolf from the herd[34] do not fit the modern definition of pet: 'Any animal that is domesticated or tamed and kept as a favourite, or treated with indulgence and fondness.' (*OED*, 1989: 9.625) or 'for primarily social and emotional reasons rather than economic purposes' (Serpell & Paul, 1994: 129). Even locusts etc., valued for their 'song' at bedside,[35] and singing or talking birds[36] had a psychological and ornamental function of their own, which differentiated their companionship from that of a tame dolphin,[37] hare,[38] pig[39] or lap-dogs.[40]

The burials and eulogies of race horses may chiefly have been a further way for the owners to perpetuate the fame and glory they had won with their champions. However, their surviving epitaphs and the funerary epigrams of war horses, hounds and watchdogs more often referred to the gratitude of the mourners for their personalized interactions with these 'service animals'.[41] Although they are not to be mistaken for pets, their eulogies are relevant to the present survey.

Not surprisingly, cat epitaphs (either inscriptional or fictional) or burials have not been identified at this point. As has been remarked by Lentacker & De Cupere (1994), 'the likelihood that we will find skeletal elements of (Roman) cats is minimal', for which there are many different reasons, including the cat's natural behaviour. An intended cat burial[42] would have been rather unlikely in the classical world, due to the cultural status of *Felis silvestris catus* (see Bodson, 1987, 1998). The domestic cat originated from Egypt where eventually household cats were regarded as sacred (Malek, 1993: 75). Whether the earliest cat burial (*c.* 1350 BC) evidenced in that country had been privately erected by the mourning Prince Djehutymose or inspired by a general trend in the Egyptian religion is still open to discussion (*ibid.*: 123–5). Whatever the answer, the thousands of mummified cats excavated in special cemeteries over the first millennium, especially from *c.* 400 BC on, were buried for religious purposes. The domestic cat was introduced to Europe, through Greece, in the sixth century BC. Despised for its native sacred status, it was reluctantly admitted by the Greeks and, later, by the Romans. Although evidence of its acceptance slowly increased during the Roman Empire, it never became as popular as dogs or even parrots.

ANIMAL MERITS

Qualities

Physical and behavioural qualities were recorded to a greater or lesser extent, depending on the animal species. Beauty,[43] cleverness and intelligence,[44] affection[45] and cheerfulness[46] were praised in lap-dogs; courage and self-devotion in the hound Tauron, who saved his master

Zenon from an assaulting boar, at the cost of his own life,[47] and faithfulness in the hound Lydia, 'savage in the woods, gentle at home'.[48] A race mare was remembered for her 'wind-footed, lightest limb'.[49] Corinna's parrot, 'gift brought from the limit of the world', was 'the loquacious image of the human voice'.[50] Lesbia's honey-sweet sparrow was cherished for its hopping here and there and its chirping to its mistress exclusively,[51] the 'singing' locust of Philaenis for its 'melodious noise'.[52]

Interactions

Merits and qualities were evoked with reference to bitter-sweet pictures of interactions. 'The wealthy house of Alkis' sounded empty since his 'shrill-voiced locust has flown to the meadows of Hades and the dewy flowers of golden Persephone'.[53] The tame dolphin would not 'dance anymore to the tune of the pierced reed . . . beside the ships'.[54] Lap-dogs appreciated as 'playmates' of children[55] and of grown-ups[56] were described at length more than any other species. Pretty Pearl was remembered lying in her master's or mistress's bosom.[57] Sweet Fly always shared her mistress's bed in the nicest manner except for her barking at any human rival, and 'beamed at her smile with gentle bites'.[58] Patrice ate out of her master's hand, drank from his cup, and wagged her tail when greeting him.[59]

Most modern dog owners might define their relationship with their dogs in the same way (Hart, 1995). Indeed, the attitudes and reactions stated in the ancient evidence related to well-known characteristics of canine behaviour, so often reported that they have become commonplace in the universal literature on dog ownership. In Greek and Roman epitaphs, which were intended to commemorate each dog's own personality in a life-like style, they must be taken at face value.

HUMAN GRIEF

Whatever their age, dog owners openly wept for their animals,[60] and expressed affliction and bereavement.[61] They lamented over the irreparable loss and railed against destiny deemed cruel in all cases,[62] yet even more so when Agathias' 'poor partridge' was killed by the housecat,[63] or when the bitch Pearl died while giving birth.[64]

Taking action to bury the animal and pay it last respects seemed to provide the mourners some relief. One of them was proud to announce that he had brought the earth for the burial mound, erected the tombstone, and written the epitaph.[65] Another addressed his dead companion:

'I am in tears, while carrying you to your last rest place as much as I rejoiced when bringing you home in my own hands fifteen years ago'.[66]

DISCUSSION

Human–animal companionship was so strongly felt by people of the earliest periods that it was not supposed to end at the master's or mistress's death. As in other societies (Malek, 1993: 45; Clutton-Brock, 1995: 10–12), animal companions were sometimes slaughtered and burnt or buried with their former owners. Patroklos' horses and dogs[67] or the Gallic ponies, all the many dogs and pet birds (nightingales, parrots and blackbirds) of Regulus' son,[68] were burnt on their pyres to share their owners' fate, although the precise intention remains speculative as regards Patroklos. The motives for these 'man-oriented' customs and treatment are not clear in all cases, but at any rate are not the reflection of sentimental attitudes towards the animals concerned. Nor were these animals the primary objects of concern when they were represented with their masters or mistresses on their own tombstones (Woysch-Méautis, 1982). This practice, which first appeared in archaic Greece (seventh–sixth century BC), implicitly recognized animal companionship as an important feature of the deceased person's life, worthy of being publicized and immortalized. Despite the conventional style observed in ancient funerary art, those representations were not deprived of their full meaning, as may be judged from the late Gallo-Roman tombstones (third–fourth century AD) of children shown with their pets.

Conversely, attention was directed to the animal companions which were perpetuated for their own sake. They were portrayed from life in paintings, sculptures or literary forms. Such was the case with Issa, Publius' Gallic lap-dog, pictured by a painter whose work is now lost and described by the Roman poet Martial (*ca.* 40–*c.* 104 AD) in a famous epigram.[69] When they died, many pets were mourned, buried and lavished with funerary portraits and epitaphs which constituted the final and supreme stage of the ancient human–animal companionship. Before examining the wording and content of their epitaphs with respect to pet-keepers' motivations, two features must be pointed out to delineate further the general framework of this multifaceted process.

First, in the opinion of the ancients, the deities cared for animals, either wild or domestic. As for the latter, Socrates' (469–399 BC) recommendation to pray to the gods for cattle and sheep, shepherds' and animal breeders' prayers for their livestock and dog companions, and travellers' prayers for their mounts provide evidence of the belief in divine attention to animal welfare (Bodson, 1980). Animals mentioned after, yet together with, one's wife, children and other relatives were not taken as substitutes for human relationships, nor were they considered as mere economic products. Livestock animals, although their health and fecundity were of primary importance in contributing to the household's livelihood in a mixed farming system, were considered as living beings (*zoa*) in an empathy-based approach (see Midgley, 1983: 112–14) and were regarded as no less privileged to supernatural protection than the other – that is, human – members of the community.

Second, animals living in human surroundings were also entitled to justice in a way which transcended philosophical debates on animal rights as well as legal proceedings against animal abuse. Stoic, Epicurians and Neo-platonists engaged in endless arguments for and against animal rationality and its consequences in terms of extending human justice to them. Whatever the motives – the animal's own nature, its kinship with human nature, or man's moral duty to be just and benevolent in all circumstances (Sorabji, 1993: 212–19) – court actions were occasionally taken against those who victimized animals, to say nothing of spontaneous, even violent public disapproval.[70] Notwithstanding the Hesiodic myth of Zeus' restricted conferring of justice on human beings, ancient popular wisdom held that people had the power to protect themselves from each other or, to put it in Lonsdale's terms (1979: 156) 'to restrain man from preying on his own kind'.[71] 'There are Furies of the dogs too', went a Greek proverb. This meant that the chtonian powers of vengeance (Furies), whose mission was pursuing unpunished criminals, would retaliate sooner or later against any offender, even if the victim were a mere dog.[72] Complementing the gods' gift of health and welfare, the benefit of divine or immanent justice was thus perceived by the general public as extending to all creatures, including animals placed under the control of humans and exposed to their possible cruelty.

Such was the conceptual background against which ancient people of all walks of life bestowed burials and memorials on the animals. To keep here to those animals which provided services, e.g. horses, hounds and watchdogs, or those which were integrated as pets in the family circle, burial was a way for the grieving owner to cope with emotional pain through action. A significant part of the process was achieved by epitaphs describing the animal companion's admirable qualities (for example song, talk, dance, swiftness) and the wide variety of its indoor and outdoor interactions. These interactions brought the animal owner unalloyed and gratefully remembered pleasure. Expressed by many references to daily life, such pleasure was further underlined by the use of words such as *athurma* and *deliciae*. The Greek *athurma*, whose literal meaning was 'plaything', 'toy' causing 'joy' and 'delight', was the common term used to refer to a pet. The Latin *deliciae*, also in current use to qualify persons one was fond of – for example children, sweetheart – defined the animal as one's 'enjoyment', 'delight', or 'favourite'. Pleasure was obviously a major motivation of ancient pet-keeping. But pets were not only seen as attractive living toys or playmates. The joy or delight they brought to their owners had an emotional dimension involving both partners and was explicity stated in chosen terms. As regards humans, it was said to be *philia*, the general Greek term for 'friendship', and justified by feelings of gratitude for the devotion of a hound such as Tauron[73] or for the 'splendid goodwill and beauty' of Divine, the bitch mourned by a girl left anonymous.[74] Conversely, the feelings identified by the owner in his or her animal were called *storge*. This term, literally 'affection', acquired the meaning 'filial' or 'brotherly affection' when applied to children (and 'parental affection' if applied to parents). In the context of pet-keeping, it compared the animals' attitudes towards their masters to the affection either of children towards their parents, or of sisters and brothers towards each other. Such had been the affection between Divine and her young mistress, while the bitch Parthenope's *storge* for her master was an unrivalled 'filial affection'. Indeed, Parthenope, implicitly considered a champion engaged in the sport-like competition of attachment and fondness, was given funerary honours as 'prize of the contest (*athlon*) of filial affection (*storge*)'.[75] In addition, dog companions were also referred to as 'foster children' (*trophimoi*) in ancient Greek language.[76] The word was used in Divine's epitaph.[77] As with Latin *alumna* ('foster child') used in Helena's epitaph (see Fig. 3.1),[78] the affective connotation of *trophimos* (*-e*) might have been further enhanced in lap-dogs' eulogies.

This terminology raises the question of the animal companions' personification. Mary Midgley's perceptive pages (1995: 344–51) make valuable reading with respect to classical evidence. No definite argument has yet been able to settle the matter of sentimental anthropomorphism in ancient pet epitaphs, although it may be inferred here and there. Indeed, as has been seen, ancient people adopted the pattern of human funerals to bury their animal companions and were therefore easily and naturally induced to refer to them using a humanized vocabulary superficially defined by Lazenby (1949: 299) as 'exaggerated pathos'. In all likelihood, the statement '*like* a human being' (my italics) in Crown's epitaph did not mean the humanization of the animal, but emphasized his mistress's pride at having fulfilled her commitment to her pet in every possible respect.[79] Undoubtedly, puppies and monkeys taken as substitutes for children attest to the animal anthropomorphization; however, none of the Greek and Roman pet epitaphs recovered thus far refers to monkeys or puppies. Nor did epitaphs include any sentence prefiguring modern statements such as 'The more I know people, the more I love my dog.' or 'Deceived by the world, never by my dog.' (Kete, 1994: 33–4; compare Thomas, 1984: 118; Ritvo, 1987: 86).

CONCLUSION

Ancient Greek and Roman pet owners' motivations identified in surviving animal epitaphs thoroughly underscored the keeper's pleasure as the primary benefit of having animal companions. This pleasure stemmed from physical, cognitive and emotional criteria generally combined with each other, to an extent that depended on the species involved, but was not limited to the owner's self-interest. All epitaphs, although mainly those relating to mammals, especially lap-dogs, suggest that the satisfaction of pet companionship was based on both partners' joy or delight, which arose from their interactions at two closely dependent stages: first, the animal's exuberant attitudes, tricks and play, and the keeper's own stimuli and responses to his or her companion; second, their reciprocal affection. For all the other feelings of gratitude, admiration or confidence, this affection emerges as the most distinctive feature of ancient pet epitaphs.

Assessing how many ancient people had pets at any given time, and how many of these, on average, were given burials is an impossible task. Nonetheless, for all their gaps and limits, Greek and Roman animal epitaphs still provide a meaningful insight into the motivations of ancient pet owners. Pet owners were people from all economic, social and intellectual backgrounds in both classical civilizations and from early stages, although evidence increases after the fourth century BC. It has yet to be established, however, whether classical pets were 'instruments of follie' (see Serpell, 1996: 43), or if, in fact, 'the Roman upper classes . . . took their affection for these animals to even more bizarre extremes than the Greeks.' (*ibid.* 46). The matter is not simple, and might not be settled without further inquiries into the ancient views on animal nature in general and pets in particular. Meanwhile, the epitaphs surveyed in this chapter suggest that strong and selfless affection for animals should no longer be considered a uniquely modern phenomenon.[80]

NOTES

A.P. = *Anthologia Palatina* (or *Greek Anthology*; see Paton, 1970).
C.I.L. = *Corpus Inscriptionum Latinarum.*
I.G. = *Inscriptiones Graecae.*

1. The adjective 'classical' is used for the sake of brevity in the sense: 'of ancient Greece and Rome'.
2. Eubulus, fr. 114 Kassel-Austin (quoted by Athenaeus 12.519a): 'For how much better, I ask you, for a human being to bring up a human being provided he have the means, than a splashing, quacking goose, or a sparrow, or a monkey, always plotting mischief.' Clement of Alexandria, *Paedagogus* 3.30 (sharply blaming the women who cared for pets instead of children, widows, orphans, and the elderly).
3. Athenaeus 12.518f–519b (the question 'Do not the women in your country bear children?' addressed by Masinissa (238–148 BC), King of Numidia, to the Sybarites, fond of lap-dogs and monkeys). Plutarch *Pericles* 1.1 (the same question by the Roman Emperor August to some wealthy foreigners 'carrying puppies and young monkeys about in their bosoms and fondling them.'). Cf. *On Tranquillity of Mind* 13 (*Moralia* 472c). Juvenal *Satirae* 6.654–5.
4. Plutarch *Pericles* 1.1.
5. For example, Preston Day, 1984.
6. *I.G.* 14.2128.2 (Rome; Roman Empire).
7. Plutarch *Solon* 7.3.
8. Compare Lord Byron's dog buried in 'the precincts of the sacred Abbey of Newstead.' Ritvo, 1987: 86.
9. Herodotus 6.103.
10. Aelianus, *On the Animals* 12.40. Such was also the case of Rhodope's dog: for reference see n. 16.
11. Vermeule & Neuerburg, 1973, no. 84. Other examples: Pfuhl & Möbius, 1979, nos. 2196, 2197, 2199. Cf. Theophrastus, *Characters* 21.9. All tombstones carved with animals were not animal gravemarkers. See, for example, Freyer-Schauenburg, 1970; Vermeule, 1972.
12. *A.P.* 7.190.4 (Anyte, third century BC): epitaph of a locust and a cicada. See below, ad n. 53 (*AP* 7.189).
13. *A.P.* 7.203.4 (Simias, end of the fourth to early third century BC): of a partridge.
14. *A.P.* 7.199.4 (Tymnes,third century BC?): of a bird; 211.4 (*idem*): of a dog.
15. Catullus 3.13–15.
16. Iplikçioglu, Çelgin & Vedat Çelgin, 1991, no. 22.9.
17. Still a convenient reference book, although the interpretation is outdated. General survey of the human–animal relationship in ancient Rome: Toynbee, 1973.
18. Plutarch *Themistocles* 10.9–10. Xanthippos, Pericles' father, lost his dog before the battle of Salamis (480 BC), when he was evacuated by boat from Athens. The dog, having been prevented from boarding with his master, swam after him and eventually died from exhaustion.
19. See above, p. 28.
20. For example, see above, n. 18.
21. Plutarch *Alexander* 32.12.
22. *C.I.L.* 12.1122a (Hadrian's epitaph of his mount Borysthenes; see *Historia Augusta. De vita Hadriani* 20.12; Cassius Dio 79.10.2).
23. Pliny the Elder, *Naturalis Historia* 10.121–123. Cf. Plutarch *Sollertia animalium* 19 (*Moralia.* 973B): on 'a certain barber' who had bred a jay.

24. *I.G.* 12.458.3 ((Mytilene; Roman Empire).
25. On cats, see below, p. 31.
26. *A.P.* 7.210 (Antipater of Sidon, second century BC).
27. *A.P.* 7.215 (Anyte), 216 (Antipater of Thessalonica, first century BC–first century AD).
28. *A.P.* 7.216.4.
29. *A.P.* 7.202 (Anyte).
30. *A.P.* 7.203 (Simias).
31. Page-Dawe-Diggle, 1981: 456–458, no. CXLVI (Tauron, Zenon's dog; mid-third century BC); Page, 1942: 462–3, no. 109.2 (second epitaph of the same). Page et al., 1981: 457, rightly refuted Gorteman, 1957: 116–18, who claimed that 'the present epitaphs may represent merely a "motif littéraire".' Other examples: Page et al., 1981: 291, no. LXIX (ascribed to Simonides); 481–2, no. CLXI (fragmentary epitaph for a dog; papyrus-text of an anthology; *ca.* 100 BC). Pfuhl & Möbius, 1979, no. 219. etc.
32. *A.P.* 7.211 (Tymnes); *C.I.L.* 9.5785 (Ricina; Roman Empire).
33. *A.P.* 7.208 (Anyte); 212 (Mnasalces, third century BC).
34. *A.P.* 9.432 (Theocritus, third century BC).
35. *A.P.* 7.197 (Phaennus, mid or end third century BC).
36. Catullus 2–3 (sparrow); Ovid *Amores* 2.6 (parrot); Pliny the Elder *Naturalis Historia* 10.121–123 (raven); Martial, *Epigrams* 7.87.8 (nightingale); *I.G.* 14.56 (Syracuse; Roman Empire; nightingale); Statius *Silvae* 2.4 (parrot). Also *AP* 7.203–206 (partridges). Compare Loughlin, 1993.
37. *A.P.* 7.214 (Archias, first century BC?).
38. *A.P.* 7.207 (Meleager, 140/20–60 BC).
39. Chamoux, 1974 (second to third century AD).
40. For example, *I.G.* 12.458 (Mytilene; Roman Empire); 459 (Mytilene; second century AD); 14.1360 (Rome; Roman Empire = Moretti, I. 1972, no. 317); 2128 ([Rome]; Roman Empire); 1647 (Rome; Roman Empire = Moretti, III. 1979, no. 1230). For example, *C.I.L.* 6.29896 (Rome; Roman Empire); 10.659 (Salerno; Roman Empire); 13.488 (Auch; second or third century AD).
41. As earliest occurrences of that personalization, see Homer *Iliad* 19.400–424; *Odyssey* 17.290–305.
42. Only one representation on a Greek human tombstone *c.* 430 BC (see Woysch-Méautis, 1982: 111, no. 70) as opposed to hundreds of dogs (*ibidem*, nos. 153–210, 256–334).
43. For example, *I.G.* 14.1647.5. (Rome; Roman Empire: 'Divine'); *C.I.L.* 6.29896.2 (Rome; Roman Empire: 'Pearl').
44. For example, *C.I.L.* 10.659.7 (Salerno; Roman Empire: 'Patrice'). Compare *C.I.L.* 9.5785 (Ricina; Roman Empire: a dog 'vigilant vehicle-keeper').
45. For example, *I.G.* 14.1647.4.
46. *C.I.L.* 10.659.13.
47. Reference above, note 31 (no. CXLVI, ll. 7–12).
48. Martial, *Epigrams* 11.69.2–3.
49. *A.P.* 7.212.2 (see n. 33).
50. Ovid, *Amores*, 2.6.37.
51. Catullus 3.6–10.
52. *A.P.* 7.198 (Leonidas of Tarentum, third century BC).6.
53. *A.P.* 7.189 (Aristodicus of Rhodes, third century BC?).1.3–4; cf. *A.P.* 7.194 (Mnasalces).4: 'Democritus' house rang with the melodious song of his locust.'
54. Reference above, note 37.
55. *I.G.* 14.1647.6–7.
56. *I.G.* 12.459.1.

57. *C.I.L.* 6.29896.7.
58. *C.I.L.* 13.488.3–10.
59. *C.I.L.* 10.659.9–13.
60. *C.I.L.* 10.659.1 (a grown-up master: see below, *ad* n. 66); *I.G.* 14.1647.8 (a young girl: see below, *ad* n. 74).
61. According to McNicholas & Collis (1995: 132–3), 'in the majority of cases the term "bereavement" is not applicable to pet loss'. They are referring to contemporary society where (p. 135) 'outward expressions of grief and sorrow are not generally condoned'. It is impossible to determine to what extent pet loss was not only 'disturbing', but also 'disruptive' (see p. 133) for ancient pet owners. From their feelings expressed in pet epitaphs, in a society where pet loss was culturally recognized (see above, § 2), the term 'bereavement' seems to be appropriate. See Rajaram *et al.*, 1993; Gosse & Barnes, 1994.
62. *C.I.L.* 13.488.4. Compare Catullus, 3.16; Ovidius, *Amores* 2.6.
63. *A.P.* 7.204 (Agathias Scholasticus, sixth century AD).5–6; compare 206 (Damocharis, sixth century AD).
64. *C.I.L.* 6.29896.11.
65. *I.G.* 14.2128.
66. *C.I.L.* 10.659.1–2.
67. Homer *Iliad* 23.171–174. For further examples of dog and horse burials in Late Bronze Age and Early Iron Age Crete and mainland Greece, see above, n. 5 (e.g. pp. 26–7).
68. Pliny the Younger, *Letters* 4.2.3.
69. Martial, *Epigrams* 1.109.
70. Pliny the Elder, *Naturalis Historia* 10.121–123.
71. Hesiod, *Theogony* 275–280. Sorabji, 1993: 117. Hesiod's account is still open to investigation since animals of a given species do not prey on their own kind (cf. the Latin proverb: 'Dog does not eat dog-flesh', quoted by Varro *On the Latin Language* 7.31). Man only behaves as a wolf against his fellowmen (Plautus, *Aulularia* 495: *lupus est homo homini*).
72. *Appendix proverbiorum* 2.20 Leutsch & Schneidewin (1839: 397).
73. Reference above, n. 31; text *ad* n. 47.
74. *I.G.* 14.1647.6.
75. *I.G.* 12.459.3 (climax of the epitaph).
76. Aelian, *On the Animals* 11.31; 16.39.
77. *I.G.* 14.1647.9–10.
78. Fig. 3.1. See above, n. 11.
79. See above, text *ad* n. 16.
80. Serpell & Paul, 1994: 136; Serpell, 1996: 23.

REFERENCES

Translations of Greek and Latin sources are quoted from the 'Loeb Classical Library', except for Greek and Latin inscriptions (my translations).

Bodson, L. (1980). Deux expressions du sentiment religieux dans la prière personnelle en Grèce. 1. La prière pour les animaux. In *L'expérience de la prière dans les grandes religions. Actes du colloque de Louvain-la-Neuve et Liège (22–23 novembre 1978)*, ed. H. Limet & J. Ries, pp. 149–72. Louvain-la-Neuve: Centre d'Histoire des Religions.

Bodson, L. (1987). Les débuts en Europe du chat domestique. *Ethnozootechnie*, **40**, 13–38.

Bodson L. (1998). Critères d'appréciation de l'animal exotique dans la tradition grecque ancienne. *Les animaux exotiques dans les relations internationales: espèces, fonctions, significations. Journée d'étude. Université de Liège, 22 mars 1997*, ed. L. Bodson, pp. 139–212. Liège: Université de Liège.

Chamoux, F. (1974). L'épitaphe du cochon d'Édesse. In *Mélanges de Philosophie, de Littérature et d'Histoire ancienne offerts à Pierre Boyancé*, pp. 153–62. Rome: École française de Rome.

Clutton-Brock, J. (1995). Origins of the dog: domestication and early history. In *The Domestic Dog: its Evolution, Behaviour, and Interactions with People*, ed. J. Serpell, pp. 8–20. Cambridge: Cambridge University Press.

Freyer-Schauenburg, B. (1970). KUÔN LAKÔNOS – KUÔN LAKAINA. *Antike Kunst*, **13**, 95–100.

Georgoudi, S. (1984). Funeral epigrams for animals. *Archaeologia*, **11**, 36–41 (in modern Greek).

Gorteman, C. (1957). Sollicitude et amour pour les animaux dans l'Égypte gréco-romaine. *Chronique d'Égypte*, **32**, 101–20.

Gosse, G.H. & Barnes, M.J. (1994). Human grief resulting from the death of a pet. *Anthrozoös*, **7**, 103–12.

Hart, L.A. (1995). Dogs as human companions: a review of the relationship. In *The Domestic Dog: its Evolution, Behaviour, and Interactions with People*, ed. J. Serpell, pp. 161–78. Cambridge: Cambridge University Press.

Herrlinger, G. (1930). *Totenklage um Tiere in der antiken Dichtung mit einem Anhang byzantinischer, mittellateinischer und neuhochdeutscher Tierepikedien*. Stuttgart: W. Kohlhamer.

Immerwahr, S.I. (1973). *Early Burials from the Agora Cemeteries*. Princeton; NJ: American School of Classical Studies at Athens.

Iplikçioglu, B., Çelgin, G. & Vedat Çelgin, A. (1991). *Epigraphische Forschungen in Termessos und seinem Territorium*, I. Vienna: Österreichische Akademie der Wissenschaften.

Kete, K. (1994). *The Beast in the Boudoir. Petkeeping in Nineteenth-Century Paris*. Berkeley: University of California Press.

Lazenby, F.D. (1949). Greek and Roman household pets. *The Classical Journal*, **44**, 245–52, 299–307.

Lentacker, A. & De Cupere, B. (1994). Domestication of the cat and reflections on the scarcity of finds in archaeological contexts. In *Des Animaux introduits par l'homme dans la faune de l'Europe. Journée d'étude. Université de Liège, 20 mars 1993*, ed. L. Bodson, pp. 69–78. Liège: Université de Liège.

Leutsch, E.L. & Schneidewin, F.G. (ed.) (1839). *Corpus Paroemiographorum Graecorum*, I. Göttingen: Vandenhoeck & Ruprecht.

Lonsdale, S.H. (1979). Attitudes towards animals in Ancient Greece and Rome. *Greece and Rome*, **26**, 146–59.

Loughlin, C.A. (1993). Psychological needs filled by avian companions. *Anthrozoös*, **6**, 166–72.

Malek, J. (1993). *The Cat in Ancient Egypt*. London: British Museum Press.

McNicholas, J. & Collis, G.M. (1995).The end of a relationship: coping with pet loss. In *The Waltham Book of Human–Animal Interaction: Benefits and Responsibilities of Pet Ownership*, ed. I. Robinson, pp. 127–43. Oxford: Pergamon.

Midgley, M. (1983). *Animals and Why They Matter*. Athens: The University of Georgia Press.

Midgley, M. (1995). *Beast and Man. The Roots of Human Nature*, revised edn. London & New York: Routledge.

Moretti, L. (1972–1979). *Inscriptiones Graecae Urbis Romae*. I–III. Rome: Istituto Italiano per la Storia Antica.

OED (1989). *The Oxford English Dictionary* (1989), Vol. XI, 2nd edn. Prepared by J.A. Simpson & E.S.C. Weiner. Oxford: Clarendon Press.

Page, D.L. (ed.) (1942). *Greek Literary Papyri. Texts, Translations and Notes*. London & Cambridge, MA: W. Heinemann & Harvard University Press.

Page, D. L. (ed.) (1981). *Further Greek Epigrams*. Revised and prepared for publication by R.D. Dawe, & J. Diggle. Cambridge: Cambridge University Press.

Paton, W.R. (ed.) (1970). *The Greek Anthology*, II. London & Cambridge, MA: Heinemann & Harvard University Press.

Pfuhl, E. & Möbius, H. (1979). *Die ostgriechischen Grabreliefs*. Mainz: Philipp von Zabern.

Preston Day, L. (1984). Dog burials in the Greek world. *American Journal of Archaeology*, **88**, 21–32.

Rajaram, S.S., Garrity, T.F., Stallones, L. & Marx, M.B. (1993). Bereavement – loss of a pet and loss of a human. *Anthrozoös*, **6**, 8–16.

Ritvo, H. (1987). *The Animal Estate. The English and Other Creatures in the Victorian Age*. Cambridge, MA: Harvard University Press.

Serpell, J. (1996). *In the Company of Animals. A Study of Human–Animal Relationships*. Cambridge: Cambridge University Press (Canto).

Serpell, J. & Paul, E. (1994). Pets and the development of positive attitudes to animals. In *Animals and Human Society. Changing Perspectives*, ed. A. Manning & J. Serpell, pp. 127–44. London & New York: Routledge.

Sorabji, R. (1993). *Animal Minds and Human Morals. The Origins of the Western Debate*. London: Duckworth.

Thomas, K. (1984). *Man and the Natural World. Changing Attitudes in England 1500–1800*. London: Penguin.

Thompson, H.A. (1951). Excavations in the Athenian Agora: 1950. *Hesperia*, **20**, 52–3.

Toynbee, J.M.C. (1973). *Animals in Roman Life and Art*. London: Thames and Hudson. (Reprint: Baltimore: The Johns Hopkins University Press, 1997.)

Vermeule, C. (1972). Greek funerary animals, 450–300 BC. *American Journal of Archaeology*, **76**, 49–59.

Vermeule, C. & Neuerburg, N. (1973). *Catalogue of the Ancient Art in the J. Paul Getty Museum. The Larger Statuary, Wall Paintings and Mosaics*. Malibu, s.n.

Woysch-Méautis, D. (1982). *La représentation des animaux et des êtres fabuleux sur les monuments funéraires grecs de l'époque archaïque à la fin du IVe siècle av. J.-C.* Lausanne: Bibliothèque Historique Vaudoise. (Cahiers d'archéologie romande 21.)

SOPHIA MENACHE

4

Hunting and attachment to dogs in the Pre-Modern Period[1]

INTRODUCTION

The phenomenon of dog-keeping in pre-modern societies raises challenging questions that have not yet received satisfactory attention in historical research (Menache, 1998). To mention but one crucial issue: was dog-keeping practicable among traditional societies, given the negative attitude towards dogs of all monotheistic religions? (Menache, 1997). This question becomes especially pertinent with regard to the Middle Ages in light of the ubiquitous presence of the Catholic Church and its widespread socio-political power. It is the premise of this chapter that dog-keeping resulted from the need of human beings to project onto their pets a variety of hopes and beliefs, needs and frustrations (Phineas, 1973–4; Perin, 1981; Thomas, 1984; Ronecker, 1994; Clutton-Brock, 1997). As a projection mechanism (Laplanche & Pontalis, 1984), dog-keeping may be considered a universal phenomenon; however, its character, scope and manifestations have been in a continuous state of change. In other words, the universal emotional basis of dog-keeping was not strong enough to make pets immune to the socio-economic, cultural and mental factors that dictated the nature and scope of pet-keeping in a given space and time.

Beginning with the wide-ranging subject of pet-keeping, this chapter goes on to examine the gradual evolution of dog-keeping in medieval Christendom, while focusing on the contribution of late medieval hunting to turning dogs into companion animals. Since hunting (with the help of dogs) was an important activity of the upper classes, it provided a leading factor in the evolution of dog-keeping, with all its affectionate connotations, among the medieval élite. This perspective will allow a brief discussion of the mental process that in the long run paved the way for dog-keeping in Western society. However, one should refrain from broad sociological/gender generalizations. In the particular context of pre-modern hunting and dog-keeping, it will be more accurate to speak in terms of the nobility and, more specifically, male groups. Although medieval women also participated in hunting, their participation was marginal, and we lack satisfactory evidence with regard to their approach to hunting and/or to the contribution of canines in the chase.[2] This is not to say that women did not play an important part in dog-keeping – indeed, some archaeologists believe that it was women who began the domestication process (Zeuner, 1954; Clutton-Brock, 1987).[3] Rather, gender should be taken into account when analysing the various factors that favoured affectionate attitudes towards dogs as companion animals. From a geographical–chronological perspective, this study centres on Western Europe from the fifth to the fifteenth century. This wide scenario was dictated by the need to follow the gradual evolution of a process that reached maturity towards the late Middle Ages. It also justified the selective method applied hereafter.

PET-KEEPING IN THE PRE-MODERN PERIOD: A CONTRADICTION IN TERMS?

Joel S. Savishinsky (1983: 114) accurately remarked that the indiscriminate definition of pet-keeping for both foraging and modern societies implies a similarity of intent that may not exist, since similar patterns of behaviour mask underlying differences of attitude and treatment. Savishinsky's remark is specially relevant with regard to traditional societies for two main reasons. In traditional societies, the socio-economic environment provided a supportive framework for each and every member, thus supposedly weakening the affective dependence of human beings on companion animals. Moreover, dogs were entrusted with guarding functions – either of members of the household or their cattle – that went beyond (and perhaps were irrelevant to) their role as companion animals. Savishinsky further emphasized the fact that in settled or the so-called civilized societies, pets became enduring, anthropomorphized members of the family. Conversely, the pets of hunters and foragers were merely ephemeral members of their communities: 'They are frequently treated as disposable, if animated toys, subjected to physical abuse and neglect, insufficiently fed, and either killed or allowed to die when their amusement value declines'(p. 114). Thus, the mere existence of dogs in different societies cannot be viewed as proof of dog-keeping.

This conclusion finds ample confirmation in the light of the affective essence attributed to pet-keeping in the past 200 years. Affection/tenderness/devotion are recognized as the most prominent human emotions lying at the core of pet-keeping. Jerrold Atlas (1982), in a panel devoted to 'The human/companion animal bond: interaction between man and animal' at a meeting of the International Psychohistorical Association, corroborated the widespread premise that pet-keeping basically involves an emotional commitment (p. 8). Whether the basis of the relationship between humans and animals is conscious, unconscious, or based on evolutionary development of the type envisaged for a Jungian collective unconscious, people expect reciprocity from animals and, consequently, enter into special, intimate primary relationships with them. The emotive basis of the human–animal relationship is perhaps best expressed in the semi-humorous saying: 'My dog is the only one who understands me' (Bunch, 1982: 9).

As indicated above, a primary methodological problem in the investigation of dog-keeping in medieval society is presented by the powerful position of the ecclesiastical establishment, on the one hand, and its reservations, one may even say open aversion to dogs, on the other. One should note that in the framework of all monotheistic religions – Judaism, Christianity, Islam – animals share an inferior role in the interplay between God, human beings, and all other creatures on earth. The complete supremacy of humankind over all animals was established by an almighty God and, as such, was absolute and did not leave room for further considerations:

> So God created man in his own image, in the image of God created he him; male and female created he them. And God blessed them, and God said unto them, Be fruitful, and multiply, and replenish the earth, and subdue it: and have dominion over the fish of the sea, and over the fowl of the air, and over every living thing that moveth upon the earth. (*Genesis* 1: 27–8)

The basic recognition of the mastery of human beings was further strengthened by the principle of *nomina res essentiant;* i.e. the names Adam gave the animals (*Genesis* 2: 19) not only suggested their character but also influenced their role and destiny on earth (Muratova, 1977; Dronke, 1983). Acknowledgement of human domination 'over every living thing that moveth upon the earth' favoured an abusive, instrumental approach to animals, whose very existence was justified by their serving the needs of human beings. Moreover, the biblical tenets did not remain confined to the geographical–cultural space of Asia Minor; one may find similar, patronizing attitudes in classical Greece, as well. Aristotle, for instance, claimed that animals were created for the sake of human beings.

Domestic animals were there to labour, the wild ones to be hunted (*Politics*, 1256 a–b). Needless to say, the biblical anthropocentric approach heavily influenced the disparaging attitudes towards animals in Christian doctrine (Biese, 1905). Theologians such as Tertullian (*c*. 160–*c*. 225) (cols. 864–5), Origen (*c*. 185–*c*. 254) (pp. 245, 254–55), Augustine (354–430) (pp. 293–4), Bede (c. 673–735) (p. 21; cols. 31–2) and, later, Peter Comestor (d.c. 1179) (cols. 1062–4) maintained that human beings held total mastery over animals, since the former were created in God's image and, therefore, were beneficiary of His wisdom. As pointed out by Jean Batany (1984), 'an ideology [i.e. Christianity] that declared the primary rule of men over animals, could not accept, but with great reluctance, any close contact between the human and animal species' (p. 41; Payson Evans, 1896).

At the theoretical level, at least, the supremacy of human beings over the animal kingdom thus obstructed – if not precluded – the approach to animals in general, and dogs in particular, as companion animals, with all its emotional, egalitarian essence. This latent contradiction is not restricted to the condition inherent in human beings 'created in God's image' and their pets; it also embraces the degree of correlation between dogmatic principles and the ubiquity of dogs in daily practice as a whole. The question thus arises: When and how did medieval people abandon their rather comfortable status of absolute ruler over nature – which was accorded to them by divine will – and, exercising their free choice and acting in open defiance of theological tenets, place themselves close to dogs, to the point that they turned canines into their 'best friends'?

Hunting – as it developed among the upper classes in the late Middle Ages – represented a crucial step in the evolution of dog-keeping. After the medieval nobility transformed canines into their main partner in the recreational killing of wild animals, this peculiar partnership did not actually change the traditional (i.e. patronizing, abusive) approach to the animal kingdom. Rather antithetically, dogs served to manifest the mastery of human beings over all living creatures, in perfect accord with the will of Providence as enacted in *Genesis*. In principle, therefore, the peculiar partnership of human beings and dogs in hunting was not based on the formers' conscious or unconscious acceptance of the dog as a companion animal or on an affectionate attitude towards canines. Rather, it was founded on the existing gap between expectations and reality: between human beings who, at least in theory, were accorded complete rule over nature by an omnipotent God and, in open contrast, individuals who were actually unable to overcome, let alone rule over, wild animals.

Dogs helped to fill this gap, both by becoming the most acquiescent slaves to the will of their masters, and by further helping the latter to defeat other animals. By making dogs the intermediaries between hunters and the killing of other animals, whether for food or for sport, hunting thus provided an important stage in the gradual acceptance of dogs in human society (Grossman, 1993). However, hunting *per se* did not provide the affectionate basis for dog-keeping. This was a by-product of the evolution of hunting itself – with all its socio-economic, cultural, and mental undertones – which reached maturity in the late Middle Ages.

THE 'PRE-HISTORY' OF HUNTING

The partnership of human beings and dogs in hunting is not unique to the Middle Ages. From their very early domestication more than 10 000 years ago, dogs were used as companions in the chase. As an early hunting partner, the dog was often buried with its master. These first seeds of an affective relationship between human beings and dogs can already be detected by the end of the Bronze Age, when most European populations ceased to eat dog meat (Haudricourt, 1962; Bökönyi, 1974; Sahlins, 1976). The gradual development into settled societies and the discovery of agriculture weakened neither the importance of hunting nor the participation of dogs in this activity. On the contrary, dogs were closely connected to hunting in Homeric narrative, where they appear as the most helpful auxiliaries of men (*Iliad* XVIII: 573; *Odyssey* XIV: 21; XVII: 291–327). Among the many representations of dogs in hunting themes during the classical period, the following should be mentioned: the wall painting 'Dogs chasing a boar', the vase painting 'Hunter with his dog', the clay plaque 'Hunter with his dog', the sculpture 'Dogs attacking wild boar', and the fresco 'Hunting scene'. Besides pictorial representations, their contribution to hunting bestowed on dogs the praise of classical authors. Xenophon (431–350 BC) approached hunting dogs as one of the most important possessions worthy of adorning an estate. He further advised that 'dogs should be large, and they should have their heads light, short and sinewy; the lower jaw muscular; the eyes upraised, black and bright; the face large and broad' (*Agesilaus*, 9. 6). By the time of the historian and geographer Strabo (*c*. 64 BC– AD 25), hunting dogs were being exported from Britain to Rome, where they were greatly valued for their size. Britain was so famous for its mastiffs that Roman emperors appointed an officer on that island with the title of *Procurator cynegii*, whose sole business was to breed and transmit canines to the amphitheatre (Gray, 1989).

HUNTING IN THE MIDDLE AGES

The settlement of Germanic peoples in the Roman Empire throughout the fifth and sixth centuries gave further impetus to hunting, this activity still serving as a very important source of subsistence. Marc Bloch, who refused to see in hunting just a 'sport', has indicated both the economic value of meat – which was a primary element on the nobility's table – and the danger presented by wild beasts to a society whose daily life was very close to the woods (Bloch, 1968). Without neglecting the economic and security factors that supported hunting throughout the Middle Ages, one may also point to a 'socialization process', which turned hunting into a behavioural pattern that suited the upper class (Godman, 1990; Hen, 1995). Again, the emotional linkage between hunters and their dogs becomes manifest in the incorporation of the latter in burial gifts. Five graves from the Merovingian period that contain dogs, two of them wearing leather collars, were found at Vendel; the dogs were probably used by the deceased for hunting (Delort, 1983; Ohman, 1983). Beyond the customary level, the growing attachment to dogs also received the force of law. Throughout the sixth century, the Salic, Alemanic and Burgundian codes carefully established heavy fines for the theft or murder of dogs, according to their classification into the well-defined categories of hunting-dogs, pets and shepherd-dogs (*Pactus Legis Salicae*, Eckhardt, 1955: 132–4; *Leges Burgundiorum*, von Salis, 1892: tit. 97; *Leges Alamannorum*, Lehman, 1966: tit. 78). The Welsh laws of Hywel Dda, believed to have been written around 945, assessed the value of dogs according to their breed, age, training and function. No less important was the social status of their owners:

> The king's greyhound, if trained, is worth one hundred and twenty pence; if untrained, sixty pence; if a year old, thirty pence; if a cub in the kennel, fifteen pence; and from the time of its birth until it opens its eyes, seven pence half-penny. The king's lapdog is worth a pound; the lapdog of a freeman, one hundred and twenty pence; that of a foreigner, four pence; and a common house dog is of the same value as the latter. . . .Whoever possesses a dunghill dog, its value is four pence. A shepherd's dog shall go in the morning before the flock, and fetch them home at night. Its value is the same as a steer, which is perfect in all its parts. (*The Ancient Laws of Cambria*, 1823)

An indirect, but still remarkable, testimony of the widespread affective attitudes towards dogs took form in the growing ecclesiastical antagonism to dog-keeping, particularly among the clergy, which may or may not have been connected to the Church's opposition to the clergy's

Fig. 4.1. Representation of the many breeds of dog that existed in the Middle Ages. All of them are wearing collars of leather or metal; some are also wearing muzzles.

participation in hunting.[4] Merovingian Church Councils refer twice to dogs. The Council of Epaône (511 AD) declared that bishops and clerics were not allowed to keep dogs (and also hawks) for hunting (canon four) (*Decreti Prima Pars*, col. 126).[5] The Council of Mâcon (585 AD) provided an additional justification for the ban while dissociating dogs from hunting: a clergyman was forbidden to possess dogs, since people might then be afraid to approach his house (*domus ecclesiae*) (canon 13) (*Sacrorum Conciliorum ... Collectio*, col. 955). Both decrees were later incorporated into the so-called *Vetus Gallica* (canon 42: 1–2), which is the first systematic collection of Canon Law in the Merovingian period. Charlemagne also reiterated the prohibition of dog-keeping for hunting purposes by ecclesiastics (802 AD) (*Capitulare*, n. 33). On the other hand, we have rich testimony about the large number of dogs that accompanied the Frankish army and the many practical problems they caused.

The ubiquity of dogs in daily life, including both in ecclesiastical and in monastical circles, is amply corroborated by contemporary art:

heads of dogs became a very popular motive in Anglo-Saxon book illumination, and they were later adopted by the *scriptoria*. The latter produced decorated manuscripts in the Franco-Saxon style, which were thought to originate in the Abbey of St Amand. It is also possible that the impressive breed of St Hubert Jura Hound was developed in the monastery of this name located in the French border region of Switzerland during the eleventh century. Conciliar edicts, therefore, did little to halt the gradual inclusion of dogs in medieval society. On the contrary, social norms and legislation provide ample evidence of the presence and acceptance of dogs in both secular and ecclesiastical circles.

HUNTING AND ATTACHMENT TO DOGS

In all instances mentioned up to now, there is clear proof of the ubiquity of dogs in medieval society, notwithstanding the antagonism manifested by both theological dogma and ecclesiastical legislation. The question still remains: to what degree was the approach to dogs accompanied by elements of affection/devotion that were to characterize pet-keeping? Hunting was an important source of alimentation, and the involvement of dogs in this activity did not free them from the instrumental perspective that characterized the approach of medieval society to the animal kingdom as a whole. Though some seeds of change are discernible during the early Middle Ages, they are more evident towards the end of the thirteenth century, when hunting lost its primary role of subsistence – at least with regard to the upper sectors of society – and acquired all the attributes of a pattern of social behaviour that was used to differentiate between the nobility and the lower classes. The dissociation between hunting as a virtuous activity and its former economic function fostered the emergence of a new pattern of behaviour, which permeated the rich treasure of medieval myth. Gradually, hunting became listed among the virtues of popular heroes; St Eustace and St Hubert, Alexander, the Emperor Octavian, King Arthur, Charlemagne, Partonope of Blois, Sigfried and Tristan are only a few of the many hunters of worth and rank (Rooney, 1993). Alfred the Great, King of Wessex (871–99 AD), was said to be an expert and keen hunter at the age of 12, and he himself undertook the instruction of his falconers, hawkers and dog-keepers (Trew, 1940). Of Edmund, his grandson, it was written:

> When they reached the woods they took various directions among the woody avenues; and lo, from the varied noise of the horns and the barking of the dogs, many stags began to fly about. From these, the king, with his pack of hounds, selected one for his own hunting, and pursued it long

> through devious ways with great agility on his horse and with dogs following. (*Cotton MSS*, British Museum, Cleopatra B. 13)

One scene in the Bayeux Tapestry shows King Harold setting out with a hawking party accompanied by five dogs, all of which are collared.

The 'ennoblement' of hunting as a behavioural pattern that suited monarchs and nobles brought about the gradual welcome of dogs as companion animals in the upper ranks of medieval society. In order, though, to serve as a projection mechanism for this class, thereby preserving social stratification, a clear statutory barrier needed to be erected to obviate the indiscriminate ownership of hunting dogs. In 1016, during the reign of Cnute the Dane, the 'council' at Winchester made it a crime for poor people to keep greyhounds:

> No mean person may keepe any greyhounds, but freemen may keepe greyhounds, so that their knees be cut before the verderors of the forest, and without cutting of their knees also if he does abide 10 miles from the bounds of the forest. But if they doe come any nearer to the forest, they shall pay 12 pence for every mile; but if the greyhound bee found within the forest, the master or owner of the dog sall forfeit the dog and 10 shillings to the king. (Ash, 1927)

According to later authorities, the law called for a 'mean person' who possessed a greyhound to be fined. Officers of various ranks were appointed, without salary, to enforce the royal will. The fines went to the royal treasury, but the greyhounds themselves were taken away to be destroyed (Ash, 1933). An exception was made in the case of 'little dogges (all which dogges are to sit in ones lap), because in them there is no daunger'. To gain this exemption, a dog had to be small enough to be able to pass through a 'dog gauge', which was in the form of an oval ring, seven by five inches in diameter; an attached swivel allowed it to be hung from a girdle.

The Norman conquest of England brought about an extension of restrictive laws in the framework of the Forest Law. The King's Forest was primarily a legal and administrative concept that referred to areas placed under a special Forest Law, designed to preserve game and keep them for hunting (Postan, 1976). Whatever the regulations by which previous kings had protected their hunting, evidence that subjects were restricted or prevented from hunting on their own land dates back to 1066 (Reynolds, 1994). From the late twelfth century, especially in the period following Henry II's drastic deforestations, the matter of the Forest became a burning political issue (see, for example, the Assize of the Forest (1184) in Stubbs, 1929). Although the fines were changed from time to time, similar

Fig. 4.2. Veterinary attention to hunting dogs, with special attention to the mouth and legs.

principles appear again in the well-known Forest Law promulgated by Edward I (1278) (Ash, 1927). In the Kingdom of France, the Capetian kings lacked the power necessary to enforce regulations of this kind, but local lords took the initiative; for example, hunting was completely forbidden to the peasants of Bigorre by the beginning of the twelfth century (*Les fors de Bigorre, c.* 13). The ban was usually accompanied by heavy fines and ever-growing levies. Elsewhere in Europe, peasants were often obliged to provide food and shelter not only for authorized hunters but also for their dogs.

In applying the force of law to restrict hunting and the ownership of hunting dogs to the upper class, medieval kings indirectly bestowed a new meaning to hunting; it was detached from its former substantial essence and, instead, attached to the noble virtues of courage and bravery, supposedly characterizing a particular, well-defined class, the knighthood. Indeed, in 1394, Hardouin de Fontaines-Guérin wrote that all nobleman must be trained to hunt with dogs and to develop a love and mastery of the sport (lines 1–3). This new social meaning created a solid basis for a fresh approach to dogs, especially hunting dogs, which were transformed into a symbol of the knightly class. When this transition was made, dogs began to be treated with tenderness, devotion, one may say even brotherhood, as was suitable for members – though of an inferior status – of the same group. In this regard, one should bear in mind Savishinsky's (1983) observations regarding the use of pets as symbols of status in cultures with established patterns of domestication, particularly stratified societies, and how pets in the Middle Ages 'become a kind of living heraldry, an animated, almost totemic symbol of class and group identity' (pp. 116–17).

The transformation of dogs into companion animals among the nobility was accompanied by ever-growing expense for their care. King John of England, for instance, bestowed the manor of Baricote in Warwickshire to Boscher, a servant, for keeping a white bitch with red ears, to be delivered to the king at the end of the year, when another dog was to be sent into his care. In Hertfordshire, Hugh Pantulf held the manor of Stanford for keeping a greyhound for the king; and three castles in Monmouthshire were given to William de Breosa for ten greyhounds and other dogs. Payments of up to 100 shillings in cash were given to servants in Hampshire for similar purposes (Ash, 1933). Such expenditures were not exceptional; rather, they reflected the sophistication of hunting practices and the ever-growing cost of this pastime. Vast sums were spent on transporting the hunting party around the country and giving opulent gifts of hunting hounds and equipment. Members of the nobility kept huntsmen in their households to maintain their kennels and foresters to maintain their parks or forests.

In parallel, there was a conscious attempt to define more clearly the substance of hunting and its various components. The term 'hunting' itself was used to mean the noble, highly formalized art of hunting the beasts of venery – the hart, hare, boar and wolf – and the beasts of chase, which included lesser types of deer and the fox (Rooney, 1993). The noblest activity was considered the chase of the stag with horse and hounds, the season for this sport lasting between 3 May and 14 September. The number of hounds at relay posts in such a chase might vary from six to 12 (Thiébaux, 1967). Packs of hounds were also kept for other sporting purposes: spaniels for putting up birds for the hawks, greyhounds for the pursuit of hares and deer, and dogs of the mastiff type for hunting boar (Salzman, 1926). After the deer was captured and its corpse disemboweled, the hounds received their share: the heart, lungs, liver and windpipe were washed, cut into pieces, mixed with blood and bits of bread, and arranged on the hide. It was advised that each huntsman cut himself a switch to keep the feeding dogs from fighting over the hide. Jacques de Brézé (*c.* 1481) relates that when the stag had fallen dead, Madame descended from her horse and spoke to each of the hounds individually; she even refused to stop for food or drink herself until the hounds had been fed (lines 40, 42).

THE MEETING POINT BETWEEN HUNTING AND DOG-KEEPING: HUNTING TREATISES

The report of Jacques de Brézé hints at a new literary genre that developed in Europe from the thirteenth century onwards: hunting treatises, some of them in the form of illuminated manuscripts. This new genre provides a faithful expression of the gradual change in attitude towards dogs, which lost its former utilitarian character, especially among the nobility. Once hunting was turned into a battlefield between knight and beast – in which the former had to prove his bravery and courage in accordance with the elaborate behaviour code of chivalry – the way was paved for dogs to become 'man's best friend'; as such, they were accorded distinctive treatment, especially affection and tenderness.

The earliest treatise on hunting with dogs is *De arte bersandi*, whose anonymous author chose Guicennas, a German knight who lived during the reign of Frederick II, as his source of inspiration (pp. 5–6). Among the many pieces of advice offered to help his readers, themselves potential hunters, the writer emphasizes proper communication with dogs: 'You should speak kindly [with your dog], taking care of avoiding any anger against him, but to be as gentle as possible, to lay your hand on his head, and to kindly caress him'. Again, during the chase itself, the hunter must be 'amiable towards him, and kindly caress his head, and give him cheese, not much, but piece by piece . . .' (pp. 20–2). In another early example of the genre, *La Vénerie de Twiti*, the author, William Twich or Twety – King Edward II's *veneur*, who died before 1328 – strongly advises speaking gently with dogs, first and foremost during hunting. Thus, in order to encourage their speed, the hunter should call: '*en avant, sire, en avant!*'. If the dogs distanced themselves too much, their speed should be slowed by calling towards them: '*hou, mon ami, hou! doucement, mon ami, doucement!*' (p. 11).[6] Expressions such as 'sir', 'my friend' and 'kindly' gradually reduced the gap between human and animal while creating the illusion, if not the reality, of a fluent communication between the two partners in the chase.

The best known of medieval hunting treatises was the *Livre de la chasse* (1387–91), written by Gaston Phébus, Count of Foix, who owned

Fig. 4.3. Representation of a dog kennel in an aristocratic estate, showing the daily removal of hay to keep the kennel clean and the humidity low.

1600 hunting dogs and had a passionate love of venery.[7] Between 1406 and 1413, large parts of his book were translated into English in the celebrated *Master of Game* by Edward, second Duke of York (Hubbard, 1949). After speaking about 'the nature of beasts of venery and of chase which men would hunt', Gaston begins his description 'of the nature of the hounds which hunt and take them':

> And first of their noble conditions that be so great and marvellous in some hounds that there is no man can believe it, unless he were a good skillful hunter, and well knowing, and that he haunted them long, for a hound is a most reasonable beast, and best knowing of any beast that ever God made. And yet in some case I neither except man nor other thing, for men find it in so many stories and [see] so much nobleness in hounds, always from day to day, that as I have said there is no man that liveth, but must think it.

To this touching declaration, Gaston encloses historical testimony of the exceptional virtues of hounds from the times of Clovis, the first Frankish king, up to his own time. Thence, the unavoidable conclusion as to the many merits of this breed (pp. 106–12):[8]

> A hound is true to his lord and his master, and of good love and true. A hound is of great understanding and of great knowledge, a hound hath great strength and great goodness, a hound is a wise beast and a kind [one]. A hound has a great memory and great sense, a hound has great diligence and great might, a hound is of great worthiness and of great subtlety, a hound is of great lightness and of great perseverance, a hound is of good obedience, for he will learn as a man all that a man will teach him. A hound is full of good sport; hounds are so good that there is scarcely a man that would not have of them, some for one craft, and some for another. Hounds are hardy, for a hound dare well keep his master's house, and his beasts, and also he will keep all his master's goods, and he would sooner die than anything be lost in his keeping.

Following a detailed description of the different breeds, their nature, inclinations, and the most common sicknesses of canines, Gaston dispenses practical advice, such as the proper size for the kennel, its ventilation and daily cleaning (pp. 140–2). He further recommends that in order to cheer the dogs on, the owner should be well acquainted with their names and call them continually by name. He himself could name individual hounds, whose appellations he recorded, such as Bauderon, Baudellette, Bloquiau, Briffault, Cliquau, Fillette, Huielle, Huiiau, Loquebaut, Mirre, and Ostine (pp. 146–9). Dogs were occasionally given a voice, as Jacques de Brézé did in *Les dits du bon chien Souillard*. In that touching poem, written before 1490, the dog Souillard declares, in a chivalrous manner, his deepest love for his masters, three of whom were distinguished personages in the court of France, including the king himself, whom he had served on earth and sea (pp. 56–8).

The Spanish novelist José Ortega y Gasset (1964) interpreted the moment when people and dogs began to hunt as a team in terms of a defeat for human reason: 'This is the point at which the dog is introduced into hunting, the only effective progress imaginable in the chase, consisting, not in the direct exercise of reason, but rather in man's accepting reason's insufficiency and placing another animal between his reason and the game' (p. 455). Whether this was a conscious or unconscious stage, hunting became the mark of medieval nobility. The close linkage between medieval nobility and its pets was clearly reflected in the minstrel's innocent mourning of Garin le Loherain, the medieval epic hero, whose murdered body was surrounded by howling hounds: 'He was a gentleman and he was much loved by his dogs' (Paris, 1863, Vol. 2, p. 244).

The socio-economic, cultural equation of **nobility = hunting = dogs** was, therefore, consciously established to the point that it permeated the code of behaviour of the upper class. Moreover, European monarchs themselves became involved in the writing of hunting treatises, an interesting and perhaps unique feature of this literary genre. The first Spanish book that dealt with dogs, *Libro de la Montería*, was written by Alphonso XI, the Avenger, King of Castile and Leon, *c.* 1349; it focuses on the crucial importance of dogs in hunting. Two other works were also written or dictated by medieval monarchs: *Geheimes Jagdbuch* (*c.* 1499) by Maximilian I of Habsburg, the Holy Roman Emperor and Count of Flanders; and *La Chasse Royale* by Charles IX, King of France. The publication of these royal treatises, though, appears to have been delayed for a considerable period, for Alphonso's treatise was published only in 1582 and Charles' in 1625 (Hubbard, 1949).

The identification of dogs with hunting was further corroborated in pictorial representations. Most paintings/sculptures of dogs show them as predatory animals, whose speed in the chase and ferocity at the kill made them ideal companions for hunters (O'Brien, 1994). Nevertheless, a distinct separation between the medieval nobility and its dogs still existed, at least in theological terms. In his long series of prayers to God to save his body and soul and allow his entrance into Paradise, Gaston Phébus, notwithstanding his unconditional love for dogs, did not ask even once for divine mercy for these creatures (Tilander, 1975, pp. 27–69). Companion animal, yes; but man's best friend still could not enter celestial Jerusalem.[9]

The hunting treatises of the fourteenth and fifteenth centuries provide a vivid reflection of changing attitudes towards dogs. They hint at a meeting point where dogs were divorced from the animal kingdom to be allowed a respectable entry into the apex of human society, among the medieval nobility. True, the final goal was not the dog itself, but the success of his master in hunting. Still, after the dog had become a condition *sine-qua-non* for this success, it was dissociated from other animals and invested with a unique place in human society. Or, as James Serpell (1995) accurately expressed it, dogs were placed in 'the no-man's land between the human and non-human worlds' (p. 254).

CONCLUSION

As a projection mechanism, the different attitudes towards dogs throughout the Middle Ages hint at the changing needs and expectations that affected this society in a given time and space. From this perspective, the discovery of canine loyalty from the thirteenth century onwards is not fortuitous. Rather, it conjoined with the maturation of chivalry and its behavioural code, in which the fidelity of the knight towards his lord played a critical role. At this stage, the way was paved for medieval society to welcome Argos, Odysseus's dog (*Odyssey*, XVII: 291–327), and all his companions to become an integral part of human society and even a source of inspiration for the average human being. The welcoming of dogs into Christendom should be considered an additional manifestation of the process of change that advanced European society as a whole into the modern era. The gradual disintegration of the feudal regime, coupled with the transition to a commercial economy – the rules of which were unknown to many – opened the doors to dogs not only among the nobility, but also among the emerging bourgeoisie and the peasants. New challenges and expectations faced Western society on the eve of the Modern Period; but human beings were no longer alone. They had paved the way for their 'best friend' to be at their side as the traditional framework of medieval society was breaking apart.

NOTES

1. This chapter is dedicated to my dog, Rocki, whose beloved memory remains with us.
2. Medieval practical manuscripts, lavishly illustrated, depict only men, hounds and their quarry. John I of Portugal's *Livro da montaria*, devoted largely to the boar hunt, alludes nowhere to women; nor does Alfonso XI of Castile, in his descriptions of week-long pursuits of the bear (Cummins, 1988).
3. By domestication, we mean the process not only of rendering animals under human control but also of attaching them to the *domus*. The verb 'to domesticate' embraces the concept of civilizing and naturalizing; namely, to introduce an animal to a place where it is not indigenous in order to share intimacy with its keepers.
4. In the Bible, hunting is practised by sinners (like Nimrod and Esau), thereby justifying its association with sin (*Genesis* 10: 9, 25: 27).
5. The *Decretum* of Gratian reinforced the biblical ban, establishing the general sinfulness of the hunter and forbidding hunting and hawking to the clergy; it further included a scheme of penances to be enjoined on the perpetrators of such crime (*Decreti Prima Pars Dist.* 86, *c.* 8–9, in *ibid.*, cols. 299–300). But Canon Law did not change rooted behaviour patterns, and the more worldly members of the clergy continued to engage in hunting to the point that the hunting cleric became a common satirical figure.
6. Although the manuscript also exists in English, the original language is difficult to establish.
7. Born on 30 April 1331, Gaston represents the archetype of a medieval noble: married to Agnes, the daughter of Jeanne de France, Queen of Navarre, and Philip, Count of Evreux, he was a very handsome man, whose sex appeal was further strengthened by his learning and knightly behaviour.
8. The English translation is from *The Master of Game by Edward, Second Duke of York* (Baillie-Grohman, & Baillie-Grohman, 1909).
9. 'For without are dogs, and sorcerers, and whoremongers, and murderers, and idolaters, and whosoever loveth and maketh a lie'. *Rev.* 22: 15.

REFERENCES

Ash, E.C. (1927). *Dogs: Their History and Development.* London: Ernest Benn Ltd.
Ash, E.C. (1933). *The Book of the Greyhound.* London: Hutchinson & Co.
Atlas, J. (1982). The human/companion animal bond: interaction between man and animal. *Psychohistory News,* **2**, 8.
Augustine. *Confessions,* trans. H. Chadwick (1991). Oxford: Oxford University Press.
Baillie-Grohman, W.A. & Baillie-Grohman, F. (eds.) (1909). *The Master of Game by Edward, Second Duke of York.* London: Chatto & Windus.
Batany, J. (1984). Animalité et typologie sociale: quelques parallèles médiévaux. *Cahiers d'études médiévales,* **2**, 41–54.
Bede. Hexaemeron, *Patrologia Latina,* vol. 91.
Bede. Vita *Sancti Cuthberti,* ed. and trans. B. Colgrave (1940). Cambridge: Cambridge University Press.
Biese, A. (1905). *The Development of the Feeling for Nature in the Middle Ages and Modern Times.* New York: Burt Franklin.
Bloch, M. (1968). *La Société Féodale.* Paris: Albin Michel.
Bökönyi, S. (1974). *History of Domestic Mammals in Central and Eastern Europe.* Budapest: Akadémiai Kiadó.
Bunch, R. (1982). The human/companion animal bond: interaction between man and animal. *Psychohistory News,* **2**, 9.
Capitulare missorum generale. ed. A. Boretius (1883). In *Monumenta Germaniae Historica, Capitularia Regum Francorum,* vol. 1. Hannover.
Clutton-Brock, J. (1987). *A Natural History of Domesticated Mammals.* London: British Museum and Cambridge University Press.
Clutton-Brock, J. (1997). Hard hunting and heavy petting. *Times Literary Supplement,* **20 June**, 6.
Cotton MSS. British Library, Cleopatra B. 13.
Cummins, J. (1988). *The Hound and the Hawk: the Art of Medieval Hunting.* London: Weidenfeld and Nicolson.
Decreti, in *Corpus Iuris Canonicum,* ed. Friedberg, 2 vols. vol. 1. Leipzig, 1897.
Delort, R. (1983). La saga du chien. *L'histoire,* **62**, 48–60.
Dronke, P. (1983). La creazione degli animali. In *L'uomo di Fronte al Mondo Animale nell' alto Medioevo.* pp. 809–42. Spoleto: Settimane di Studio del Centro Italiano sull'alto Medioevo 31 (1985).
Eckhardt, K.A. (ed.) (1955). *Pactus Legis Salicae.* Göttingen: Musterschmidt Verlag.
Gaston Phébus. *Livre de chasse,* ed. Gunnar Tilander (1971). *Cynegetica* 18. Karlshamn.
Gaston Phébus. *Livre des oraisons – Les prières d'un chasseur,* ed. Gunnar Tilander (1975). *Cynegetica* 19. Karlshamn.
Godman, P. (1990). The poetic hunt. In *Charlemagne's Heir,* ed. P. Godman & R. Collins, pp. 565–89. Oxford: Oxford University Press.
Gray, E.A. (1989). *Dogs of War.* London: Robert Hale.
Grossman, L. (1993). *The Dog's Tale: a History of Man's Best Friend.* London: BBC Books.
Guicennas. *De arte bersandi: Le plus ancien traité de chasse de l'Occident.* ed. Gunnar Tilander (1956). *Cynegetica* 3. Upsala.
Hardouin de Fontaines-Guérin. Trésor *de vènerie* (sic!), ed. H. V. Michelant (1856). Paris.
Haudricourt, A.G. (1962). Domestication des animaux, culture des plantes et traitement d'autrui. *L'homme,* **2**, 40–50.
Hen, Y. (1995). *Culture and Religion in Merovingian Gaul A.D. 481–751.* Leiden: Brill.
Hubbard, C.L.B. (1949). *An Introduction to the Literature of British Dogs.* Ponterwyd.
Jacques de Brézé. *La Chasse – Les Dits du Bon Chien Souillard.* ed. Gunnar Tilander (1959). *Cynegetica* 6. Lund.

La Vénerie de Twiti: Le plus ancien traité de chasse écrit en Angleterre. ed. Gunnar Tilander (1956). *Cynegetica* 2. Upsala.
Laplanche J. & Pontalis, J.B. (1984). *Vocabulaire de la psychanalyse.* Paris: Presses Universitaires de France.
Lehman, K. (ed.) (1966). *Leges Alamannorum. Monumenta Germaniae Historica, Leges Nationes Germanicarum,* Vol. 5–1. Hannover.
Les Fors de Bigorre. ed. J. Fourgous & G. de Bezin (1901). Bagnères.
Menache, S. (1997). Dogs: God's worst enemies? *Society and Animals,* **5**, 23–44.
Menache, S. (1998). Dogs: a story of friendship. *Society and Animals,* **6**, 67–86.
Muratova, X. (1977). Adam donne leurs noms aux animaux: l'iconographie de la scène dans l'art du moyen âge – les manuscrits des bestiaires enluminés du XIIe et du XIIIe siècles. *Studi Medievali,* **18**, 367–94.
O'Brien, C. (1994). *Dogs in Art.* London: Studio Editions.
Ohman, I. (1983). The Merovingian dogs from the boat-graves at Vendel. In *Vendel Period Studies,* ed. J.P. Lamm & H.A. Nordstrom, pp. 167–81. Stockholm.
Origen. *Contra Celsum,* ed. H. Chadwick (1965). Cambridge: Cambridge University Press.
Ortega y Gasset, J. (1964). A veinte años de caza Mayor del Conde de Yebes. In *Obras Completas de José Ortega y Gasset,* Vol. 6, *Brindis y Prólogos,* 6th edn, pp. 450–81. Madrid: Alcazor.
Paris, P. (ed.) (1983). *Etude sur les chansons de geste et sur le Garin le Loherain de Jean de Flagy.* Paris: C. Dauniol.
Payson Evans, E. (1896). *Animal Symbolism in Ecclesiastical Architecture.* London: W. Heinemann.
Perin, C. (1981). Dogs as symbols in human development. In *Interrelations between People and Pets,* ed. B. Fogle. pp. 68–88. Springfield, IL: Charles C Thomas.
Peter Comestor. *Historia Scholastica, Patrologia Latina,* Vol. 198.
Phineas, C. (1973–4). Household pets and urban alienation. *Journal of Social History,* **7**, 338–43.
Postan, M.M. (1976). *The Medieval Economy and Society.* Harmondsworth: Penguin Books.
Reynolds, S. (1994). *Fiefs and Vassals: the Medieval Evidence Reinterpreted.* Oxford: Clarendon Press.
Ronecker, J.P. (1994). *Le Symbolisme Animal: Mythes, Croyances, Légendes, Archetypes, Folklore, Imaginaire.* Saint-Jean-de Braye: Dangles.
Rooney, A. (1993). *Hunting in Middle English Literature.* Cambridge: Boydell.
Sacrorum Conciliorum Nova et Amplissima Collectio. ed. G. Mansi (reprint Graz, 1960–61), Vol. 9.
Sahlins, M. (1976). La pensée bourgeoise: Western society as culture. In *Culture and Practical Reason,* ed. M. Sahlins. pp. 70–9. Chicago and London: University of Chicago Press.
Salzman, L.F. (1926). *English Life in the Middle Ages.* Oxford: Oxford University Press.
Savishinsky, J.S. (1983). Pet ideas: the domestication of animals, human behavior, and human emotions. In *New Perspectives on Our Lives with Companion Animals,* ed. A.H. Katcher & A.M. Beck. pp. 112–31. Philadelphia: University of Pennsylvania Press.
Serpell, J. (1995). From paragon to pariah: some reflections on human attitudes to dogs. In *The Domestic Dog: Its Evolution, Behaviour and Interactions with People,* ed. J. Serpell. pp. 245–56. Cambridge: Cambridge University Press.
Stubbs, W. (1929). *Select Charters and Other Illustrations of English Constitutional History.* Oxford: Clarendon Press.
Tertullian. *Liber de Resurrectione Carnis, Patrologia Latina,* Vol. 2.
The Ancient Laws of Cambria (1823). Trans. William Probert London.

Thiébaux, M. (1967). The mediaeval chase. *Speculum*, **42**, 260–74.
Thomas, K. (1984). *Man and the Natural World: Changing Attitudes in England, 1500–1800.* Harmondsworth: Penguin Books.
Trew, C.G. (1940). *The Story of the Dog and His Uses to Mankind.* London: Methuen.
von Salis, L.R. (ed.) (1892). *Leges Burgundiorum. Monumenta Germaniae Historica, Leges Nationes Germanicarum,* Vol. 2–1. Hannover.
Zeuner, F.E. (1954). Domestication of animals. In *A History of Technology*, ed. C. Singer, E.J. Holmyard & A.R. Hall. pp. 327–31. New York and London: Oxford University Press.

ERICK L. LAURENT

5

Children, 'insects' and play in Japan

INTRODUCTION

From spring until late autumn each year in the countryside of Japan, many children (mostly boys) try to catch insects (*mushi*) using an assortment of tools. Most rural schools still ask their eight- to ten-year-old pupils, as homework during the summer vacation, to prepare a collection of insects.[1] Books for children that deal with *mushi* are numerous and the information presented is very detailed. Children also seem to be the main target of the activities of the Japanese 'insectariums' (Yajima, 1990). Insects seem to be the first animals with which a child plays and by which he learns concretely about nature. These facts tend to show that in Japan *mushi* belong mainly to the child's world.

This chapter deals with the cultural status of the insects, or rather the 'ethnocategory' known as *mushi* to be specific (Laurent, 1995, 1998), in the world of the child in Japanese culture. The following questions will be addressed: What is the importance of *mushi* in the child's world, especially in teaching children about nature? What is the status of *mushi* for the child – toy, playmate, or even pet?

METHODS

In order to answer these questions, three kinds of research were undertaken. Firstly, the author acted as a participant observer during fieldwork in rural areas.[2] Here, children's activities involving *mushi* were recorded: catching them, playing with them, observing them, listening to them, breeding them, singing songs about them, and collecting them for homework. These activities constitute the main data for this research.

Secondly, an analysis of the content of children's educational books about nature in general (in which the part devoted to *mushi* is sizeable) or about *mushi* in particular was carried out.

Lastly, interviews at the 'Japan Pet Fair' with representatives of companies selling *mushi* as well as tools and equipment for breeding were conducted, along with an analysis of the contents of their leaflets.

RESULTS

Mushi in Japanese culture – economic aspects

Economic importance of mushi *in Japan*

The economic importance of *mushi* in Japan can be seen through the selling of *mushi* and the assortment of implements to catch and breed them, in department stores and post offices. The species concerned are mainly of two types: (1)'autumn singing insects', crickets (*matsumushi*,[3] *suzumushi*[4]), or, less often, the singing grasshopper (*kirigirisu*[5]) bought to bring an atmosphere of summer or autumn into the house; and (2) the rhinoceros beetle (*kabutomushi*[6]) or stag beetle (*kuwagatamushi*[7]) for children to play with and breed.

The preparation and selling of edible *mushi* also represent a not insignificant part in the local economy of some prefectures in central Japan (Nagano, Gifu), as these *mushi* are now considered typical souvenirs from those regions. Even if the catching of some of these species sometimes involves children, it can be considered an exception and these activities are never viewed as 'play' as such.

The silkworm is another insect that used to bring wealth to relatively poor rural areas of Japan. It has, however, become less important nowadays in terms of the number of people involved, as well as of the income generated. Children used to play an important part in the various activities of breeding, sometimes even being allowed one or two days' holiday from school in order to help with the work.

Historical perspective of the autumn singing insects

Since the Heian period (794–1185 AD) and still today in the countryside, 'autumn singing insects' in cages are considered a welcome present to be brought into the house and listened to.

The first reference in the literature to the selling of 'autumn singing insects' (crickets) seems to date back to 1685 in Kyôto.[8] These crickets were sold in big square baskets suspended by a pole put on the shoulders, and on top of which were placed smaller cages sold together with the *mushi* (Matsuura, 1983). At first, the *mushi* were caught directly from nature, but as their popularity increased and their marketing progressed, crickets began to be bred by the end of the eighteenth century (Sakamoto, 1957, 1976; Matsuura, 1983; Kanô, 1990). Around 1820, the first strolling cart *yatai* appeared, pushed along the roads by fishermen or peasants.[9] From the Meiji period (1867–1912), the market tended to be held in permanent places called '*mushiya*', '*mushi* shops', where several species of *mushi* (e.g. 'autumn singing insects', fireflies and jewel beetles), as well as cages[10] and small tools related to their catching and breeding were sold. Lafcadio Hearn (1898, 1987) gives a price list for 1897 of the 12 most popular species, as well as a list of the best places to hear the different species of *mushi*. The decline in the selling of *mushi* began in the 1930s and the *mushiya* (or such shops) almost completely disappeared after World War II (Kanô, 1990).

The selling of mushi *in present-day Japan*

The species of the 'autumn singing insects' sold in contemporary Japan are the singing grasshopper *kirigirisu* and the bell cricket *suzumushi*.[11] Each year, *suzumushi* can be found in nearly every post office[12] in Japan (Fig. 5.1). From July onwards, they can be sent all over the country in special packages.

General department stores, however, remain the best places to buy *mushi*. Every summer, department stores nationwide devote a significant portion of their seasonal goods display to *mushi*. About half of this space contains several types of insecticide,[13] the remaining half being devoted to *mushi* themselves, breeding tools and apparatuses. The income from the selling of the *mushi* alone for the entire country climbed to 5 billion yen[14] for the year 1992 (*Asahi Shimbun*, 8.7.1993).

Let me give, as an example, the list of *mushi* and goods related to *mushi* sold in July 1996 in one of Kyôto's biggest department stores.[15]

a. Rhinoceros beetles and stag beetles.
Collecting material:[16] nets (one size, three different colours) (¥800); plastic boxes in the shapes of cameras or radios (to be slung across the shoulder) (¥480).
Breeding material: set for breeding beetles (terrarium containing pieces of wood, 'mattress',[17] food, trough) in three sizes (¥500–1000–1800); plastic terrariums in six different sizes, with hemstitched covers (¥250–880–1280–1980); two types of 'mattress' (¥400) – 'insect mattress' and 'food-mattress'; pieces of wood of different sizes (from ¥200 to 700); three types of food: powder (¥700 for

(*a*)

(*b*)

(*c*)

Fig.5.1. Examples of post office advertising about *suzumushi* crickets.

three packs of 25 g), red and orange jelly (¥480 for ten packs of 10 g), liquid in plastic bottles called 'insect-water' (¥700) or syrup (*mitsu*) (¥300) specifically designed for rhinoceros beetles only.

Living *mushi*: rhinoceros beetles (¥400–1300–1600), stag beetles (¥400–1000–1800–2300).

b. For *suzumushi* crickets.[18]

Collecting material: none.

Breeding material: plastic cages with hemstitched covers in two sizes (¥350–700); wooden cages (¥650); two types of sets for breeding crickets, decorated with bridges, moss and plants (terrarium containing pieces of wood, litter, food, trough) (¥1100–4300); 'mattress' (¥350); pieces of wood of two sizes (¥350–380);[19] two types of food: powder (¥200–250), liquid in plastic bottles called 'insect-water' (¥300) or syrup (¥300).

Living *mushi*: adults (¥120–180), larvae (¥300 for 10, ¥550 for 20, ¥800 for 30).

To a lesser extent, rural areas seem to follow the same pattern. With the arrival of summer, each village's shop display abounds with butterfly nets, rhinoceros beetle cages, and packs of 'mattress' for stag beetles (Fig. 5.2).

The selling of rhinoceros beetles in department stores seems to have originated in the 1960s, the period of a so-called 'rhinoceros beetle boom' among children, especially boys, which was extensively covered in the mass media of the time (Kaneko *et al.*, 1992). From then on, the rhinoceros beetle as well as the stag beetle, which seems to have experienced a similar popularity boom, won first place among the *mushi* sold on the Japanese market. Every summer, several millions of them are sold.[20] In order to satisfy this need, rhinoceros beetle breeders started to appear. People began to breed rhinoceros beetles in order to sell them to post offices or department stores. For instance, the 'Midori' city park of Yûra in Wakayama prefecture started in 1987 to breed rhinoceros beetles on a 300 m^2 area, and it sells thousands of them each year to the department stores of the city. The money is used to restore the park (*Hidaka Nippô*, 3.7.1991). In some places, the soil is even warmed in order to activate the adult moulting so that the adults can be sold sooner.[21]

I interviewed a breeder who lives with his wife in the town of Takatô (Nagano prefecture). He had been cultivating mushrooms since 1967, but began to 'convert' to rhinoceros beetle breeding in 1985, on the advice of a friend who said he could make a lot of money from that. Even though there was only one place in the nearby city of Ina where his *mushi* were

(a)

(b)

Fig. 5.2. Display of goods related to *mushi* (Kyôto, 1992).

(*c*)

(*d*)

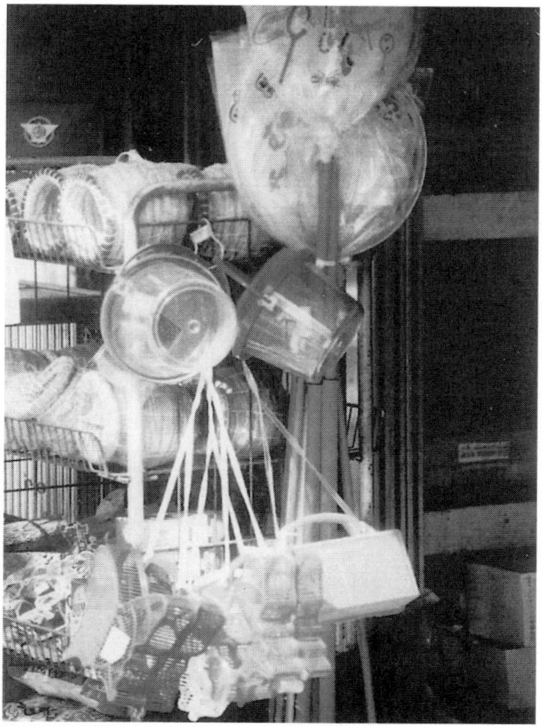

sold, the business was prosperous and he has expanded the area devoted to rhinoceros beetles year by year. The orders come from the local post office,[22] the department stores of Ina, as well as from people that come directly to the farm.

Technically speaking, one could say the breeding of the rhinoceros beetle is a 'semi-natural' process. Rhinoceros beetles (1000 per 3.3 m^2) are placed in fenced enclosures full of humus, bits of wood and leaves. They lay eggs at the end of August, and the adult moulting takes place the following summer. In between, the natural process is left to take place by itself. The conditions of the substratum (humidity, thickness, and so on) are carefully monitored but this does not require constant care. 'It's not a very hard job, quite suitable for old people' says the Takatô breeder. However, human intervention is needed in order to prevent disease transmission and genetic malformations. Thus, every two or three years, 'new blood' is introduced by bringing new, wild beetles into the breeding population. In this way, 4000 rhinoceros beetles are bred for the market each year.

More so than for the breeding of mammals or birds, the success of breeding insects is uncertain and cannot on its own bring a steady income. 'Entomo-zootechny' is not yet very well developed as a branch of knowledge, and individual breeders have to adapt their methods to each new situation they face.[23] In addition, insects do not have a long history of domestication (apart from bees and silkworms) with human intervention to improve the species. They are thus quite fragile, dependent on strict *eco-preferanda*, and sensitive to small environmental changes or imposed behavioural variations.

Teaching about nature and the seasons

Each species of *mushi* is linked with (sometimes used as a symbol for) a particular period of the year, and related to a particular time of the day, as well as to particular methods for catching them and particular modes of play.

One can find in academic literature concerning folklore, lists of traditional seasonal games in nature linked to animals and plants[24] (see also Iijima, 1991). From these lists (hereafter referred to as A and B), it is clear that the games that concern *mushi* parallel the seasons.[25] In the spring, there is 'butterfly catching' (List B). In the summer, there is observation and play with the aquatic *mushi* 'whirligig beetle and true water beetle' (List A), and the 'snail' (A), 'firefly catching' (A, B), 'cicada catching' (A, B), play with '*onimushi*' (literally 'demon-*mushi*', rhinoceros beetle and stag beetle) (A), '*onimushi* fighting' (A), 'dragonfly catching' (A, B), 'play with dragonfly' (A); play with the 'ant-lion' (A), 'chase of the *batta* locust' (A), the 'collecting of *tsuchibachi* wasp's larvae' (A), and the 'making of *mushi* cages in straw' (A). In the autumn, there is 'listening to singing insects' (A), observation and play with the 'red dragonfly' (A), the 'collecting of the *inago* locust' (A), 'spider' fighting (A), observation of the 'mantis' (A), and the 'collecting of wasp nests' (B). In the winter, nothing is mentioned concerning *mushi*.

From my personal observations and an analysis of recent educational books concerning *mushi* (see below), during the late 1980s and the early 1990s, it is apparent that some of the activities have simply disappeared, or disappeared as children's play. For instance, one can no longer speak of the collecting of wasp larvae or locusts or even the listening to singing insects as typical child activities.[26] The chase of the *batta* locust as well as the making of *mushi* cages[27] seem to have almost disappeared (Laurent, 1997).

Some of the activities, such as the traditional catching of dragonflies or spider fighting, were still popular some years ago. Activities such as observing and playing with aquatic *mushi*, snails, and rhinoceros beetle fighting, and cicada and butterfly catching still occur to some extent in the countryside but much less so in urban areas.

One cannot speak of the creation of new games as far as *mushi* are concerned. However, new species have entered the universe of children's play, possibly influenced by educational books or television programmes, for example the woodlouse, ladybird and weevil. Moreover, the popularity of certain species has changed; nowadays the rhinoceros beetle and the stag beetle seem to be the most appreciated by Japanese children. These preferences are certainly influenced by fashion but are also determined by the availability of the species in nature or in shops. Lastly, the status of certain species has also changed. For example, catching a firefly is not as easy as it used to be some years ago, because it is now officially protected. In certain places, firefly catchers are reported to the police and may receive a relatively heavy fine.

Educational books and personal experience

In Japan, there are many educational books that aim to introduce *mushi* to children. They use a mixture of cartoons, pictures, didactic drawings and diagrams. They explain how to breed them, why (that is, in which ways) they are interesting, or sometimes they simply show the child some games involving certain species of *mushi*. Recent educational

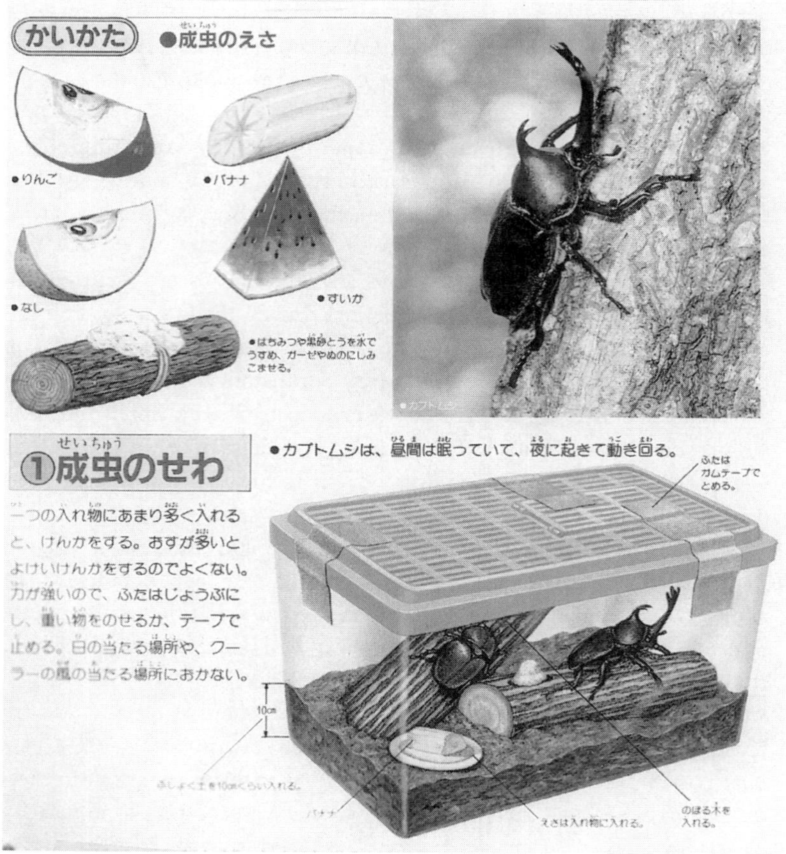

Figs. 5.3 & 5.4. Educational books about *mushi*: the case of the rhinoceros beetle.

books concerning *mushi* are almost always constructed according to the same general pattern.[28] To follow are some examples of these types of book.

For each of the 36 *mushi* described in *The Mushi, Life and Breeding* (Miyatake, 1983, 1989), first the ecology and development are explained, then the different methods (often very specific) of catching the *mushi* are made known. Finally, the method for breeding the *mushi* at home is presented in detail (material required, type of food, etc.). The only *mushi* presented which are not insects are spiders. For instance, the headings concerning the rhinoceros beetle (Figs. 5.3 and 5.4) are as follows: 'Way of life', 'Ways of breeding' ('Care of the adult', 'Care of the eggs', 'Care of the larva', 'Care of the chrysalis'). As can be seen, the text concerning this

Figs. 5.3 (*cont.*)

mushi has a very practical aim: breeding. The majority of the chosen species (83.3%) follow this trend, but the example of the cricket (*kôrogi*) shows us that this is not always the case. The headings are 'Way of life' and 'Observation'.

All the 31 *mushi* covered in *Collecting Mushi* (Saitô, 1990) are insects. For each of these, similar information is given, the accent being on information for breeding (83.9% of the items, less practical though than in the previous book), and colour photographs.

In the *Great Dictionary of Play* (Masuda, 1989, Vol. 2: 375–80), intended mainly for teachers and educators, we find in the section entitled 'Catching' (*toru*), technical instructions concerning the collecting of different sorts of plants, animals and minerals. Among the activities presented, two are concerned directly with *mushi*: 'Catching the cicadas' and

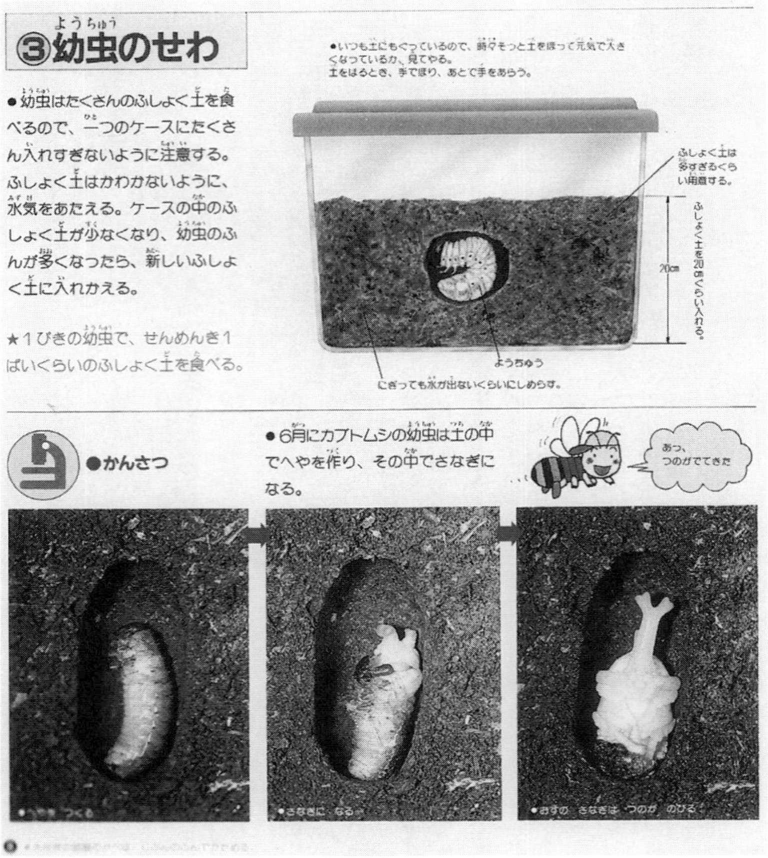

Fig. 5.4.

'Catching the dragonfly'. Others, such as catching spiders, locusts, fireflies and butterflies, are only mentioned without explanation. In the section entitled 'Breeding' (*kau-sodateru*), we find three species of *mushi*[29] whose methods of breeding are explained: the cricket *suzumushi*, the snail and the rhinoceros beetle. Two others are briefly referred to because their breeding method is very similar to that of the previous species: the *kôrogi* cricket and the stag beetle.

Through these interactions with *mushi*, the child undergoes a psychological experience that can be interpreted as the foundation of 'being taught about nature'. For instance, raising and observing *mushi* (as homework during summer) always means keeping records, in the form of a diary, including personal comments and sketches. In this regard, the leaflet from the post office that advertises the selling of *suzumushi* crickets

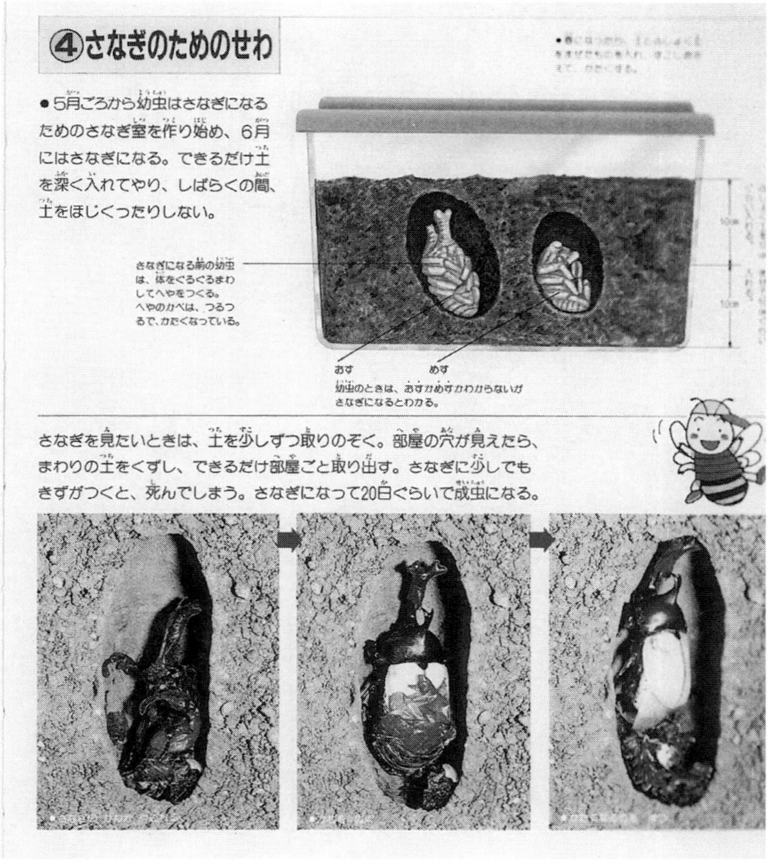

Fig. 5.4. (*cont.*)

(see Fig. 5.2) tells children: 'while observing, let's write a diary with pictures!' *Mushi* are usually looked after every day by the child and kept until they die. Their death, however, seems to give rise to a relative indifference from children.

As we have seen, children can engage in many activities with *mushi*. Below are details of a few of these.

Dragonfly catching The first evidence of dragonfly catching (*tombo tori*) dates back to the middle of the Edo period (i.e. the first half of the eighteenth century) (Uemura, 1990). Dragonflies can be caught by hand or with a net, but the traditional way of catching them is with a tool called a *buri* in western Japan or a *toriko* in Tôkyô (Fig. 5.5). The tool is made of two small balls, stones or shells wrapped in cloth or paper (preferably red) and tied to a silk thread. The tool is thrown in the air towards a dragonfly; its

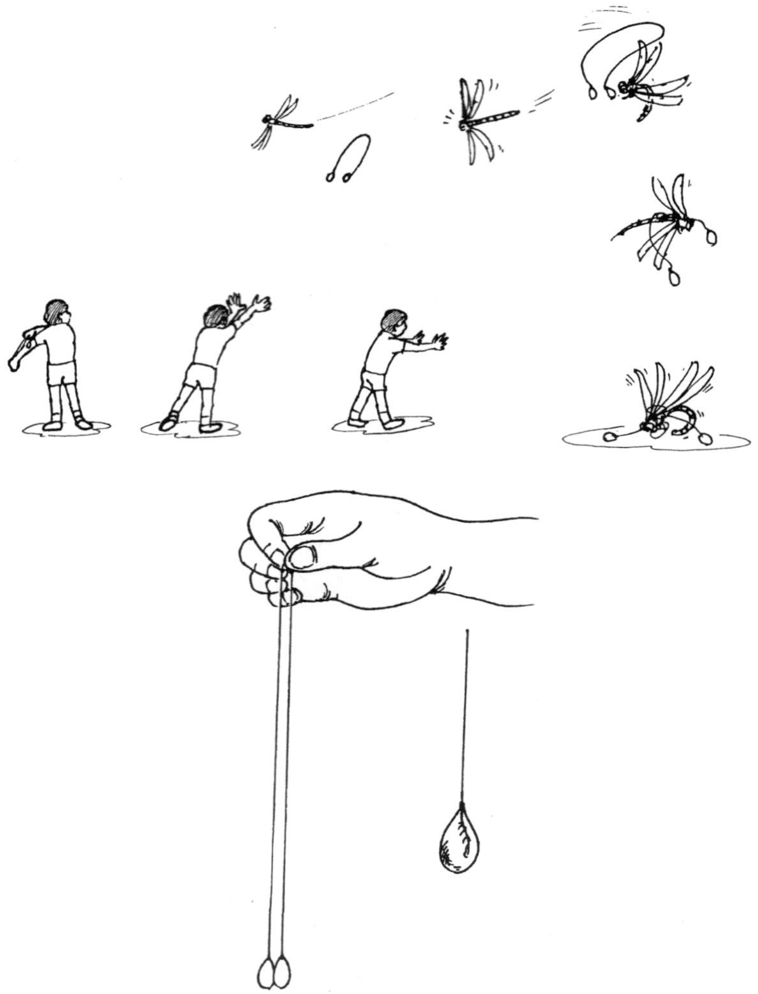

Fig. 5.5. Dragonfly catching with a *buri* or *toriko* (drawing by Haruji Deguchi).

wings get tangled in the thread and it falls to the ground. The right hand holds the thread in the middle and the left hand holds the balls. So that the balls are not hurled first when thrown, both hands are to be projected in the air at the same time, the right (i.e. the thread) being the first.[30] All the difficulty lies therein. The tool is to be thrown about 1 m ahead of the dragonfly so that it flies into it. Although very popular amongst boys until the 1960s in the Japanese countryside, dragonfly catching does not seem of much interest to present-day children.

Spider fighting Traditionally, spider fighting,[31] at an organized level, occurred as a popular gambling amusement among fishermen along the coasts of Kyûshu, Shikoku as well as the Pacific coast of Honshu from Hiroshima to Chiba. Nowadays, fights can only be seen as religious rituals or annual events, occurring on fixed days in Kagoshima and Kôchi prefectures (Kawana &Saitô, 1985).

Another type of spider fighting,[32] was very popular nationwide until the 1960s or 1970s as a boys' game. Boys make spiders fight by throwing one in another spider's web or by putting two spiders together in the same small-scale arena demarcated by pieces of wood. The spiders were abandoned after the fight and were never bred. We can find the same type of game in 1930's Yokohama in Otani (1958): every 10- or 12-year-old boy used to possess his own spiders that were sometimes exchanged for sweets. The spiders were kept in small boxes and known individually by the boy-owner. The fights usually continued until one of the spiders died.

Activities with the rhinoceros beetle and the stag beetle These are certainly for children the two most popular and best-selling *mushi* in contemporary Japan. Two major traits that seem to account for their popularity are their inoffensive character as well as their spectacular appearance. The black and horny rhinoceros beetle is indeed one of the biggest coleoptera. One of the first references to play with the rhinoceros beetle is found in Kurimoto (1811), who wrote about the breeding of the larvae, and about children pulling small carts into which adult rhinoceros beetles had been placed. The same type of game still occurs in the Japanese countryside today. More often, little boys make the rhinoceros beetles fight each other for a small piece of watermelon, or make them pull weights.

Catching cicadas In *Megane*, Toson Shimasaki (1913, 1974: 77–80) mentions the catching of a *minmin* cicada[33] by hand. The mischievous child speaks to the cicada in these words: 'If you don't fly away, I'm gonna catch you'. Then, the cicada flies away making a loud noise with its wings, to which Shimasaki applies the onomatopoeia *bata bata bata*. From spring to autumn, the cicadas are perhaps the *mushi* most often caught in Japan. Being everywhere, even in the centre of cities, clinging to tree trunks or walls, they do not constitute a very difficult prey animal for children equipped with nets and cages, and who can climb trees. What seems of most interest to them is the cry[34] cicadas make when caught, as well as the tricks used for catching them.

Activities with other species The great number of names given by children to a particular species can be seen as an indicator of their interest in that species. In this respect, the snail seems to be one of the most popular. Its common and/or regional names (e.g. *dedemushi, katatsumuri*)[35] – whose syllabic redundance tends to show their belonging to the world of children – are explicit references to a snail's habit of protruding and retracting its horns, to the shape of its shell, and its sluggish appearance. Some of the games that children play with snails and other *mushi* in present-day Japan include: the snail, locust or woodlouse race; whirling *kanabun*[36] from the end of a rope; ant-lion and mantis or spider feeding.

Children's songs: mushi *as playmates*

Fireflies Most Japanese children's songs about *mushi* are related to fireflies (Iijima, 1991). They also seem to be the most ancient ones (Minami, 1961, 1983). However, in 1913, Kawaguchi already complained that firefly catching had become silent. The most famous song seems to be the generically titled '*hotaru koi*' ('Come, firefly!'), very often sung by Japanese children during the summer nights when firefly catching occurs. Though one can find many regional versions,[37] here follows what I would consider to be the most common form:[38]

> Fi-, fi-, firefly come!
> Fi-, fi-, firefly come!
> Water there is bitter,
> Water here is sweet.
> Fi-, fi-, firefly come . . .

Snails (dendenmushi) All Japanese children know this song about snails:[39]

> Snail, snail
> Where is your head?
> Stick out your horns, stick out your spear, stick out your head.

Butterflies (chôchô) This song of the butterfly is also very popular:[40]

> Butterfly, butterfly,
> You stop on a colza leaf
> You grow tired of the colza
> You stop on a cherry
> From cherry flower to cherry flower
> Stopping and having fun
> Having fun and stopping.

Autumn singing insects Some songs about autumn singing insects use the onomatopoeia of their 'voice'. For instance, the following song about a cricket:[41]

> The *suzumushi* cricket sings
> Rin rin rin rin rin rin rin

Singing through the long autumn nights
How interesting, the voices of *mushi*!
The *matsumushi* cricket also started to sing
Chinchiro chinchiro chinchirorin
Singing through the long autumn nights
How interesting, the voices of *mushi*.

Calling songs Among children's songs about *mushi*, some are sung to the animals, in order to catch them (firefly, dragonfly . . .) or to play with them (snail . . .). Apart from the cases of the firefly, dragonfly and ant-lion, these are not really used in this sense any longer. Having lost their function, they became just songs of *mushi*, learnt at school.

Kawaguchi (1913: 339), remembering his childhood, writes that 'fireflies would easier come toward us if we sang rather than kept silent'. The following poem expresses a similar sentiment. Jôsô[42](1661–1704):

Beckoning voices
Have stopped.
Fireflies, by the thousands.

Here follows a song, collected in Wakayama prefecture (Ekawa village), to call the *minomushi*[43] out of its sheath, sung while rolling it between the fingers:[44]

Grandfather, grandmother [*twice*]
Grandfather is back, open the door!
Grandfather, grandmother [*twice*]
Grandmother is back, open the door!

Another song from Ekawa village voices a clear threat to the *mushi* if it does not come out: '*Minomushi*, get out, come. If you don't, I will shorten your bottom!' In the same respect, Ishigami (1983) gives (from Okayama prefecture): '. . . we will peck at your buttocks'.

In Hase village, children used to sing to call the ant-lion out of its pit: 'Ant-lion, ant-lion, come, get out. If you don't come, I'll cut you with scissors'. And a similar version can be found in literature (*Hase bunkazai senmon iinkai*, 1973): 'Ant-lion, ant-lion, come, get out. If you don't come out now, you will never come out'.

In Ekawa, there is also a song to call the singing *kirigirisu* grasshopper: 'Grasshopper, well really! If you don't come, I'll kill you!'.

We find a variation of the song of the snail, together with a taunt, in Tanabe (Wakayama prefecture): '*Denden mushi mushi*, if you don't get out, I'll stab your buttocks'; or '. . . we'll break your cauldron'[45] (Minakata, 1913–18, 1971).

Calling songs are also recorded to try to get bats out of caves (Kawaguchi, 1913), or to call a line of ants (Minakata, 1913–18, 1971).

As opposed to folktales (*mukashibanashi*),[46] in which mainly mammals are featured (for instance the fox, racoon dog,[47] dog, monkey, mouse and/or rat, and wolf), in children's songs we instead find mostly plants and *mushi* as main characters (Iijima, 1991). This could be interpreted as a sign that *mushi* are important in the child's world. They seem to be perceived as a familiar animal, not (too) dangerous, belonging to the everyday and with whom communication seems possible. *Mushi* are even called by the child to play with him. And if *mushi* hide, the child threatens them, as one would do with a sulky child refusing to come and play with the others. On the other hand, mammals remain animals of legends and moral tales, often bearing a supernatural power (e.g. the fox and racoon dog). They often have a marked positive (exemplified in moral virtues) as well as a negative (when feared for its power) side to their nature. Thus, they are viewed quite differently from the relative neutrality with which *mushi* are tinged in children's songs.

It is important to mention here the great success of the current insect (mainly firefly and dragonfly) protection movement among children, especially in primary schools. Things have certainly changed: during the 1950s, children were involved in a fight against 'pests'. Some villages' authorities (Ekawa for instance) used to give each child a piece of paper limed with glue on which were drawn squares. The children were supposed to stick a 'pest' in each square, after which they brought the paper full of *mushi* to the town hall where they would receive a coin to buy sweets.

Status of the *mushi* in the child's world: toy, playmate or even pet?

A matter of definition

We find a complex situation in the Japanese language (and culture) as far as the concept of 'pet' is concerned. In present-day Japan, the word '*petto*' (from the English 'pet') is widely used to refer to pets and companion animals[48] (dogs and cats are the best examples). There exists another expression, *aigandôbutsu* (literally 'animals to love and play [or take pleasure] with'), that seems at first sight to be used as a perfect synonym for *petto*. It appears, however, that, as far as *mushi* are concerned at least, several species ('autumn singing insects', firefly, rhinoceros beetle, to give but a few examples) are widely referred to as *petto* in the literature (Hearn, 1898, 1987; Makino, 1967; *Petto yôhin daihyakka*, 1987; Nomura, 1994; see also Laurent, 1998), but never as *aigandôbutsu*. Such introduced expressions are numerous in the Japanese language, and it is well known that they undergo a history of their own once they have been adopted. More research is certainly needed to clarify the difference between the two concepts, but I have the impression, from my fieldwork interviews and from reviewing the literature, that the real synonym for pet is actually *aigandôbutsu*; *petto* being a group whose contents appear less well defined and more culturally marked. What must be kept in mind, however, is that the fact of belonging to the category *petto* does not necessarily imply being a pet in the English sense of the term; although we could consider the 'ethnocategory' '*petto*' as a fairly good approximation for 'pet'.

The mushi *as petto?*

According to Serpell and Paul (1994: 129), 'the word [pet] is generally applied to animals that are kept primarily for social or emotional reasons rather than for economic purposes. . .'. Similarly, according to Bonduelle and Joublin (1995), one of the most salient characteristics of pets, as opposed to domestic animals, is their not having a direct commercial utilization. Serpell and Paul (1994: 133) refer to pet-keeping as 'treating individual animals with indulgence and fondness'. According to these characterizations, *mushi* could reasonably be thought of as belonging to the group 'pets'. Their commercialization can be viewed as a secondary phenomenon, something that has emerged from the combined effects of Japanese people's fondness for *mushi*, and the distancing of urban Japan from nature. As we saw, some *mushi* are bred to be sold, kept and fed in cages in houses and gardens. Until recently, *mushi* were only considered domestic animals in rare cases, for example the bee and silkworm. However, if we consider the rhinoceros beetle, the stag beetle and the autumn singing insects, we could ask ourselves whether to include *mushi*, as far as Japanese culture is concerned, in the category of *petto*, or even pets.

In order to answer that question, let us first take a look at the '4th Japanese Pet Fair'.[49] The marked presence of *mushi* puzzled me at first. My surprise was not only due to their presence, but also to the fact that they were the only real animals at the fair![50] Four stands sold *mushi* as well as related tools and apparatuses. Of the four spokespersons, three were absolutely positive: 'Yes, *mushi* are *petto* !'. One of them explained to me that different species of *mushi* correspond to each period in a child's development. First, a child begins to play with ladybirds, and progresses slowly up to the dragonfly catching that requires skill and know-how, passing

Figs. 5.6, 5.7 & 5.8. Leaflet of a company selling *mushi* and goods related to *mushi*.

through play with rhinoceros beetles and cicada catching (among others).

Examining the leaflets the companies distributed would convince anyone of the truth of those words. For instance (Figs. 5.6, 5.7 and 5.8), in one of these, *mushi* appear very humanized, cute and smiling, inviting the child to play.[51] The stress is put on outdoor activities, such as sports. The three species that monopolize the advertising space are the rhinoceros

(a)

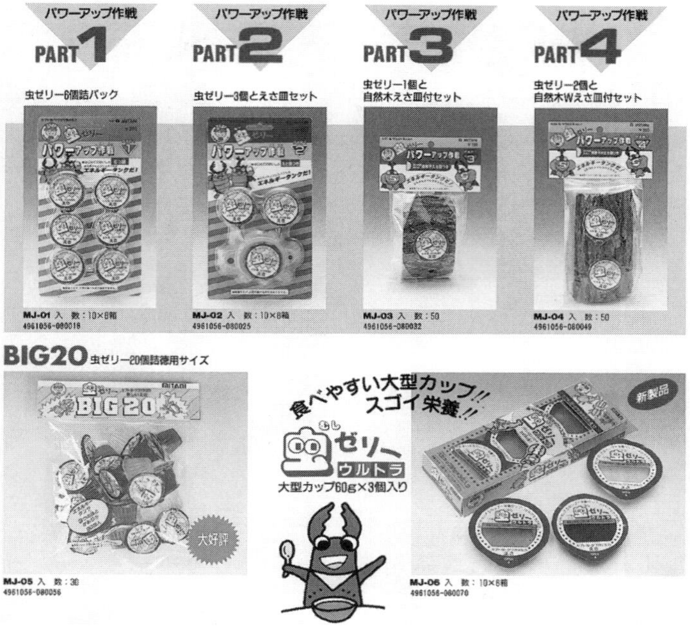
パワーアップ作戦 PART1
虫ゼリー6個詰パック
MJ-01 入 数：10×8箱
4961056-080018
パワーアップ作戦 PART2
虫ゼリー3個とえさ皿セット
MJ-02 入 数：10×8箱
4961056-080025
パワーアップ作戦 PART3
虫ゼリー1個と
自然木えさ皿付セット
MJ-03 入 数：50
4961056-080032
パワーアップ作戦 PART4
虫ゼリー2個と
自然木Wえさ皿付セット
MJ-04 入 数：50
4961056-080049
BIG20 虫ゼリー20個詰徳用サイズ
大好評
MJ-05 入 数：30
4961056-080056
食べやすい大型カップ!!
スゴイ栄養!!
ゼリー
ウルトラ
大型カップ60g×3個入り
新製品
MJ-06 入 数：10×8箱
4961056-080070
4

(b)

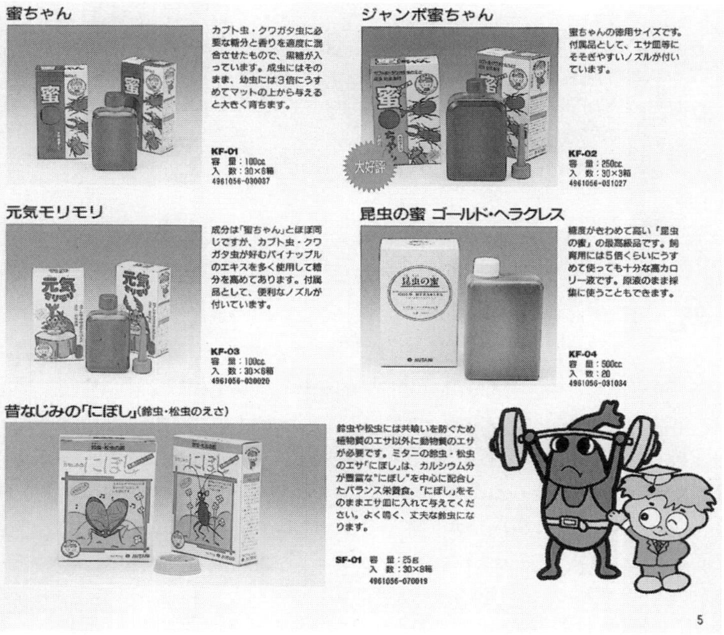
蜜ちゃん
カブト虫・クワガタ虫に必要な糖分と香りを適度に混合させたもので、黒糖が入っています。成虫にはそのまま、幼虫には3倍にうすめてマットの上から与えると大きく育ちます。
KF-01
容 量：100cc.
入 数：30×6箱
4961056-030037
ジャンボ蜜ちゃん
大好評
蜜ちゃんの徳用サイズです。付属品として、エサ皿等にそそぎやすいノズルが付いています。
KF-02
容 量：250cc.
入 数：30×3箱
4961056-031027
元気モリモリ
成分は「蜜ちゃん」とほぼ同じですが、カブト虫・クワガタ虫が好むパイナップルのエキスを多く使用して糖分を高めてあります。付属品として、便利なノズルが付いています。
KF-03
容 量：100cc.
入 数：30×6箱
4961056-030020
昆虫の蜜 ゴールド・ヘラクレス
糖度がきわめて高い「昆虫の蜜」の最高級品です。飼育用には5倍くらいにうすめて使っても十分な高カロリー液です。原液のまま採集に使うこともできます。
KF-04
容 量：500cc.
入 数：20
4961056-031034
昔なじみの「にぼし」(鈴虫・松虫のえさ)
鈴虫や松虫には共喰いを防ぐため植物質のエサ以外に動物質のエサが必要です。ミタニの鈴虫・松虫のエサ「にぼし」は、カルシウム分が豊富な"にぼし"を中心に配合したバランス栄養食。「にぼし」をそのままエサ皿に入れて与えてください。よく鳴く、丈夫な鈴虫になります。
SF-01 容 量：25g
入 数：30×8箱
4961056-070019
5

Fig. 5.7.

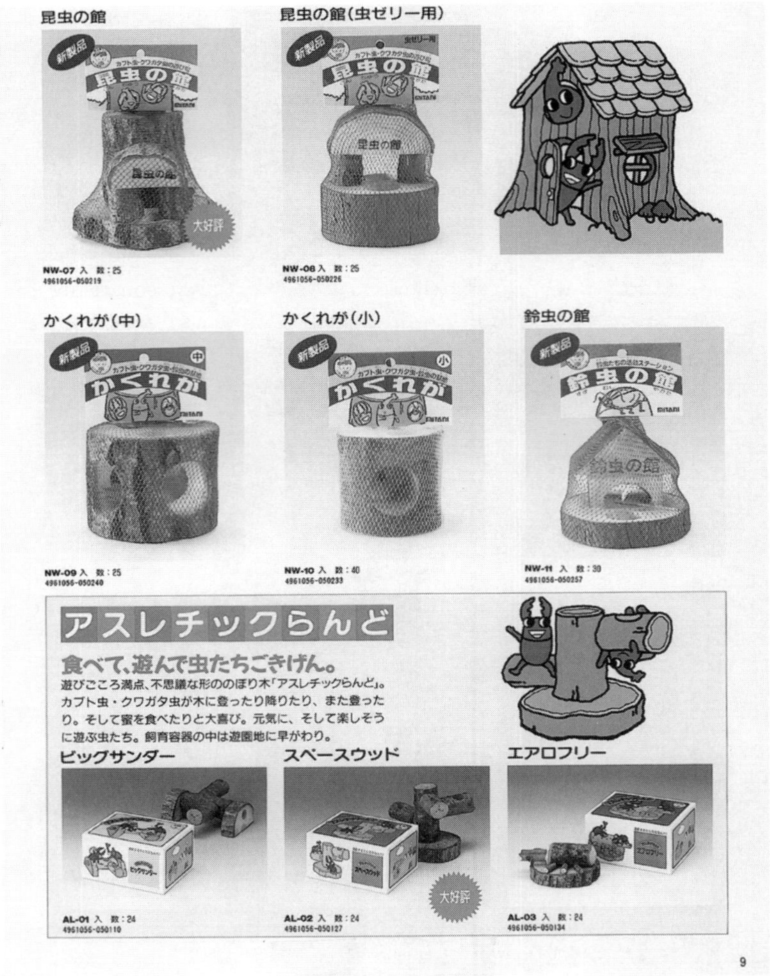

Fig. 5.8.

beetle, the stag beetle and the *suzumushi* cricket. The catalogues contain aqua-terrariums of various sizes that are specifically designed for certain species of *mushi*; mats that supposedly 'suppress odors and absorb liquids'; species-dependent food (powder, jelly, honey, tonic drinks); small pieces of wood used as troughs; nests, some of them having been designed for *mushi* to exercise in;[52] butterfly and cicada nets; and a deodorant spray[53] that supposedly 'evacuates unpleasant bodily odors of insects as well as odors of rotting food'.

The way *mushi* are presented at this fair, one feels they are almost human (eating, resting, exercising, sometimes having unpleasant odours). One also feels they invite the child to play; that both belong to, and move in, the same kind of world. The *mushi* performs the same actions (plays, partakes in everyday activities) as the child. The world of the child is thus clearly transposed to the world of the *mushi*. In this sense, the *mushi* cannot be considered a toy. He or she must be seen as a playmate for the child. This idea is reinforced by the contents of educational books and children's songs about *mushi* (as reported earlier).

It is also important to remember that children constitute the majority of those who visit Japanese insectariums, such as the ones linked to zoological gardens (as in Tama near Tôkyô), as well as the independent ones (like Kashihara in Nara prefecture). Such places offer various activities with *mushi* designed specifically for children: outdoor-play activities, initiations into nature, video programmes, and didactical conferences (see also Yajima, 1990).

Moreover, it is rather difficult to determine from interviews whether *mushi* are perceived as *petto* by children or adults. This new category is still rather vague, and several species of *mushi* are certainly 'borderline cases' in popular perception. My experience as a participant observer leads me to believe that some species of *mushi* (rhinoceros beetle, stag beetle, spider, firefly, for instance) do fit as *petto*, owing to the special link that exists between them and children, and through (mutual) interactions, in which play and emotional bonding are important elements.

CONCLUSION

The common perception that *mushi* are toys, on the same level as wooden or plastic ones with which a child plays, must be challenged. Considering the analysis of the contents of songs and educational books, and considering as well the explicit references made to *mushi* as *petto*, *mushi* should be thought of as playmates rather than toys or tools. The concept of *mushi* as a toy or a tool is not, however, absent from the complex image of the *mushi* in the Japanese child's world. One can indeed witness a certain reification, but it seems rather recent, and apparently generated mainly from the marketing of certain species of *mushi*. Children do not gather or collect *mushi* so often any more; they buy them. *Mushi* have become commodities, with a price, put in shopping carts in department stores, next to fruit juice or laundry detergent. In this regard, one of the people in charge at the insectarium of the zoological garden of Tama feels invested with a mission to make a bridge between the world of the 'toy-*mushi*' and 'real nature'. It is therefore considered important to show living insects to children even though they are one of the most difficult animals to present to people (due to short lifespan, small size, lack of sponsors). It is also considered important for the child to touch the insects and to move amidst them flying. Thus, the insectarium is seen as a kind of bridge. After all, this is already one of the roles which zoological gardens perform with larger animals.

It has been argued, nonetheless, that to some extent *mushi* could be considered as *petto*. Even if one cannot accept the view of *mushi* as pets on the same level as cats or dogs (they have no names, there is no mourning when they die...), they are more to Japanese children than just wild animals on the one hand, more than toys, and even more than mere playmates on the other hand. We witness indeed a special relationship between some species of *mushi* and children, involving pleasure and play, that refers to a special link between nature and humans.

It is mainly through nature, and more particularly thanks to the *mushi*, that the Japanese child learns about the passage of the seasons.[54] The child knows from an early age, for instance, that the fireflies and rhinoceros beetles appear at a very specific period of the year and that they are not eternal beings. In its traditional formulation, the *mushi* teach the child something the books cannot: they give the child concrete material for experimental dialogue with nature: the ways of catching, treating, raising, observing *mushi* depend on the species involved and the child therefore learns about biological diversity. An insect's relatively short lifespan also shows the child ontogenic development and the cycles of life. Feeding and keeping *mushi* require personal observation, thinking and even experiments. But the most important point lies in the fact that *mushi* stand on a borderline, neither pets nor wild animals, giving them a particular place inside 'close nature' (*mijikana shizen*), as well as a special function as a link between distant nature and humans. For this reason, one could assert that the *mushi* constitute for Japanese children one of their most important natural elements, something they will never completely forget.

Traditionally, the transmission of knowledge concerning *mushi* used to occur orally, mainly from (grand-)father to (grand-)son. An evolution seems to have appeared in this respect quite recently. Nowadays, knowledge concerning nature tends to be transmitted through illustrated books and television. Thus, the transmission has become much more 'mediate'. Often, the child first sees the book or television, then talks to his (grand-)father before going to experiment by himself in nature. It has also become necessary to supply the *mushi*, to sell them, because in most cases the urban child cannot play with them directly any more. Does it not prove, after all, the importance of the *mushi* in Japanese culture – the role *mushi* have in Japan? They could have been simply forgotten; but, on the contrary, they are marketed, 'mass-mediatized', and, in the process, are given an importance they may have never had before.

Finally, one can witness a decrease in interest in *mushi* by children. This is most probably due to the increase in the number of games and other attractions targeted at children (e.g. television, computer, magazines). The marketing and, more precisely, the recent publicity campaigns centered on *mushi* can indeed be explained by the need to stay competitive with these other products, which are thought of as more appealing to the child. That *mushi* do succeed in imposing themselves in such an aggressive market can, however, be interpreted as a sign of the vitality of their role as far as the child is concerned (as educational tools, for the purpose of play or even as a socio-affective object) in Japanese culture and society.

NOTES

1. One can witness a divergence between the Japanese countryside, where children still have direct contact with insects, and urban Japan, where this tends to be less true. However, in the parks, temple or shrine gardens and along the river banks of Kyôto, a lot of children still do catch insects.
2. Mainly during two periods of six months, from May until October 1989 in Hase village (Nagano prefecture), and from May until October 1991 in Ekawa village (Wakayama prefecture). Also in Kyôto city since 1988 and Gifu prefecture since 1994.
3. 'Pine cricket', *Dionynus marmoratus* de Haan (family Eneopteridae).
4. 'Bell cricket', *Homœogryllus japonicus* de Haan (family Phalangopsiidae).
5. *Gampsocleis buergeri* de Haan (family Tettigoniidae).
6. *Allomyrina dichotoma* L. (family Scarabeidae).
7. Family Lucanidae, especially *Dorcus curvidens* Hope.
8. In the *Yôshûbushi, The Chronicles of the Yamashiro Province* (southern Kyôto) (cited in Kanô, 1990).
9. They were called *yamadashi*: '(came) out of the mountain'.
10. In Tôkyô, Japanese cages for insects were first manufactured at the beginning of the nineteenth century (Hearn, 1898, 1987).
11. If one can speak of trends, *suzumushi* seem to be the most popular at the end of the twentieth century, as far as the general public is concerned. For instance, in a public park in the town of Niihama (Ehime prefecture), a shelter for *suzumushi* was built in 1991, in which many of them were released in order to create a sound mood. On the other hand, the *kantan* crickets (*Œcanthus longicauda* Matsumura) seem to attract the favours of a more restricted audience. They cannot be bought but should rather be listened to in nature; there is in Tôkyô a 'Society for the Listening to the *Kantan* Cricket' (*Asahi Shimbun*, 4.10.1992).
12. The phenomenon, at a national level, seems to be limited only to this species. One can even find *suzumushi* and cages given as a present in certain post offices for each deposit of a certain amount of money. In 1988, 140 of these gifts had

been offered (*Takatô Nippô*, 28.7.1989). Several species of insects are also sometimes given in train stations as 'summer gifts'.

13. There exist about 30 different types, whose specificity lies in the principle of action (chemical, electrical or mechanical), on the one hand, and in the species aimed at, on the other.
14. One pound sterling is about 200 yen, as of June 1999.
15. The surface is clearly divided, as is nearly always the case, into goods related to rhinoceros beetles and stag beetles on the one hand, and to *suzumushi* crickets on the other.
16. Although these items are sold with the beetles, they are in no way specific to these insects. Nets, for instance, are extensively used by children to catch butterflies or cicadas.
17. *Matto* is the abbreviated form for *mattoresu*, derived from the English mattress.
18. In the shop, their voices compete with dogs barking, cats meowing, and birds squawking; but not all to their disadvantage, it should be said.
19. The shopkeeper could not answer in what way the pieces of wood specific to crickets were different from those sold for beetles.
20. Older people do not seem to understand this sudden love of *kabutomushi*. These were not so appreciated in traditional Japan, sometimes being called 'demon-*mushi*' (*onimushi*). 'It doesn't sing. It cannot be eaten. It's not even beautiful. No interest at all . . .' (70-year-old female informant).
21. Yamaguchi city (information gathered from the 7 pm News of the NHK, national broadcasting network, on June 29, 1992).
22. They asked him to start breeding *suzumushi* crickets as well. So far he has refused, not having the knowledge for doing so, but he said he might try very soon.
23. 'Certain larval stages require rain, others not. It's difficult, there are no written documents available. I have to follow my instincts', so says my informant.
24. *Ibaraki minzoku gakkai* [Folklore Society of Ibaraki] (1970) cited in Iijima (1991). *Kodomo no saiji to asobi* [Seasons and children's plays]. Anonymous (1948) cited in Iijima (1991). *Gunmaken Minakamigun Minakamichô Fujihara no denshôasobi* [Orally transmitted games in Fujihara, Gunma prefecture, Minakami district, Minakami town]. I shall call these lists A and B, respectively.
25. We can see that the elements contained in list A (from Ibaraki) are more numerous. It does not seem to be due to a regional disparity or a chronological evolution. It is just that the number of items in list A are more numerous in regard to all the species considered.
26. I would rather characterize the first as an activity of middle-aged men, the next as one of middle-aged women, and the last as belonging to a very general population.
27. Plastic cages or sometimes wood ones can now be bought very easily.
28. I will just give two precise examples: (A) Miyatake (1983, 1989); (B) Saitô (1990).
29. Without considering the tadpole (*otamajakushi*), usually viewed as a *mushi* in Japanese folk animal categories, but put here in the group 'fish' (*sakana*).
30. Otherwise, because the dragonfly is above the balls whereas the thread is below, the wings will not get tangled in the thread.
31. Three types of spiders are used: *haetorigumo*, the 'chasing' spider (with no web) from the Salticidae family, bred in the Edo period for the entertainment of the nobility, particularly *nekohaetori* (*Carrothus xanthogramma* (Latreille)); *jigumo*, from the Atypidae family, living in the ground, and making no web; as well as the most often used *koganegumo* or *kenkagumo* (literally 'fighting spider'), from the family Argiopidae, which makes a web.

32. *Jôrogumo* or *kenkagumo* (*Nephila clavata* Koch) are more specifically used for this type of fight.
33. *Oncotympana maculaticollis* Motschulsky.
34. Effectively piercing and metal-like, and different according to species, it can in no way be compared to the nearly inaudible sounds made by European cicadas.
35. *Dedemushi, dendenmushi, katatsumuri, maimai, maimaitsuburi, tsurugame, namekujira, minamushi, tsunodashi, gegebo,* and so on.
36. The word refers to the only coloured species in the scarabeid family, mainly the greenish ones which are morphologically close to *Rhomborrhina sp.* The children seem to be particularly interested in the sounds the insects make while flying: the more it tries to escape, the more noisy it becomes. A similar game was popular in Belgium and France some decades ago, with a June beetle. The Japanese June beetles (*Melolontha sp.*) are part of the larger folk understanding of the category *kanabun*.
37. Minami (1961, 1983) has found 34 of these.
38. 'Ho, ho, hotaru koi! Ho, ho, hotaru koi! Acchi no mizu wa nigai zo, kocchi no mizu wa amai zo. Ho, ho, hotaru koi. . .'.
39. 'Dendenmushi mushi, katatsumuri. Omae no atama wa doko ni aru? Tsuno dase, yari dase, atama dase.' The first written mention of this song seems to be found in *Hinamikiji*, published in 1676 (cited in Yanagita, 1930, 1990). There is also a song with very similar words from the thirteenth–fourteenth century, 'The princess who loved insects' (*Mushi wo mezuru himegimi*).
40. After Okumoto (1990), the melody of this song comes from a popular Spanish song introduced from France at the beginning of Meiji era, the words having been written in 1874 by a schoolmaster named Nomura Akitari: 'Chôchô chôchô, na no ha ni tomare. Na no ha ga daitara, sakura ni tomare. Sakura no hanano, hanakara hana he . . . Tomare yo asobe yo, asobe yo tomare.'
41. 'Are suzumushi ga naiteiru. Rin rin rin rin rin rin. Aki no yonaga wo nakitôsu. Aa omoshiroi mushi no koe. Are matsumushi mo nakidashita. Chinchiro chinchiro chinchiro rin. Aki no yonaga wo nakitôsu. Aa omoshiroi mushi no koe.'
42. 'Yobu koe wa taete hotaru no sakan kana' (cited in Katô, 1978).
43. Name of the larva (caterpillar), as well as the adult, wingless female of the psychids moths. The larva makes a sheath of small branches, leaves and/or pine needles in order to protect itself during its development. The sheaths are suspended from trees, and children like to play with them, trying to make the caterpillar come out.
44. 'Jîjî bâbâ jîbâ, Jîsan kaettekita, doa akerô. Jîjî bâbâ jî bâ, Bâsan kaettekita, doa akerô.'
45. That is, apparently, the shell.
46. Literally 'old stories' – tales and legends that constitute a genre of popular literature.
47. *Tanuki: Nyctereutes procyonoides viverrinus* Temminck.
48. The recently introduced expression '*konpanion animalu*' seems to be known only to specialists.
49. 'Dai yonkai japan-petto-fea' (Kôbe, 23–24 March 1991).
50. One of the most striking peculiarities of this fair was indeed the absence of animals. They were only represented as fluffy or plastic models in cages.
51. Anthropomorphism is very common in Japanese advertising, and it is out of the question to take it as a clue for *mushi* being *petto*. It is rather their place and function in the depicted child's activities that allow us to think of them as *petto*; that is the way they are anthropomorphized.
52. The text that accompanies the picture says: '*Mushi* who climb and go down from trees are just like children in good health having a lot of fun'.

53. On the label of which is a drawing of a stag beetle and a rhinoceros beetle under a shower of flowers. One can easily imagine the contents as being odoriferous.
54. The existence of a '*mushi* market' (though not new if we consider the 'autumn singing insects' sold since the seventeenth century) for children, especially in urban settings where the actual presence of *mushi* is declining, is a strong sign of their importance in Japan.

REFERENCES

Asahi Shimbun (4.10.1992). Yama ai ni naku mushi no jo'o? [The queen of the *mushi* who sing between the mountains], p. 1.

Asahi Shimbun (8.7.1993). Sodateru tanoshimi konchû ga bûmu [Boom in the breeding and plays with insects], p. 8.

Bonduelle, P. & Joublin, H. (1995). *L'animal de Compagnie.* Collection 'Que sais-je?' Paris: Presses Universitaires de France.

Hase bunkazai senmon iinkai (1973). *Inadani. Hasemura no Minzoku* [Folklore of Hase – Valley of Ina]. Hase: Hase bunkazai senmon iinkai.

Hearn, L. (1898, 1987). Insect-musicians. In *Exotics and Retrospectives*, pp. 37–80. Rutland and Tôkyô: C.E. Tuttle.

Hidaka Nippô (3.7.1991). Kabutomushi 200 hiki wo shoshukka. [First sending of 200 rhinoceros beetles], p. 3.

Iijima, Y. (1991). Kodomo to shizen [Children and Nature] *Kodomo no Minzokugaku* [Ethnography of children], pp.152–86. Tôkyô: Shin'yôsha.

Ishigami, G. (1983). *Nihon Minzokugo Daijiten* [Great Dictionary of Japanese Popular Language]. Tôkyô: ôfûsha.

Kaneko, H., Konishi, M., Sasaki, K. & Chiba, T. (1992). *Nihonshi no Naka no Dôbutsu Jiten* [Dictionary of Animals in Japanese History]. Tôkyô: Tôkyôdô shuppan.

Kanô, Y. (1990). Naku mushi no bunkashi [Cultural chronicle of singing *mushi*]. In *Shirîzu Shizen to Ningen no Nihonshi. 5. Mushi no Nihonshi* [Japanese History of Nature and People. 5. Japanese History of *Mushi*], ed. D. Okumoto, pp. 56–65. Tokyô: Shinjinbutsu ôraisha.

Katô, M. (1978). *Mushi.* Sendai: Kinkôdô shuppanbu.

Kawaguchi, M. (1913). Hokaku to yobikoe. [Catchings and appeals]. *Kyôdo Kenkyû*, **1** (6), 339–41.

Kawana, T. & Saitô, S. (1985). *Kumo no Kassen. Mushi no Minzokushi* [Spider Fights. Notes of Folklore concerning *Mushi*]. Tôkyô: Miraisha.

Laurent, E. (1995). Definition and cultural representation of the category *mushi* in Japanese culture. *Society and Animals*, **3**, 61–77.

Laurent, E. (1997). Les contenants pour *mushi*, leur signification dans la culture japonaise. *Ebisu*, **15**, 61–88.

Laurent, E. (1998). *Les Mushi dans la Culture Japonaise.* Lille: Presses Universitaires du Septentrion. 2 volumes.

Makino, S. (1967). *Petto* [Pets]. Osaka: Kyôikusha.

Masuda, Y. (1989). *Asobi no Daijiten* [Great Dictionary of Play], 2 volumes. Tôkyô: Shoseki.

Matsuura, I. (1983). *Naku Mushi no Kansatsu to Kenkyû* [Observations and Research on Singing *Mushi*]. Tôkyô: nyûsaiensu sha.

Minakata, K. (1913–18, 1971). Kishû zokuden [Popular legends of the Kishû region]. In *Minakata Kumagusu Zenshû* [Complete Works of Minakata Kumagusu], Vol. 2, pp. 322–64. Tôkyô: Heibonsha.

Minami, K. (1961, 1983). *Hotaru no Kenkyû* [Research on Fireflies]. Tôkyô: Saienchisutosha.
Miyatake, Y. (1983, 1989). *Mushi. Kurashi to Kaikata* [The *Mushi*, Life and Breeding]. Osaka: Hikari no kuni.
Nomura, J. (1994). *Shodôbutsu no Kaikata* [Methods of Breeding Small Animals]. Tôkyô: Seibidôshuppan.
Okumoto, D. (1990). *Mushi no Nihonshi* [Japanese History of *Mushi*]. Tôkyô: Shinjinbutsu ôraisha.
Otani, T. (1958). Kumo no asobi [Playing with spiders]. *Hidebachi*, **10**, 20–1.
Petto yôhin daihyakka (1987). [Encyclopedia of Products Related to Pets]. Tôkyô: yasei-sha.
Saitô, A. (1990). *Mushi Atsumare* [Collecting *Mushi*]. Osaka: Hikarinokuni.
Sakamoto, T. (ed.) (1957, 1976). *Fûzoku Jiten* [Dictionary of Habits], pp. 692–3. Tôkyô: Tôkyôdô shuppan.
Serpell, J. & Paul, E. (1994). Pets and the development of positive attitudes to animals. In *Animals and Human Society*, ed. A. Manning & J. Serpell, pp. 127–44. London and New York: Routledge.
Shimasaki, T. (1913, 1974). *Megane* [The Glasses]. Tôkyô: Horupu shuppan.
Takatô Nippô (28.7.1989). *Suzumushi purezento chokinsêru jisshi.* [Present of *Suzumushi* Crickets to each Saver].
Uemura, Y. (1990). Tombo wo tsutta koro [The days dragonflies were chased]. In *Shirîzu Shizen to ningen no nihonshi. 5. Mushi no nihonshi* [Japanese History of Nature and People. 5. Japanese History of *Mushi*], ed. D. Okumoto, pp. 121–8. Tôkyô: Shinjinbutsu ôraisha.
Yajima, M. (ed.) (1990). *Konchû no Kuni. Tama Dôbutsu Kôen Konchû Seitaien wo Tsukuru* [Insects Country. Building an Insect Ecological Park in Tama Zoological Garden]. Tôkyô: Keyaki shuppan.
Yanagita, K. (1930, 1990). Tengyûkô [Reflections concerning snails]. In *Yanagita Kunio Zenshû* [Complete Works of Yanagita Kunio], Vol. 19, pp. 7–177. Tôkyô: Chikuma shobô–Chikuma bunko.

NORINE DRESSER

6

The horse *bar mitzvah*: a celebratory exploration of the human–animal bond

INTRODUCTION

I confess. It took years before I located the Long Island, New York woman who gave the horse *bar mitzvah*. Even though the New York NBC-TV sports department confirmed that in December 1993 they had featured several sound bytes on the celebrated steed and his owner, by the time I contacted them, the station no longer had any backup information. All I knew was that at the ceremony, the 13-year-old horse wore a *yarmulke* (Jewish skull cap). Despite the missing information, I clung to the name because of the seemingly absurd juxtaposition of those words. I was rewarded for holding on to the phrase when over time I discovered a cat *mitzvah* and a bark *mitzvah* for a 13-year-old Labrador retriever whose 20 dog guests received monogrammed hats to wear at a barbecue reception.

The horse *bar mitzvah* presents an incongruous pairing of a solemn religious ceremony with an unexpected participant. The idea is startling, risky, yet useful in making people consider human–animal relationships from a different perspective. The purpose is to focus on rites of passage and other rituals, especially those taking place at holiday times. Such celebrations reveal a complex relationship between humans and animals. Folklorists often examine rites of passage, customs, beliefs and ceremonies to understand human behaviour and culture. I am no exception. This has been the foundation of my research, teaching, and writing for over 25 years.

Why do people impose human celebrations on other species? Is it merely an expression of human imperialism? Or is it because urban America has no ready-made outlets for expressing love for other creatures? To investigate this phenomenon, I conducted interviews at two different Blessing of the Animals ceremonies. I observed a Halloween 'Boo at the Zoo', National Pet Cemetery Day, and a graduation at Guide Dogs of

America. I conferred with grief counsellors working with zoo employees following the Philadelphia Zoo fire in which 23 primates died, questioned hosts of a cat *mitzvah* and a dog wedding, a cat owner who sat Shiva (a Jewish mourning custom) for her deceased pet, participants at a reptile exposition, basset hound picnic, and a lobster wedding. This is only a fragment of the data base used to document a variety of human rituals for animals.

CHIMPANZEES

The genesis for this study occurred in 1994, when an Elderhostel course, 'Human Communication with Chimpanzees', lured my husband and me to the Chimpanzee and Human Communication Institute at Central Washington University in Ellensburg, Washington, for two weeks of intensive study. There we were introduced to the famous chimpanzee, Washoe, and her four chimpanzee companions. In a cross-fostering experiment, Washoe was raised by human parents, Drs R. Allen and Beatrice T. Gardner, who had obtained the chimp from the wild in 1966 when she was 10 months old. Washoe lived with the Gardners in their home, almost like a human offspring. She ate human food with utensils, but at night slept in a locked trailer.

Because of chimps' superior manual dexterity, the Gardners experimented to discover if Washoe, their 'adopted child', could learn to communicate with 'her parents' through signing. They raised her as if she were a deaf human child. The Gardners never spoke in front of Washoe in any language other than the American form of sign language (ASL). They never used any rote training in transmitting the language. Instead, Washoe acquired language as human infants do – by observing and imitating her 'parents' signing to each other and to her. Later, Washoe taught an adopted infant chimpanzee, Loulis, how to sign. Subsequently, they were joined by three more chimps cross-fostered by the Gardners, and currently all five live together at the Central Washington University Institute where they sign to one another, to the staff, and to themselves.

As part of their continuation of being treated as humans, the chimps celebrate birthdays and all major American holidays. The rationale is that because the chimps were raised as members of a human family, they enjoy and actively anticipate these rituals.[1] On Thanksgiving, the chimps eat turkey dinners. For Christmas, they have a decorated tree and receive gifts. At Easter time, the chimps search for hidden plastic eggs stuffed with raisins and other goodies. On birthdays, they receive such gifts as slingshots, whistles, mirrors and combs. On Halloween, they wear masks, eat pumpkin seeds, and dress up in clothes and scarves. When Loulis lost his baby tooth, the staff's version of the tooth fairy deposit was to give him a treat of newly dried pears.

In providing human festivities for chimps, the laboratory simultaneously fulfills its United States Department of Agriculture (USDA) requirement to provide environmental enrichment programmes (also called behavioural enrichment) for primates.[2] Environmental enrichment attempts to narrow the gap between captivity and the wild through the design of habitats and the development of techniques that stimulate the animals' natural behaviours. One method consists of eliminating boredom by enhancing play activities and creating challenges in obtaining food.

Clearly, the institute's reason for animal celebrations appears to have links to science and law. On the other hand, how does one explain the need for horse, cat and dog owners to have *bar mitzvahs* and other human celebrations for their pets? Pet owners have no governmental body demanding extra activities, so why are these rituals so popular? Are they merely satirical occasions and excuses for merry-making, or is there some deeper meaning?

CAT *MITZVAH*

On June 7, 1994, a California couple gave a cat *mitzvah* for their 13-month-old kitten Fifi Katz. Fifi's 'mom' is Jewish and the 'dad' is a former Jesuit priest. As to whether any of their guests were offended by a religious ceremony for a cat, the 'mom' reported that of the 90 guests who received printed invitations, only one person was reluctant – she did not like cats.

Guests brought gifts of fresh roses, cat food dishes, games, fishing poles and a cat video of birds and butterflies, with a barking dog to spice up cat-viewer interest. One guest planted a tree in Israel in Fifi's honour. Someone else brought a white lace-trimmed religious head covering and tied it onto the unwilling kitten.

The hosts served vegetarian Greek food, and their guests danced on the beach to live bouzouki music. Later, the hostess and her stepmother sang two parodies from 'Fiddler on the Roof'.[3] Friends and family pretended to weep during the 'Sunrise, Sunset' song:

> Is this the kitty that I carried?
> Is this the little cat at play?
> I don't remember growing older.
> When did they?

When did she get to be a beauty?
When did she get her first fur ball?
Wasn't it yesterday when they were small?

To add to the mirth, the singers changed the last line of the well-known 'Sunrise, Sunset' chorus to, 'One season following another, laden with happiness and fleas'.

Although no rabbi was present, a Jewish obstetrician/gynaecologist gave a blessing and speech urging that we treasure in life that which gives us pleasure. Clearly, Fifi was precious to her owners. She had given them endless moments of joy which they chose not to overlook.

Fifi's 'father' explained that, of course, the event was for fun and that most people had a very good time. On the other hand, there was a serious side to it, as well. He emphasized that to him, traditions were beautiful. Even though the cat *mitzvah* was a 'tongue-in-cheek' occasion, he deeply respected traditions. The true purpose of the occasion was to celebrate the beauty of creation as manifested in a particular little animal and, at the same time, to realize our own at-homeness in the universe. He claimed that it was a celebration of people's relationship with nature and the cosmos.

HORSE *BAR MITZVAH*

When I finally tracked down Rachel, the sponsor of the horse *bar mitzvah*, she told me that gratitude and love for her animal companion were the reason for honouring him with a human religious ceremony. She claimed that her horse, Sonny, had enabled her to accomplish more in life and had given her more love than she had ever received either from her family or former spouse.

Rachel,[4] a special education teacher, views Sonny as a magical agent. As soon as she took ownership of the 12-year-old gelding pacer, he gave her life new direction; she became a writer, horse trainer, a foaling assistant, and publicity director of a race track. At the time of Sonny's purchase, she had given him a 12-year birthday party, but when wondering what to do for his thirteenth, one of her students, then studying for his own *bar mitzvah* (the 13-year celebration marking a Jewish boy's transition into manhood), suggested a *bar mitzvah* for Sonny.

Delighted with the suggestion, in May 1993 Rachel invited two dozen family members and friends from the track and school to celebrate at the stables. Over Sonny's stall she hung a banner announcing, 'Today, I am a *mensch* (an upright, honorable person)', which is a paraphrase of the human post-*bar mitzvah* pronouncement, 'Today I am a man'. Sonny wore *tzitzit* (a fringed religious garment that goes over the head like a *poncho*) and a *yarmulke* (skull cap) held in place with hair pins.

Some of Rachel's guests had never been to a real *bar mitzvah* and were unfamiliar with the traditional Jewish foods that Rachel served: gefilte fish with horseradish, whitefish and carp, pickled herring, and lox and bagels. She poured traditional Manischewitz red sacramental wine. After Sonny sampled it, he lifted his head, licked his lips, then with gusto lapped up a quart. Later, the wine cast an uncharacteristic calm over the steed.

Next, Rachel brought in a cake decorated with the words, 'Congratulations to my son, the accountant'. This related to the traditional Jewish saying, 'My son, the doctor'. Then Rachel created her own version of a candle-lighting ceremony common at many *bar mitzvah* receptions. When called, a family member came up to the cake to be recognized for the role he or she played in the life of the *bar mitzvah* and to light one of the cake's 14 candles, one for each year plus one for good luck. Friends who had helped 'dose' the horse when he was ill were included. Subsequently, Rachel's nephew David, serving as stand-in rabbi, read aloud in Hebrew from a prayer book. Rachel tried to coax the horse to *daven* (move his head up and down) by luring him with a can of soda pop, but he was too interested in the ceremony and just wanted to look into David's book. When the nephew closed the book, Sonny nudged him, causing Rachel to comment, 'Look, this horse likes Jewish. *Mazel Tov* (Congratulations)!' Then she kissed Sonny's face.

Just like *bar mitzvah* boys, Sonny received gifts, a fountain pen, the traditional gift for human *bar mitzvahs*, as well as horse presents: carrots, sugar, apples, a tote box, and a bouquet of flowers, which Sonny first sniffed then ate. Although she had originally planned the event as a kind of joke, after the guests had left and Rachel was alone cleaning up, she was overwhelmed by what had transpired. 'He [Sonny] kind of understood what it meant. It was like *bar mitzvahing* a son', she said proudly. The horse *bar mitzvah* was a memorable event not only for Rachel, but for the sports community as well, becoming the subject of newspaper articles and television sports news.

Two years later, while Rachel was riding Sonny out on the trail, Sonny suffered a fatal accident. His unexpected death came as a severe blow to Rachel, and she buried him on the farm where he was boarded, planting an apple tree and chrysanthemums next to his grave with an engraved marker reading: 'My Beloved Sonny, 1980–1995'. The farm manager erected a fence around the grave and added a ring where riders can tie up their horses on their way to the trail and stop to pay their respects. Two years after Sonny's death, Rachel tearfully maintains, 'He was the light of my life'.

DOG WEDDING

Dr Cynthia Wayne (Surrogate Mother)
and
Fair Isle Highland Lassie (Birth Mother)
request your presence at the Canine Nuptials
(A Dog Wedding!)
of
My Heather of Queens
to
Scotty MacKenzie
Sunday, June 23, 1996, 2:30 p.m. (people time)
The Ferndale Gardens, Newton, Massachusetts
Kids Welcome BYOC (Bring Your Own Chair) No cats, please!

When I first met folklorist Cynthia Wayne in Lafayette, Louisiana, in October of 1995, she was already bubbling with plans for the June wedding. Her excitement never waned throughout the eight-month betrothal period, and although she cannot remember when she thought up the idea of marrying her beloved West Highland Terrier, Heather, to her other Westie, Scotty MacKenzie, it was an idea she seized upon with glee.

During the pre-nuptial period, enthusiasm built and reached its height when, a few weeks prior to the event, Cynthia's cousin created a Web Page to honour the wedding. It showed a picture of Heather telling Scotty that she was a 90s' girl, 'So forget about that obey stuff.'

Forty guests of all ages and 20 Westies gathered in spectacular Massachusetts showcase gardens with brilliant peacocks strutting by and swans gliding on the pond. There, next to a red and pink impatiens-covered wishing well, Cynthia added her own party touches: pink, grey, and white balloons, paper wedding bells, pink streamers decorated with dog bone stamps.

As guests arrived, children stamped the guests' hands with a dog bone stamp, handed out copies of Heather's Web Page and wedding programmes containing Heather's biography, photograph, and a description of how Heather met Scotty.

Amidst laughter and the whirring of cameras, a kilt-clad bagpiper led the wedding procession, tailed by Scotty, the groom, in his red and white tartan vest with big bow and brass button, black top hat accented by a tartan hat band and fastened with a small Westie pin. Best Person, Marcie, led him on leash, followed by three Westie bridesmaids in pink ruffs and the bride's mother in a fetching pink and white dress.

Cynthia entered next wearing a red and white Stewart dress kilt with white sweater, Westie earrings and pin. The bride had on a white see-through gown of curtain material decorated with pink flowers and mother-of-pearl buttons, a white hat with matching pink flowers, and a snap-on garter – but not for long. Heather, ordinarily very meek, shook until her complete outfit came off. Then she took offense at one of the bridesmaids and, in an unbridely fashion, began barking at her. Guests howled with laughter. When Scotty made advances on one of the bridesmaids, one friend wryly observed, 'This guy shouldn't be getting married. He's got a roving eye'.

Once the bridal party was in place, Cynthia welcomed her guests and blessed the Westies and began the vows. To Scotty she intoned, 'By accepting this pepperoni, will you promise to love and obey Cynthia and Heather?' Forgetting Heather's Web Page oath about refusing to obey, Cynthia asked, 'Heather, by accepting this pepperoni, will you promise to love and obey Cynthia and Scotty?' Then Cynthia broke a biscuit and pronounced, 'By accepting this biscuit, you agree to this union'. Scotty agreed, but Heather, in her atypical cantankerous mood, refused.

Best Person Marcie lifted a Barq Root Beer-filled glass in a toast to the newlyweds and Cynthia proclaimed, 'Everybody, this is Mr Scotty MacGregor Wayne and Mrs Heather Wayne MacGregor'. All the guests cheered and applauded, and the piper began the recessional leading the assemblage to the food – for the dogs, dog toast in the shape of Westies topped with liverwurst and cheese spread served on a doily-covered silver tray. Human guests drank Barq Root Beer and ate people biscuits in the shapes of West Highland terriers and dog bones decorated with coloured and chocolate sprinkles. They also had a pink and white iced wedding cake.

After the festivities, with the bagpiper in tow, Cynthia, Marcie and the newlyweds proceeded to a nearby rehabilitation centre to visit Cynthia's mother, a post-stroke patient. When the older woman heard the bagpipe music procession nearing her room and saw her daughter and dogs in their wedding get-ups, with Marcie carrying the leftover wedding cake, parading up the rampway, she and all the other patients lit up with anticipation and joy.

Most people enjoyed the idea of the dog wedding before and during the celebration. Mae, a wedding guest, said that when Cynthia first told her of the plans, she thought, 'I don't believe she's doing it', but over time Mae herself became more and more intrigued with the idea. By the time of the big day, she had become a willing participant. Other guests were skeptical too, asking one another, 'Have you ever been to a dog wedding before?' However, by the end of the gloriously sunny afternoon, all had been charmed in the gorgeous garden setting. Afterwards, Mae commented that she had a smile on her face for the rest of the day, while one enthusiastic 74-year-old woman exclaimed, 'It was a gas!' Only the woman at the fabric shop where Cynthia had purchased the remnant for the wedding gown expressed any negativity. It was 'really stupid'.

Cynthia, who has a zesty sense of humour, couldn't resist joking about the wedding and making puns. For example, she insisted on being photographed paying the piper. In sending a wedding e-mail update she wrote, 'Well, here's the latest poop'. She scoffed at the suggestion that the dogs were living in sin. When people asked, 'How will they consummate the marriage?', she answered, 'They've been sleeping together since August, but Heather had a hysterectomy and Scotty has clipped testicles'.

On the pink and white wedding invitation Cynthia had labelled herself as the 'surrogate mother', so there is no question as to how she interpreted her role. Cynthia is single and this provided her with the opportunity to participate in what, to outsiders, often seems like a marvellous activity: giving a wedding for one's daughter. In reality, human weddings are more often a time of financial burden and emotional stress.

Like mothers of human brides, Cynthia had the anticipation of celebrating and hosting a big event, but unlike human weddings, costs were nominal. Also, unlike human weddings, the main actors in the ritual had no input. Cynthia could create an event in any fashion she desired. Clearly, Cynthia and all her guests had a wonderful time. Everyone was geared for laughter, and no one was disappointed. Judging by Cynthia's attitude and those of her friends, the major motivation for this wedding was to have fun. From one friend's perspective, 'Cynthia created an opportunity for the people she loves to come together and share in her love of her animals – her family'. From a wider perspective, Cynthia used the dog wedding as an avenue to express love and affection between people and people, and people and animals. While some might criticize forcing animals to wear clothes, their discomfort did not last very long and, at the ceremony, Heather put a quick end to it. At its conclusion, the dog nuptials delighted everyone at the wedding and rehabilitation centre.

BLESSING OF THE ANIMALS

Every Saturday before Easter, thousands of pet owners flock to El Pueblo State Historic Park in downtown Los Angeles, California, where

Fig. 6.1. The owner enjoys the spectators' interest in her sheep before being blessed by Cardinal Roger M. Mahoney at the annual Blessing of the Animals, Los Angeles, California. Because the event takes place in the gaily decorated plaza of El Pueblo, birthplace of Los Angeles, participants frequently don colourful Mexican attire and flower-trimmed hats. Sometimes they dress their animals in hats and costumes, adding humour to this spiritual occasion.

they wait patiently in line with their ribbon- and flower-festooned animals to be blessed with prayer and a sprinkling of holy water by the Cardinal in the annual Blessing of the Animals ceremony. This tradition has its roots in the sixteenth century when farmers brought their animals to church to be blessed for fertility and health. The blessing is society's recognition to the animal kingdom of the many benefits the animals provide: food, clothing, companionship.

In the 1995 ceremony, I interviewed pet owners about their motivations for bringing their animals. One woman believed that her canary had changed after being blessed the previous year. She claimed that before being blessed, the bird had been neurotic, but afterwards it had calmed down considerably. She said her family now had a loving feeling about it; that is why she wanted to have it blessed again.

A man, pulling his Chinese breed of turtle in a wagon, stated that he had attended for three years without the turtle. That did not seem right, so last year he promised the turtle he would bring him, but he did not. It was too much trouble. Subsequently, the turtle fell seriously ill.

Fig. 6.2. A girl cradles a rabbit after being blessed by the minister inside a Unitarian Church, Palos Verdes, California. Blessing of the Animals ceremonies occur across the USA in a variety of religious venues. At this event, someone portraying Noah, wearing a long white beard and draped in a sheet, stood before a painted backdrop of an ark and assisted the minister as she sanctified each owner and animal.

Although the owner wanted to believe that the turtle's illness had nothing to do with the broken promise to his pet, he was not convinced. Consequently, like a *manda* (vow) needing to be fulfilled, this year he finally had the turtle blessed.

Asking participants why they brought their pets to the blessing then allowed me to ask questions about other celebrations for their animals. Commonly, people described elaborate birthday parties for their pets, especially dogs, and most gave presents to their pets at Christmas. One man apologized for having no home celebrations for his turtles and cat but admitted that he and his wife always wished the pets a 'Happy Birthday' and 'Merry Christmas'. A woman describing birthday festivities for her Belgian Schipperke said, 'We have no grandchildren and I don't see any reason why we can't lavish love on an animal and take good care of it'.

DEATH RITUALS

The major difference between observing death customs for humans and animals is that there are no rules for animals. Many places do not even restrict where to dispose of the ashes. All that is required are acts which bring solace to the deceased animals' human companions and caretakers. The final step can be as ritual-less as calling animal rescue to pick up the remains or having the pet put to sleep at an animal hospital. At the same time, humans have the opportunity to invent new animal death rituals, such as creating obituaries in cyberspace.[5] More frequently, however, they incorporate the animals into human funeral customs taken from their own religious/cultural backgrounds. For example, a California pet cemetery owner reported that if a pet owner's family is Asian, they frequently burn incense; if the family is Jewish, they will not place a marker on the grave until one year has passed.

In the Midwest, Dr Brigit Barnes (not her real name), a veterinarian who runs a mobile veterinary hospital, acknowledges the emotional pain both she and pet owners experience when a sick animal must be put to sleep. Because Dr Barnes sees pets and families in their own homes, the non-clinical environment makes her more aware of the family's emotional needs. Her compassion for animals and their owners led her to create a euthanasia ceremony to assist them in coping with this dramatic last act. Dr Barnes performs the ceremony not only for dogs and cats, but for pot-bellied pigs, rats, hamsters, any household pet. Ceremonies are not identical. She considers each family's religious beliefs when creating the ritual to ensure that it is congruent with the family's spiritual views.

Fig. 6.3. Christmas trees ornament graves at Pet Haven Cemetery, Gardena, California. Two weeks before Christmas, mourners bring small trees to decorate the graves of their beloved dogs, cats, horses, goats, rabbits, pot-bellied pigs, snakes, monkeys and rats in this cemetery that houses the remains of over 30 000 animals. Pet lovers visit each other and reminisce about past holidays when their animals were still a vital part of the family.

Consequently, her euthanasia ceremony flourishes and word about its comfort for humans and animals has spread; other veterinarians now refer clients to her for this farewell ritual.

Dr Barnes believes that animals have souls too and honours their spirits as they depart. As caring as she is for the animals, her concern extends to their owners. For example, she recognizes how difficult it is for a child to lose a hamster. Instead of counseling the child to go to the pet store to buy a replacement, she acknowledges the bond between the hamster and its young owner and encourages the expression of grief. Similarly, when senior citizens are about to lose their pets, she is sensitive to the series of losses they have endured and comprehends the emotional devastation of perhaps losing the last living creature to which they are connected.

After deciding about euthanasia and the disposal of the remains, the doctor and family set up an evening appointment when family members are home from school and work. When the doctor arrives, the pet is generally lying in its favourite chair or spot on the floor. It is an intensely emotional moment. Family members cry, yet everyone knows that putting the animal to sleep is the greatest kindness they can perform.

First, Dr Barnes medically assesses the situation, then gives the animal a sedative to relax it. After that has taken effect, she gently moves it to a white patchwork quilt that she regularly uses for this purpose. Then, while the pet is still awake and comfortable, she lights a special white candle and dedicates it to the pet. She reads a prayer, and afterwards family members say their farewells, perhaps with a poem, maybe a blessing, the placement of a flower next to the animal, or a child reading a goodbye message. Subsequently, Dr Barnes gives the final medication. As soon as it has taken effect, a family member lovingly folds the quilt over the pet. While still surrounding it, one of them reads a passage from *Good-bye, My Friend: Grieving the Loss of a Pet* (Montgomery & Montgomery, 1991). Dr Barnes writes a dedication to the pet and gives the book to the family to help them during the bereavement period.

For the animals, they are calmed by being at home with familiar family smells. Dr Barnes has noted that sometimes prior to the final shot, dogs will even wag their tails. For the owners, the satisfaction of saying goodbye to their beloved pet in a personal setting brings them solace. They have not just dumped it into an animal emergency room and hurried home. They have been faithful to the end. For the doctor, she knows that she has given the pets a proper send-off, that she has honoured their spirits. She has brought comfort to the owners by customizing the ceremony in accordance with their religious beliefs and social needs.

CONCLUSION

Why do people impose human celebrations on animals? There appear to be several levels of meaning. In part, it may be related to urban America's detachment from nature and an innate need to reconnect. Elsewhere in the world, existing rituals express ties to the animal kingdom. For example, Hindus in India are more closely linked to animals through the worship of animal deities such as the elephant-headed Ganesha. Days set aside to pay tribute to animal gods provide a ready-made outlet for expressing reverence for animals. In the United States, Native Americans pay respect to animals through dances to the eagle or buffalo or through animal celebrations such as the Inupiat Eskimo's Festival of the Whale.

Urban American culture has no parallel. The only traditional ceremony mainstream Americans have that honours non-human creatures is

Fig. 6.4. A cat sports a birthday tiara. The owner of the cat, Magenta, of San Antonio, Texas, observes her pet's birthday with a quiet celebration. On pets' birthdays, owners may place party hats on their animals and give them special food treats, birthday cakes and presents. Sometimes they invite human and animal guests to the festivities.

the centuries-old Blessing of the Animals. However, that occasion touches too few and, even when it does, it appears insufficient to fill the needs of contemporary urbanites. Almost everyone interviewed at the blessing ceremonies admitted that they included their pets in human rituals as well. Hence, it appears that if no animal-specific rituals are available or are in limited supply, we incorporate the animals into those originally intended only for humans.

The prevalence of such rituals may represent a change in American domestic life. According to a 1996 Dow Jones report (Palmer, 1996: 29), seven out of ten American pet owners thought of their pets as children. In addition, pet owners spent money on their pets as if they were their children, treating their pets as family members. A Gallup Poll found that 24% of Americans celebrated their pets' birthdays and 17% celebrated with gifts or birthday cakes (Pet Product News, 1995: 27). Judging by the ubiquitous sight of Christmas stockings filled with treats for cats and dogs, pet owners also offer gifts to their pets at holiday time.

The Dow Jones report provides a rationale. Because the average American family size may be shrinking and adult children are moving long distances from their parents, family pets fill an emotional void.

Fig. 6.5. A Christmas carrot wreath adorns the stable wall, Los Angeles, California. Kimi's owner, like other horse lovers, decorates his steed's stall for the holidays. Owners hang Santa decorations, 'Merry Christmas' signs, and stockings stuffed with candy canes, sugar cubes, apples, carrots and horse cookies.

Several attendees at the Blessing of the Animals corroborated this point of view, expanding upon it by admitting that their animals brought them *greater* pleasure than their children. They offer what children do not: obedience, loyalty, unconditional love.

In his book *Collective Search for Identity*, sociologist Orrin E. Klapp (1969) discusses the role of rituals in contemporary society and enumerates the benefits of such ceremonies, some of which are elaborated upon below. Of course, he is speaking about human rituals, but do they differ in function when animals are key actors? Folklorist Jay Mechling (1989) investigates animals in human roles when he considers them as *bona fide* participants in dyadic folklore, relationships of two players based on common understanding of jokes or games. Consequently, he opens up the possibility of legitimizing animals as peers with human actors when analysing ritual behaviour. Because he has removed the border between species in jokes and games, he has made it a simple step to include animals in celebrations.

Human–animal ceremonies allow people to express sentiments non-verbally that may be difficult to articulate, sentiments such as love and commitment. The man who brought his sick turtle to the Blessing of the Animals demonstrated this. Further, Dr Barnes's euthanasia ceremony releases deep emotions. Through the lighting of the candle, the physical surrounding of the dying animal and covering it with the white quilt, family members express love and respect for their departing pets.

Like ordinary human rituals, human–animal celebrations provide social identity and communitas – a feeling of oneness – as exemplified by the gatherings at the Blessing of the Animals ceremonies. At the dog wedding of Heather and Scotty, Cynthia constructed a unique, though temporary, community among her guests. One of her friends recognized this when she observed, 'Cynthia created an opportunity for the people she loves to come together and share in her love of her animals'.

Human–animal celebrations promote playfulness and creativity. The cat *mitzvah* song parody and the pretend weeping of guests are but one example taking place at that event. Rachel's horse *bar mitzvah* with its satire on common *bar mitzvah*, expressions such as 'Today I am a *mensch*', and Cynthia's decorations and puns are excellent examples of rituals providing opportunities to exercise creativity for the purpose of having fun. While the stage is set by the hosts, through their gifts and own behaviour, guests enter into the spirit of play. Through playfulness, participants therapeutically release emotions and engage in mental activities not ordinarily permitted in daily life, thus offering a break from the mundane.

While some might find it sacrilegious to have human religious ceremonies like *bar mitzvahs* and marriage ceremonies for animals, a precedent exists. In thirteenth-century France, during the Feast of the Donkey, a live ass was brought into church every January 1 (Cohen, 1994: 63).[6] Similarly, since the sixteenth century, during the Palio of Siena, Italy, horses have been brought into the church and sprinkled with holy water. Moreover, if the horses defecate in church, the community interprets it as a sign of good luck (Dundes & Falassi, 1975: 96). Notwithstanding that the Siena palio and French donkey feast are a part of role reversal celebrations, they still attest to the longevity of human–animal religious traditions. Additionally, in Germany, a donkey held the place of honour in the Palm Sunday procession, and a Mass for horses was common, especially in Bavaria (Cohen, 1994: 63). Motivations may not be consistent; nonetheless, that such traditions have endured for hundreds of years indicates that human–animal celebrations satisfy human needs.

Finally, I believe that including animals in religious and secular ceremonies does not transpire because of human imperialism. Instead, the data reveal that the human construct of reality is not limited only to things human, and that we include animals in celebrations as an act of love for our earthly companions. These actions illustrate how meaningful animals are to us as a bridge to both the natural and supernatural worlds. This seems essential for pet owners, either at a subconscious level or as overtly expressed in the statements of kitten Fifi Katz's 'father', 'Through a particular little animal. . . we realize our own at-homeness in the universe'.

Fifi's 'father' articulates what most of us only intuit: there is something spiritually uplifting about interacting with other species.[7] By so doing, it heightens awareness of our interrelatedness within a larger universe. In an unknown cosmos we acknowledge our togetherness with other creatures. Through these rituals, whether they elicit laughter or tears, humans and animals establish a profound, at times heart-rending bond with one another. Human–animal celebrations affirm life.

NOTES

1. After one Halloween, for which the chimps sign PUMPKIN DAY, Tatu asked for BIRD MEAT, their designation for Thanksgiving. After Thanksgiving, Tatu began signing SWEET TREE, a reference to Christmas (Urlie, 1996).
2. Here are some of the minimum USDA standards for laboratory chimpanzees. They must be kept in single animal cages no smaller than 5 feet by 5 feet. They must have food available and receive water twice daily. They must be looked at once a day. The physical environment must promote the psychological well-being of the animal. It is this last sentence that falls into the category of environmental enrichment (Beach, 1986). One method is to provide a challenge for obtaining food. For example, ordinarily the chimps gulp down their daily liquids, but at celebrations they face the challenge of sipping boxed juices from straws or removing frozen Kool-Aid from inside balloons.
3. Music by Jerry Bock and lyrics by Sheldon Harnick, New York: Crown Publishing, 1964.
4. Names of all informants have been changed.
5. Virtual Pet Cemetery <http://www.lavamind.com:80/pet.html> encourages bereaved pet owners to post obituaries for their pets on their Web page.
6. This celebration was a part of the topsy-turvy Feast of Fools when class distinctions were abolished: masters served slaves; men wore women's clothes; people freely gambled at the church altar, and all restraints of law and morality were unleashed.
7. I recall three times during the 1980s when I assisted artist Jan Steward in painting traditional Indian design motifs onto live elephants. In each instance, I commented afterwards how I felt emotionally touched by my physical closeness to these huge creatures. I was thrilled, felt privileged to paint their toenails in greens, blues and yellows, draw bold-coloured flowers onto their bulky sides and cautiously paint around their eyes. Simultaneously, they communicated with me by passing their inquisitive trunks all over my body. After each painting session I reported to others how it had been a mystical experience.

REFERENCES

Beach, K. (1986). A few notes on USDA requirements for captive primates. *Friends of Washoe*, **5**, 6.

Cohen, E. (1994). Animals in medieval perceptions: the image of the ubiquitous other. In *Animals and Human Society: Changing Perspectives*, ed. A. Manning & J. Serpell, pp. 59–80. London and New York: Routledge.

Dundes, A. & Falassi, A. (1975). *La Terra in Piazza: an Interpretation of the Palio of Siena*. Berkeley: University of California Press.

Klapp, O.E. (1969). *Collective Search for Identity*. New York: Holt, Rinehart & Winston.

Mechling, J. (1989). 'Banana Cannon' and other folk traditions between human and nonhuman animals. *Western Folklore*, **48**, 312–23.

Montgomery, M. & Montgomery, H. (1991). *Good-bye My Friend: Grieving the Death of a Pet*. Minneapolis: Montgomery Press.

Palmer, J. (1996). Well, aren't you the cat's meow? *Barron's*, **1 April**, 29–34.

Pet Product News (1995). **49**, 27.

Urlie, M. (1996). Tatu the time keeper. *Friends of Washoe*, **17**, 3.

JAMES A. SERPELL

7

Creatures of the unconscious: companion animals as mediators

Primitive man must tame the animal in himself and make it his helpful companion; civilized man must heal the animal in himself and make it his friend.

Carl Jung

INTRODUCTION

Ever since the publication of Erika Friedmann's ground-breaking study of recovery rates among pet-owning and non-owning heart-attack sufferers, the emphasis of research on human–companion animal relationships has moved increasingly towards identifying quantitative, physical and/or physiological benefits of pet ownership (Friedmann *et al.*, 1980, 1983; Katcher *et al.*, 1983; Siegel, 1990; Serpell, 1991; Anderson *et al.*, 1992; Friedmann & Thomas, 1995; Friedmann *et al.*, Chapter 8). Associated with this trend has been an increasing tendency to seek mechanistic explanations for some of the apparent medical benefits that have been observed. To date, at least two plausible mechanisms have been proposed involving either the immediate, physiologically de-arousing effects of tactile or visual contact with pet animals (Katcher *et al.*, 1983), and the ability of these animals to provide their owners with a form of stress-reducing or stress-buffering social support (Siegel, 1990; Serpell, 1996; Collis & McNicholas, 1998). Given the need to overcome the natural skepticism and conservatism of the medical profession, this emphasis on the quantifiable benefits of pet-keeping is entirely understandable and unsurprising. Nevertheless, it has produced in its wake a tendency to overlook or ignore some of the earliest writings on human–companion animal relationships that focus more on what I shall call the *psycho-spiritual*[1] benefits of pet ownership. The purpose of this chapter is to revisit and re-examine some of these early ideas.

THE BACKGROUND

Much of the early literature on animal-assisted therapy, or the so-called 'human–animal bond', contains allusions to the mediating properties of companion animals. These mediating powers fall into three principal areas, which I shall call, respectively, the *social lubricant*, the *animal ambassador*, and the *animal within*.

The *social lubricant* effect refers to the supposed ability of pets to catalyse social relationships between people. Psychiatrists and psychotherapists working in the 1960s and 1970s were among the first to recognize this apparent 'ice-breaking' characteristic of companion animals, and they evidently employed it to good effect when trying to establish therapeutic relationships with abnormally withdrawn or antisocial patients (see Levinson, 1969). Similar effects have also been documented among institutionalized elderly (Corson & O'Leary Corson, 1980), people walking their dogs in public parks (Messent, 1983), and among wheelchair-bound, handicapped children and adults accompanied by service dogs (Eddy *et al.*, 1988; Mader *et al.*, 1989). The mechanism(s) underlying this social lubricant effect of pets is far from clear, although, based on their work with disturbed adolescents and institutionalized elderly, Corson and O'Leary Corson (1980: 107) referred to pets as 'nonverbal communication mediators' and claimed that they offered withdrawn or otherwise isolated individuals 'a form of nonthreatening, nonjudgemental, reassuring nonverbal communication and tactile comfort and thus helped to break the vicious cycle of loneliness, helplessness and social withdrawal'. The key qualities of pets, in other words, would appear to be their special combination of human and non-human traits – their uncritical friendliness and willingness to interact socially with people, combined with their furriness and their inability to speak – in short, their intermediacy.

The second area of mediation, the *animal ambassador* idea, refers to the capacity of pets to serve as a sort of moral link with other categories of animals, and with the broader category of 'nature' of which other animals are perceived to be an integral part. As Michael Fox (1981: 38) once put it, 'keeping a companion animal can help one mature through understanding, to appreciate the intrinsic worth and basic rights of a fellow earth being'. Recently, this idea has been developed more fully by Serpell and Paul (1994), and it is tentatively supported by empirical findings that demonstrate statistical correlations between humane attitudes to animals in adulthood and positive relationships with pets in childhood (Paul & Serpell, 1993). There are, of course, other explanations for these correlations, but one plausible interpretation is that sympathetic feelings developed towards companion animals during the childhood years subsequently generalize to encompass other categories of animals, or nature itself, later in life. Pets, so the argument goes, are able to fulfil this ambassadorial role because of their ambiguous, *intermediate* position on the boundary between human and animal, culture and nature (Serpell & Paul, 1994; Serpell, 1995).

Finally, there is also a clear suggestion in the early literature that companion animals can help to connect or reunite people with something fundamental within themselves – a sort of unconscious *animal within* – and it is primarily this mediating role of pets that provides the focus of the present discussion. In his book *King Solomon's Ring*, Konrad Lorenz hints at this third area of mediation when he describes the pleasure he derives from the company of his dog as 'closely akin to the joy accorded me by the raven, the greylag goose or other wild animals that enliven my walks through the countryside; it seems like a re-establishment of the immediate bond with that unconscious omniscience we call nature' (Lorenz, 1952: 126). Lorenz's tone is somewhat vague and metaphysical, but in the writings of Boris Levinson the *animal within* idea was more clearly articulated. In *Pets and Human Development*, for example, he stated that:

> One of the chief reasons for man's present difficulties is his inability to come to terms with his inner self and to harmonize his culture with his membership in the world of nature. Rational man has become alienated from himself by refusing to face his irrational self, his own past as personified by animals. (Levinson, 1972: 6)

The solution to this growing sense of alienation was, according to Levinson, to restore a healing connection with our own, unconscious animal natures by establishing positive relationships with real animals, such as dogs, cats and other pets. He argued that pets represent 'a halfway station on the road back to emotional well-being' (Levinson, 1969: xiv) and that 'we need animals as allies to reinforce our inner selves' (Levinson, 1972: 28–9). In other words, the process of empathizing with, and relating successfully to, our pets involves tuning into and accepting our own repressed animality.

Levinson's notion of the *animal within* is clearly derived, to a large extent, from psychoanalytic theories of the unconscious, and the symbolic roles that animals are thought to play within it. According to

Sigmund Freud's ideas concerning the origins of neurosis, infants and young children are essentially similar to animals, insofar as they are ruled by instinctive cravings or impulses organized around basic biological functions such as eating, excreting, sexuality and self-preservation. As children mature, their adult caretakers endeavour to 'tame' or socialize them by instilling fear or guilt whenever the child acts too impulsively in response to these inner drives. Children, in turn, respond to this external pressure to conform by repressing these urges from consciousness. Mental illness results, or so Freud maintained, when these bottled-up animal drives find no healthy or creative outlet in later life, and erupt uncontrollably into consciousness (Shafton, 1995).

Freud interpreted the recurrent animal images that surfaced in his patients' dreams and 'free associations' as metaphorical devices by means of which people disguise unacceptable thoughts or feelings. 'Wild beasts' he argued 'represent passionate impulses of which the dreamer is afraid, whether they are his own or those of other people' (Freud, 1959: 410). Because these beastly thoughts and impulses are profoundly threatening to the ego, they are locked away in dark corners of the subconscious where their terrifying howls and shrieks can be safely ignored, at least during a person's waking hours. During sleep, however, the cage doors are unlocked, and these creatures of the unconscious rampage through the psyche like foxes in a chicken coop. To Freud and his followers, the aim of psychoanalysis was to unmask these frightening denizens of the unconscious mind, reveal their true natures, and thus, effectively, to neutralize them.

Although empirical support for Freud's overall thesis – let alone his interpretation of dreams – is unimpressive (Foulkes, 1982), the validity of the Freudian concept of the dream animal is at least tentatively confirmed by the results of recent surveys. In his exhaustive quantitative analysis of children's dreams, Foulkes (1982: 81) not only detected an extremely high prevalence of animal figures in the dreams of children between the ages of three and seven, but also found evidence that dream animals were associated with 'problems of impulse control'. Among five to seven year olds, animal dreams were reported more frequently by children diagnosed as being behaviourally impulsive and socially immature, and an increasing prevalence of animal characters was associated with a decreasing level of self-participation in dreams (implying that the animal figures substitute for the missing self). High rates of animal dreaming were also associated statistically with dreams involving aggression. Studies of adult dreams have produced comparable findings. According to one such survey:

> dreams with animal figures are more likely to be short, take place in an outdoor setting which is unfamiliar or distorted, have a great deal of activity, often of a violent nature, or be the scene of a calamity . . . As the emphasis upon animal figures increases and a greater predominance of animal figures occurs, all of the previous dream parameters become proportionately more intensified. Animal dreams are not exclusively negatively toned; sometimes the dreamer attempts to respond in an accepting or supportive role toward the animal figure, but almost without exception, if the animal figure initiates any response to the dreamer, it is some form of threat or hostility. (Van de Castle, 1983: 170)

In Freud's view, the interpretation of animal symbols in dreams depended entirely on the personal history and experience of the dreamer. He rejected the Jungian idea of 'archetypes' or universal animal symbols. Again, ethnographic survey results seem to support Freud's position. Dogs, horses and cats predominate in the dreams of American college students; marine or aquatic animals dominate the dreams of Pacific islanders; Australian Aborigines tend to dream about such things as kangaroos, wallabies, crocodiles, snakes and sting rays, while birds are relatively common in the dreams of all three groups (Van de Castle, 1983). In other words, the animal figures that find their way into our dreams tend to be the same animal figures we are most likely to encounter in everyday life.

In addition to their role in dreams, animals are also one of the universal raw materials of myths, folktales and fables the world over (Bettleheim, 1976; Sax, 1990; Doniger, 1995). Freud, like Lévi-Strauss (1966), would doubtless have accounted for the universality of this animal symbolism by arguing that animals are essentially 'good to think' – that, for the purposes of disguising uncomfortable or threatening thoughts and emotions, no other category of things could serve as well. Such a view was also consonant with the already long-established Calvinist and Hobbesian tendency to represent all antisocial or improper human conduct as the product of mankind's fundamentally 'brutish' or beast-like nature (Myers, 1998). Levinson, however, took the argument an important stage further than this by suggesting that the popularity of animal symbolism is basically innate – 'an expression of the unconscious affinity between man and animals' – rather than simply a convenient means of dressing up things we are afraid to confront in the flesh. Relations with animals played such a prominent role in human evolution, he felt, that they have become integral to our psychological well-being (Levinson, 1972: 15). In this sense, Levinson's ideas appear to anticipate those of various New Age spiritual ecologists (e.g. see Merchant, 1992), as well as E.O. Wilson's 'biophilia hypothesis' — the postulate that humans possess a biologically based, inherent predisposition to attend to, and affiliate with, life and lifelike processes (Wilson, 1984; Kellert, 1993).

THE 'SAVAGE' MIND

One obvious implication of Levinson's hypothesis is that this sense of unconscious affinity with animals is an ancestral trait; an inherited legacy of the days when humans lived closer to 'nature' than they do now. It therefore seems appropriate, at this point, to examine the validity of this claim by exploring the status of animals in the mental lives of subsistence hunter–gatherers whose lifestyles and worldviews are thought to lie closest to the original or ancestral human pattern.[2]

A remarkable degree of consistency in attitudes and beliefs about animals exists (or existed until recently) among hunter–gatherer societies from regions as far apart as Siberia, Amazonia or the Kalahari Desert of southern Africa. Briefly summarized, these beliefs include the notion that animals are fully rational, sentient and intelligent beings, in no way inferior to humans, and that the bodies of animals, like those of people, are animated by non-corporeal spirits or 'souls' that survive the body after death. While it is recognized that certain skills are needed in order to be a good hunter, it is also believed that no amount of skill or ingenuity will succeed if the animal quarry is unwilling to submit to being killed. Game animals must therefore be treated at all times with proper respect and consideration in order to earn their goodwill. Failure to treat the animal respectfully may cause either the animal's spirit or that of its spiritual guardian to demand some form of posthumous restitution. Types of spiritual retribution that may result from disrespectful behaviour include the infliction of illness, injury, madness or death on the hunter or other members of his family or clan, or loss of success in future hunting (Hallowell, 1926; Benedict, 1929; Speck, 1977; Martin, 1978; Campbell, 1984; Nelson, 1986; Guenther 1988; Wenzel, 1991; Erikson, Chapter 2). As Ingold (1994: 15) has observed, 'The hunter hopes that by being good to animals, they in turn will be good to him. But by the same token, the animals have the power to withhold if any attempt is made to coerce what they are not, of their own volition, prepared to provide . . . Animals thus maltreated will desert the hunter, or even cause him ill fortune'.

As well as being potential sources of sickness or misfortune when offended, animal spirits also provide hunter–gatherers with the means to heal illness and improve bad luck. This 'therapeutic' communication with the spirit world is typically achieved through the medium of dreams, visions or hallucinations during which the relevant animal spirits appear to the visionary, explain the spiritual source of his or her problem, and provide ritual guidance as to the best method of effecting a 'cure'. Most people are thought to be capable of experiencing these spirit dreams or visions from time to time, either by chance or through their own personal efforts. Certain individuals, however, are believed to possess extraordinary visionary powers that enable them to enter the spirit world at will. Known in anthropological circles as shamans, these individuals are credited with having an unusual affinity with the spirits of animals and nature, and a special, almost symbiotic relationship with one or more animal helping-spirits. Shamans can reputedly adopt the material form of these animal 'familiars' in order to travel about incognito, and the familiar is also the medium through which the shaman is able to enter the world of spirits and influence events at a distance from his or her body (Eliade, 1964). Some of these ideas about familiars are summarized in the following early account of Shamanism among the Penobscot Indians of New England:

> Every magician [shaman] had his helper which seems to have been an animal's body into which he could transfer his state of being at will. The helper was virtually a disguise, though we do not know whether the animal was believed to exist separately from the shaman when not in the shaman's service or whether it was simply a material form assumed by the shaman when engaged in the practice of magic . . . Direct information from Penobscot informants says that the *baohi'gan* [familiar] could be sent to fight or to work for his master the shaman. It could be sent on any mission whatsoever according to the shaman's will. We are told, too, that the owner remained inert while his *baohi'gan* was away. (Speck, 1918: 249–51)

The apparent parallels with Levinson's notions of the animal within are quite striking. In both cases, we have the concept of an unconscious domain populated by powerful animal figures. We have the idea of animal helpers mediating between people and their unconscious worlds. And we have the notion that this process of animal mediation can be used for healing purposes. However, there are also some equally striking differences. If we assume, for the sake of argument, that the hunter–gatherers' 'spirit world' is essentially synonymous with our own 'unconscious' – just different ways of representing the same thing – then one obvious difference between us and them is that hunter–gatherers are habitually engaged in a process of positive negotiation and dialogue with their inner animals. There seems to be little evidence of repression in the Freudian sense of the term. Hunter–gatherers, moreover, use the information obtained from this dialogue with the unconscious, not only to help them cope with the ordinary vicissitudes of life, but also to guide them in their dealings with nature and real animals. Consider, for example, the following description of the ways in which Bushman hunters attempt to influence the progress and outcome of their hunts:

> Throughout the hunt the hunter would monitor his every thought, emotion and action, in order to sustain the bond of connectedness with the animal by which he felt he could steer the hunt towards an auspicious conclusion . . . The bond of sympathy was something set up in the hours or days preceding the hunt, when the hunters would attune themselves spiritually to one animal species or another and, in the process, attempt to gather whatever presentiments they could about the impending hunt: the animals they might encounter, the direction they could come from, the likely dangers, the duration of the hunt. These presentiments . . . activated the hunter's entire body; they were felt at his ribs, his back, his calves, his face and eyes. His body would be astir with the 'antelope sensation', at places on his body corresponding with those of the antelope's. (Guenther, 1988: 199)

From this cultural distance, it is probably impossible for us to comprehend fully the experiences of these Bushman hunters. It appears that they not only represent aspects of their unconscious worlds in animal terms, just as we often do, but that also, at some unconscious or semi-conscious level, their thoughts and feelings actually resonate with the pulse of nature, and with the lives of the animals on whom they depend. For anyone educated in a post-Aristotelian, western tradition, it is extremely hard to imagine the profound level of identification with animals that this degree of resonance implies. Yet it is unlikely that hunter–gatherers would go to such extraordinary efforts to establish this metaphysical communion with other species if it served no useful purpose; if it did not also improve their success as hunters, and their capacity to deal with life's misfortunes. In other words, an ability to tune into and resonate with the lives of other organisms via internalized *animals within* may, indeed, be part of our biological heritage; an evolved mental skill that once enhanced our survival as hunters and gatherers.[3]

PARADISE LOST

So what became of this skill, this hypothetical, archaic ability to connect with the animal within? According to Levinson, we lost touch with it through the process of becoming civilized and urbanized, and our lives are now greatly impoverished as a result. But, while the increasing physical separation of people from nature associated with urban life has probably helped to attenuate this ability, it is possible that the process of atrophy began much earlier in human history than Levinson believed.

Subsistence hunters need to understand and identify with the animals they depend on for food because a good hunter is essentially one who learns to 'think like' his prey – to empathize with it. Hence, the marked hunter–gatherer tendency to view animals as near equals or even kinsmen, and the attendant moral conflicts associated with hunting and devouring them (see Serpell, 1996). At the same time, hunters do not ordinarily interact socially with their prey and, except at the moment of the animal's death, they exercise little or no control over it. The animal remains an independent being with a mind of its own, and it is possible for the hunter to convince himself that, if the animal allowed itself to be killed, it did so of its own free will (Ingold, 1994; Serpell, 1996).

Such an egalitarian moral ideology was incompatible with the ecological shift from hunting to animal domestication that began some 12 000 years ago. It would require, after all, a supreme feat of self-deception for a farmer or herdsman to claim that his animals were free agents. The domestic animal, almost by definition, is totally dependent for survival on its human custodian. It has no free will, as such. Moreover, because they live together in what is, to some extent, a combined social group, it is not unusual for farmers and herdsmen to establish social bonds with their animals and *vice versa*. The moral dilemma is therefore far more intense for the farmer than for the hunter, since killing or harming the animal in this context effectively constitutes a gross betrayal of trust (Serpell, 1996).

Farmers, herdsmen and others who benefit from the exploitation of domestic species have dealt with this ethical dilemma using a variety of coping strategies, but perhaps the most pervasive and durable was the idea that humans are both morally separate from and superior to all other animals (Serpell, 1996). As Ingold (1994: 16) has recently noted, the ideological difference between hunters and herdsmen primarily involves a shift from human–animal relations based on *trust* to those based on *domination*:

> In the world of the hunter, animals, too, are supposed to care, to the extent of laying down their lives for humans by allowing themselves to be taken. They retain, however, full control over their own destiny. Under pastoralism, that control has been relinquished to humans. It is the herdsman who takes life-or-death decisions concerning what are now 'his' animals . . . He sacrifices them; they do not sacrifice themselves to him. They are cared for but they are not themselves empowered to care. Like dependents in the house-

hold of a patriarch, their status is that of jural minors, subject to the authority of their human master. In short, the relationship of pastoral care, quite unlike that of the hunter towards animals, is founded on a principle not of trust but of domination.

From the typical subsistence farmer's perspective, wild animals are basically 'vermin' – annoying or menacing creatures that pose a continual danger to his life and livelihood by devouring either his crops or his livestock. Similarly, nature, in the sense of uncultivated wilderness, is a profoundly threatening concept to most farmers; the antithesis of all they hold most dear. The only animals that are considered good or beneficial are the domesticated ones that conform to the harmless and highly subordinate roles of servants or slaves (Thomas, 1983; Serpell, 1996).

In other words, the change in relations between humans and animals associated with the switch from hunting to farming produced a fundamental shift in our mental and moral taxonomy. The hunter's biological worldview, based on an egalitarian sense of continuity and connection between people and animals, was gradually displaced by a hierarchical, dualist schema that emphasizes discontinuity and difference. And the position that animals have come to occupy in the mental lives of many 'civilized' humans seems to reflect this change in status quo. In the process of distancing ourselves from other species we have also separated what we perceive to be the most human and animal-like aspects of our personalities, relegating the latter to inferior, subordinate positions where they can be mastered and controlled like domestic beasts.

Interpreted in this light, the fear of the animal within, and the repression that Freud and his followers associated with this fear, can perhaps be viewed as the product of a longstanding western tendency to deny affinity and moral obligation towards the rest of the animal kingdom. Whereas it had once been a source of therapeutic mediation and healing knowledge, the animal within became instead the potentially malevolent and destructive embodiment of our own alienation from nature and other animals.

CONCLUSION

To what extent, then, can relationships with companion animals help to heal this psychic rift? How, if at all, can they help to reconnect us with the lives of other organisms, or with nature itself? The answer must surely depend on both the animal and our relationship with it. The mediating power of pets evidently rests on their liminal, intermediate properties – their ambiguous mix of human and non-human characteristics. Eliminate this ambiguity and, presumably, you also risk eliminating their therapeutic potential. If we are willing to accept and appreciate the *animal* as well as the *human* attributes of our pets by allowing them the freedom to express at least most of their natural behaviour, they may indeed provide us with a means of overcoming what Searles (1960: 122) once called 'the existential loneliness' of our species. But if we deny and suppress our pets' animality in the way that we seem to repress and deny our own, it is hard to see what benefit either partner could derive from this relationship.

Sadly, since about the middle of the nineteenth century, the trend in pet-keeping has been towards progressively modifying, limiting and curtailing the abilities of companion animals to express the 'animal' aspects of their natures. Selective breeding has deformed their bodies to the point where some can no longer function biologically without human intervention, and, where breeding has failed, we mutilate them surgically to conform to our own arbitrary, and often bizarre, standards of physical beauty. In the interests of convenience or 'public health and safety', their behaviour and their freedom of expression have been severely restricted, and those that unwittingly transgress the boundaries of 'civilized' conduct are often swiftly disposed of (Serpell, 1995). In short, we have turned many of our pets into mere cultural artifacts, incapable of existence outside the human domain. By doing this, it could be argued, we not only turn a blind eye to the welfare of these animals, we simultaneously destroy their capacity to mediate on our behalf.

ACKNOWLEDGEMENTS

My thanks to Nicholas Humphrey, Clinton Sanders, and an anonymous referee for helpful comments on earlier drafts of this chapter.

NOTES

1. The term 'spiritual' is used here without necessarily implying the involvement of supernatural powers or agencies.
2. Anthropologists and archaeologists are understandably cautious about using living or recent hunter–gatherers as a source of insight concerning the attitudes and beliefs of our pre-agricultural ancestors. Among the very few that still survive, most living hunter–gatherers have been more or less acculturated by their more aggressive agricultural neighbours, and nearly all of them have been substantially marginalized economically. Nevertheless, as surviving exemplars of a particular – and once universal – mode of economic subsistence, recent

hunter–gatherers can hardly be ignored as a reference point, particularly when the same ideological themes are found to be shared in common by many otherwise ethnically diverse populations.

3. Recently, the archaeologist Steven Mithen (1996) has developed the idea of a separately evolved 'natural history intelligence' in humans that comes close to explaining the possible origins of hunter–gatherers' apparent psychic affinity with other animals.

REFERENCES

Anderson, W.P., Reid, C.M. & Jennings, G.L. (1992). Pet ownership and risk factors for cardiovascular disease. *Medical Journal of Australia*, **157**, 298–301.

Benedict, R.F. (1929). The concept of the guardian spirit in North America. *Memoirs of the American Anthropological Association*, **29**, 3–93.

Bettleheim, B. (1976). *The Uses of Enchantment: the Meaning and Importance of Fairy Tales*. London: Thames & Hudson.

Campbell, J. (1984). *The Way of the Animal Powers*. London: Times Books.

Collis, G.M. & McNicholas, J. (1998). A theoretical basis for health benefits of pet ownership: attachment versus psychological support. In *Companion Animals in Human Health*, ed. C.C. Wilson & D.C. Turner, pp. 105–22. Thousand Oaks, CA: Sage Publications.

Corson, S.A. & O'Leary Corson, E. (1980). Pet animals as nonverbal communication mediators in psychotherapy in institutional settings. In *Ethology and Nonverbal Communication in Mental Health*, ed. S.A. Corson & E. O'Leary Corson, pp. 83–110. Oxford: Pergamon Press.

Doniger, W. (1995). The mythology of masquerading animals, or, bestiality. *Social Research*, **62**, 751–72.

Eddy, J., Hart, L.A. & Boltz, R.P. (1988). The effects of service dogs on social acknowledgements of people in wheelchairs. *Journal of Psychology*, **122**, 39–45.

Eliade, M. (1964). *Shamanism: Archaic Techniques of Ecstacy*, trans. W.R. Trask. New York & London: Routledge.

Foulkes, D. (1982). *Children's Dreams: Longitudinal Studies*. New York: John Wiley & Sons.

Fox, M.W. (1981). Relationships between the human and nonhuman animals. In *Interrelations between People and Pets*, ed. B. Fogle, pp. 23–40. Springfield, IL: Charles C Thomas.

Freud, S. (1959). *The Interpretation of Dreams*, trans. J. Strachey. New York: Basic Books.

Friedmann, E., Katcher, A.H., Lynch, J.J. & Thomas, S.A. (1980). Animal companions and one-year survival of patients after discharge from a coronary care unit. *Public Health Reports*, **95**, 307–12.

Friedmann, E., Katcher, A.H., Thomas, S.A. Lynch, J.J. & Messent, P.R. (1983). Social interactions and blood pressure: the influence of animal companions, *Journal of Nervous and Mental Disease*, **171**, 461–5.

Friedmann, E. & Thomas, S.A. (1995). Pet ownership, social support, and one-year survival after acute myocardial infarction in the Cardiac Arrhythmia Suppression Trial (CAST). *American Journal of Cardiology*, **76**, 1213–17.

Guenther, M. (1988). Animals in Bushman thought, myth and art. In *Hunters and Gatherers 2: Property, Power and Ideology*, ed. T. Ingold, D. Riches & J. Woodburn, pp. 192–202. Oxford: Berg.

Hallowell, A.I. (1926). Bear ceremonialism in the Northern Hemisphere. *American Anthropologist*, **28**, 1–175.

Ingold, T. (1994). From trust to domination: an alternative history of human–animal relations. In *Animals & Human Society: Changing Perspectives*, ed. A. Manning & J.A. Serpell, pp. 1–22. London: Routledge.

Katcher, A.H., Friedmann, E., Beck, A.M. & Lynch, J.J. (1983). Looking, talking and blood pressure: the physiological consequences of interaction with the living environment. In *New Perspectives on Our Lives with Companion Animals*, ed. A.H. Katcher & A.M. Beck, pp. 351–9. Philadelphia: University of Pennsylvania Press.

Kellert, S.R. (1993). The biological basis for human values of nature. In *The Biophilia Hypothesis*, ed. S.R. Kellert & E.O. Wilson, pp. 42–69. Washington, DC: Island Press.

Levinson, B. (1969). *Pet-oriented Child Psychotherapy*. Springfield, IL: Charles C Thomas.

Levinson, B. (1972). *Pets and Human Development*. Springfield, IL: Charles C Thomas.

Lévi-Strauss, C. (1966). *The Savage Mind*. Chicago: Chicago University Press.

Lorenz, K. (1952). *King Solomon's Ring*. London: Methuen.

Mader, B., Hart, L.A. & Bergin, B. (1989). Social acknowledgments for children with disabilities: the effects of service dogs. *Child Development*, **60**, 1529–34.

Martin, C. (1978). *The Keepers of the Game*. Berkeley: University of California Press.

Merchant, C. (1992). *Radical Ecology*. New York & London: Routledge.

Messent, P.R. (1983). Social facilitation of contact with other people by pet dogs. In *New Perspectives on Our Lives with Companion Animals*, ed. A.H. Katcher & A.M. Beck, pp. 37–46. Philadelphia: University of Pennsylvania Press.

Mithen, S. (1996). *The Prehistory of the Mind*. London: Thames & Hudson.

Myers, O.E. (1998). *Children and Animals*. Boulder, CO: Westview Press.

Nelson, R.K. (1986). A conservation ethic and environment: the Koyukon of Alaska. In *Resource Managers: North American and Australian Hunter–gatherers*, ed. N.M. Williams & E.S. Hunn, pp. 211–28. Canberra: Institute of Aboriginal Studies.

Paul, E.S. & Serpell, J.A. (1993). Childhood pet keeping and humane attitudes in young adulthood. *Animal Welfare*, **2**, 321–37.

Sax, B. (1990). *The Frog King: On Legends, Fables, Fairy Tales and Anecdotes of Animals*. New York: Pace University Press.

Searles, H.H. (1960). *The Nonhuman Environment*. New York: International Universities Press.

Serpell, J.A. (1991). Beneficial effects of pet ownership on some aspects of human health and behaviour. *Journal of the Royal Society of Medicine*, **84**, 717–20.

Serpell, J.A. & Paul, E.S. (1994). Pets and the development of positive attitudes to animals. In *Animals and Human Society: Changing Perspectives*, ed. A. Manning & J.A. Serpell, pp. 127–44. London: Routledge.

Serpell, J.A. (1995). From paragon to pariah: some reflections on human attitudes to dogs. In *The Domestic Dog: its Evolution, Behaviour and Interactions with People*, ed. J.A. Serpell, pp. 245–56. Cambridge: Cambridge University Press.

Serpell, J.A. (1996). *In the Company of Animals*, 2nd edn. Cambridge: Cambridge University Press.

Shafton, A. (1995). *Dream Reader: Contemporary Approaches to the Understanding of Dreams*. Albany, NY: SUNY Press.

Siegel, J.M. (1990). Stressful life events and use of physician services among the elderly: the moderating role of pet ownership. *Journal of Personality and Social Psychology*, **58**, 1081–6.

Speck, F.G. (1918). Penobscot shamanism. *Memoirs of the American Anthropological Association*, **6**, 238–88.

Speck, F.G. (1977). *Naskapi*, 3rd edn. Norman: University of Oklahoma Press.

Thomas, K. (1983). *Man and the Natural World: Changing Attitudes in England, 1500–1800*. London: Allen Lane.

Van de Castle, R.L. (1983). Animal figures in fantasy and dreams. In *New Perspectives on Our Lives with Companion Animals*, ed. A.H. Katcher & A.M. Beck, pp. 148–73. Philadelphia: University of Pennsylvania Press.
Wenzel, G. (1991). *Animal Rights, Human Rights: Ecology, Economy and Ideology in the Canadian Arctic*. London: Belhaven Press.
Wilson, E.O. (1984). *Biophilia*. Cambridge, MA: Harvard University Press.

Part II
The nature of the relationship

ERIKA FRIEDMANN, SUE A. THOMAS AND TIMOTHY J. EDDY

8

Companion animals and human health: physical and cardiovascular influences

INTRODUCTION

Contributions of animals to meet basic human health needs include functioning as food, clothing, beasts of burden, working and assistance animals, in addition to medical uses such as vaccine production or as human surrogates for the development of medical procedures and products (Lierman, 1987). Animals also have well-documented detrimental effects on basic human health, including zoonoses, allergies and as sources of injury (Schantz, 1990; Plaut, Zimmerman & Goldstein, 1996; Centers for Disease Control and Prevention, 1997).

In the last two decades researchers have begun to assess the possibility that animals could have more subtle positive effects on human health, perhaps by satisfying what Maslow (1970) termed higher order needs. The cardiovascular system was a logical starting point for evaluating the effects of owning pets on human health since cardiovascular disease was one of the first chronic non-psychiatric diseases upon which psychological and social factors were found to have significant impact (Jenkins, 1976).

LONG-TERM EFFECTS

Four epidemiological studies have demonstrated the potential of pets to influence the cardiovascular health of their owners. In the first epidemiological study, pet owners were found to be more likely to be alive one year after discharge from a coronary care unit than non-owners (Friedmann *et al.*, 1980). The relationship of pet ownership to survival was independent of disease severity and other sources of social support (see Friedmann, 1995).

Two subsequent epidemiological comparisons, using larger samples of coronary heart disease patients, suggested that different types of pets might provide different benefits. Among 369 post-myocardial infarction patients with ventricular arrhythmias who participated in the Cardiac Arrhythmia Suppression Test (CAST), both pet ownership and amount of social support tended to predict one-year survival, independent of the physiological severity of the cardiovascular disease, and of other demographic and psychosocial factors (Friedmann & Thomas, 1995). Dog owners were approximately 8.6 times as likely to be alive after one year as those who did not own dogs ($p=0.02$). Social support was also a significant predictor of survival ($p=0.05$). The effects of dog ownership and social support on survival were independent of each other and of the physiological severity of the illness. No parallel independent effect of cat ownership was observed. Not only did cat ownership not predict survival, cat owners were actually more likely to die than non-owners ($p=0.03$). Cat ownership was also associated with lower social support. The difference in the effects of dog and cat ownership on survival is difficult to interpret because the effect of cat ownership on one-year survival was confounded by both the amount of social support received and the over-representation of women in the cat-owning group. In the CAST, women were more likely to die than men ($p<0.01$).

In another study, pet ownership was not related to six-month survival incidence of angina, or changes in psychological health among 454 patients who had experienced myocardial infarctions (Rajack, 1997). Furthermore, owning a pet was not related to readmission to hospital for angina or other cardiovascular causes, or to experience of further cardiac problems within six months after myocardial infarction. However, cat owners were more likely ($p=0.027$) to be readmitted to hospital for further cardiac problems or angina than people who did not own pets. This finding again must be interpreted cautiously; one significant result from multiple statistical comparisons raises the possibility of a random effect. And because Rajack (1997) did not control for physiological severity of the illness, it is difficult to compare these findings with those of the other case-control studies. These epidemiological studies (Friedmann & Thomas, 1995; Rajack, 1997) suggest that dog and cat ownership may have different effects on health status. These suggestions require additional research for elucidation.

Pet ownership may also protect people from developing coronary heart disease as well as influencing the survival of individuals who have experienced myocardial infarctions. A recent large-scale, cross-sectional, epidemiological study addressed the differences between pet owners and non-owners in physiological and behavioural variables, known as risk

Table 8.1. *Symptoms included on the checklist of minor health problems. Participants were asked to underline or check items they had experienced in the month prior to receiving the questionnaire*

Headaches	Painful joints
Hay fever	Difficulty concentrating
Difficulty sleeping	Palpitations or breathlessness
Constipation	Trouble with ears
Trouble with eyes	Worrying over every little thing
A bad back	Indigestion or other stomach trouble
Nerves	Sinus trouble or catarrh
Colds and flu	Persistent cough
General tiredness	Faints or dizziness
Kidney or bladder trouble	Trouble with feet

Source: Serpell (1991).

factors, which are associated with increased likelihood of developing coronary heart disease (Anderson, Reid & Jennings, 1992). Among 5741 people attending a cardiovascular disease-screening clinic in Melbourne, Australia, risk factors for coronary heart disease were significantly greater among the 4957 non-pet owners than among the 784 pet owners. For men, plasma levels of cholesterol ($p < 0.01$) and triglycerides ($p < 0.01$), and systolic blood pressure ($p < 0.01$) were significantly lower for pet owners than for non-owners. For women, differences in risk factors between pet owners and non-owners occurred only for those who were most susceptible to coronary heart disease – women in the menopausal and post-menopausal age groups.

The effect of pet ownership on other aspects of physical health has been addressed to a lesser degree. In a longitudinal case-control study of people who adopted pets from an animal shelter, adults who acquired a cat ($n = 24$) or dog ($n = 47$) and control participants ($n = 26$) were monitored for 10 months (Serpell, 1991). Pet adopters experienced significant decreases (dog owners: $p < 0.001$; cat owners: $p < 0.01$) in minor health problems (see Table 8.1 for a list of health problems included) one month after adopting the pet. Dog owners maintained this decrease over the ten-month duration of the study ($p < 0.05$) but cat owners did not. In this study, dog owners both appeared to walk slightly more at baseline, and reported a fourfold to fivefold increase in frequency and duration of walking ($p < 0.001$) at ten months. Members of the control group also increased their walking, although to a lesser extent ($p < 0.05$). The physiological benefits associated with acquiring a dog could have been the result of increased physical activity engendered by walking the animal. Differences in walking may also reflect other differences in lifestyle and availability of time for walking and caring for a dog. Differences in health status between pet adopters and controls may also have been confounded by the higher incidence of stressful life events among the former during the year prior to pet adoption (Serpell & Jackson, 1994).

Since the above studies address differences between pet owners and non-owners, they suffer from self-selection bias. People choose to be pet owners, and the types of pets to own, rather than being randomly assigned to pet-owning and non-owning groups. Thus, factors which determine pet ownership might also be those that determine better cardiovascular health.

Since double-blind studies of animals' impacts on health are not feasible, we must rely on the combined evidence from multiple studies. Epidemiological evidence generally supports some health benefits from pets, but the physiological mechanisms responsible for these effects remain to be determined. Whether pet ownership itself or prolonged exposure to animals is sufficient for some benefits is unclear. Epidemiological studies are also inappropriate for determining which kinds of companion animals are most effective. While recent studies suggest different health effects of dogs and cats (Serpell, 1991; Friedmann & Thomas, 1995; Rajack, 1997), there have not been sufficient studies of gerbil, hamster, fish, rabbit, bird or other pet owners to begin to evaluate the health effects of these less common pets.

SHORT-TERM EFFECTS

The remainder of this review focuses on the short-term physiological effects of animals on humans. These short-term effects, measured on the time scale of minutes rather than months or years, may or may not be the bases for the long-term effects demonstrated in epidemiological studies.

Although the epidemiological studies cited above include pets of all types, a majority of the studies of the short-term impact of animals on human physiology have been limited to the effects of dogs. This is largely a matter of convenience because dogs are popular as pets, and are easy to handle.

The immediate short-term effects of animals on human cardiovascular physiology have been investigated using a variety of research designs. The vast majority of these utilize experimental techniques in which the physiological effect of an image of an animal or an animal stimulus is measured. This review will examine the physiological responses to three different types of exposure to animals: (1) when participants are instructed or directed to look at or observe animals; (2) when participants are in the presence of animals but are given no explicit instructions; and (3) when participants are touching or interacting with animals.

Effects of looking at or observing animals or pictures of animals

Studies of the impact of looking at or observing animals include assessments of the effects of a variety of animals. The studies have documented the direct impact of animals on people's responses to scenes and the people in them, as well as examining physical indicators of parasympathetic nervous system arousal during and/or immediately after watching animals (i.e. Öhman, Fredrikson & Hugdahl, 1978; Katcher *et al.*, 1983a; Lockwood, 1983; Rossbach & Wilson, 1992; Eddy, 1995, 1996). Only one study addressed the ability of looking at or observing animals to reduce people's physiological responses to stressors (Katcher *et al.*, 1983a).

People have long thought that animals can influence perceptions of situations and the people in them. Artists, publicists and advertisers have often used friendly domestic animals to impute safety, credibility, and other trustworthy characteristics to people who accompany them (Lockwood, 1983). Research findings support the positive effects of looking at animals on some of people's moods and perceptions (Lockwood, 1983; Rossbach & Wilson, 1992).

The effect of observing animals on perceptions of otherwise ambiguous social interactions was confirmed in a study using a modified version of the Thematic Apperception Test (Lockwood, 1983). A series of scenes, with one or two people in a natural setting, were printed in two forms: one with just people present, and the other with the identical scene plus one or more animals present. The animals represented included common pets, such as dogs, as well as common wild animals such as squirrels and pigeons. A group of young adults ($n = 68$) rated most of the scenes and the people in them as significantly more friendly, less threatening, and happier when animals were included.

A few studies have addressed the cardiovascular effects of watching animals. In a single-case report, the blood pressure and heart rate of a 26-year-old male snake owner was lower during a six-minute period of watching his pet than during the preceding six minutes when he sat alone and relaxed (Eddy, 1996). In a similarly designed study of people watching familiar chimpanzees (Eddy, 1995), a caretaker and research assistants

($n = 9$) relaxed and then watched the chimps in their large play area from behind a barrier while their blood pressures and heart rates were recorded. Blood pressure and heart rate while watching the chimps tended to be slightly, but not significantly, lower than during the relaxation period (Fig. 8.1). While there were few participants in this study, the repeated measures of the same participants increased the generalizability of the findings.

As part of a study of peoples' stress responses, Katcher *et al.* (1983*a*) found decreases in blood pressures of normotensive ($n = 20$) and hypertensive participants ($n = 15$) in response to watching an aquarium with fish in it, an aquarium which had plants in it but no fish, and to staring at a blank wall. Interestingly, the *duration* of the decreases was greatest when participants were observing the aquarium which had fish in it. The potential moderating effects of observing the aquarium on the physiological responses which occur during stressful activities were also examined in the same study. Participants' blood pressure responses to the stressor of reading aloud at the conclusion of the experimental observation period were less pronounced after watching the fish than after watching the other stimuli. On the basis of this comparison, Katcher *et al.* (1983*a*) concluded that watching fish swim in an aquarium could reduce the magnitude of people's physiological responses to stressful situations.

These studies of people observing fish and chimpanzees indicate that observing animals from a safe position often encourages people to relax. A soothing silence was not required for relaxation.The encouragement to continuously renew attention to an external focus, perhaps due to the movements of the animals, facilitates relaxation and requires no prior training.

Effects of being in the presence of an animal

In six experimental studies, an animal was present in the room with the subject but the subject was not directed to focus on the animal. Five of these studies examined the effect of the animal as a moderator of stress responses (i.e. Friedmann *et al.*, 1983; Friedmann, Locker & Thomas, 1986; Grossberg, Alf & Vormbrock, 1988; Allen *et al.*, 1991; Rajack, 1997).

In one study, both dog owners ($n = 10$) and non-owners ($n = 10$) indicated significantly lower state anxiety (an indicator of physiological arousal) on a psychological checklist ($p < 0.05$) and behaved significantly less anxiously ($p < 0.05$) in a high-stress environment when the experimenter was accompanied by her dog (Sebkova, 1977). Of particular interest was the finding that the participants paid more attention to the

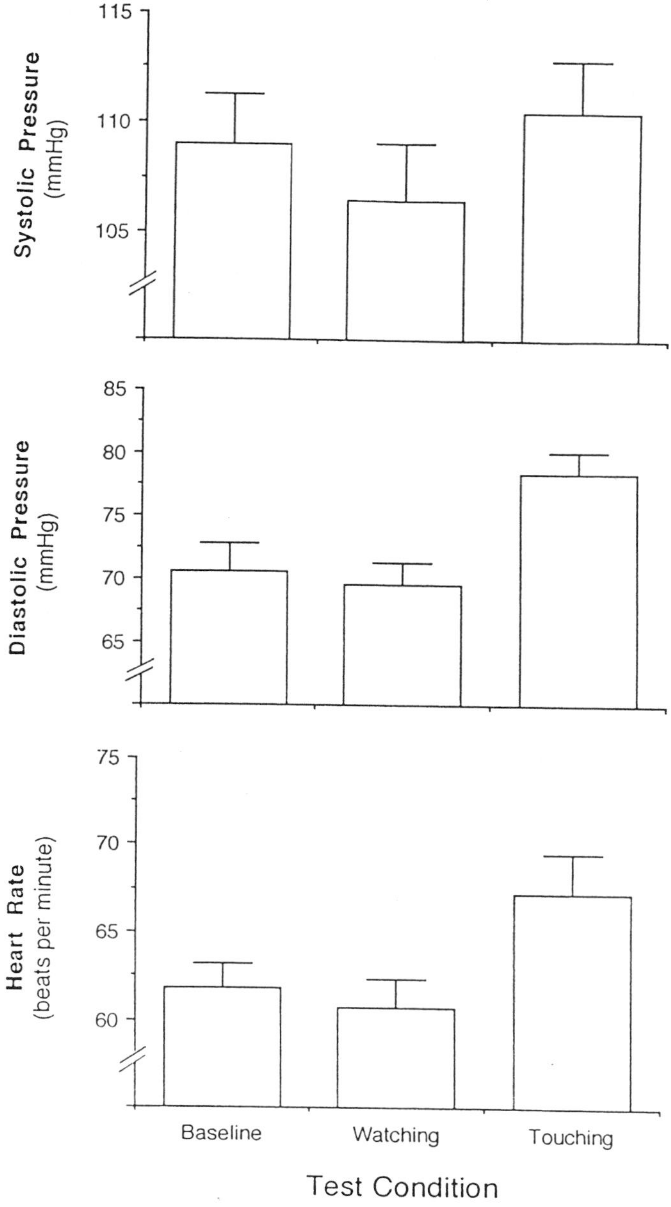

Fig. 8.1. Average systolic and diastolic blood pressures and heart rates of the caretaker ($n=1$) and research assistants ($n=9$) while relaxing (baseline), watching the chimpanzees through a mesh screen (watching), and touching/tickling the chimpanzees through a mesh screen (touching) (Eddy, 1995).

investigator's dog in a high-stress than in a low-stress situation. This suggests that the relaxing external focus of attention, or feelings of safety, provided by a non-threatening animal might be particularly important in stressful situations.

The function of a dog as a buffer to the cardiovascular effects of stress has also been evaluated (Friedmann *et al.*, 1983, 1986; Grossberg *et al.*, 1988; Allen *et al.*, 1991; Rajack, 1997). Non-stressful speech was used to evaluate whether the presence of a friendly dog would modify the cardiovascular responses of 38 nine- to 15-year-old children (Friedmann *et al.*, 1983). The children's blood pressures and heart rates were measured both with and without an unfamiliar, but friendly, dog accompanying the researcher. Blood pressures during the entire experiment were lower for the children who had the dog present at the beginning of the experiment than for those who had the dog present for the second half of the experiment only. The presence of the animal also attenuated the blood pressure increases that occurred while the children were reading aloud ($p < 0.05$). In a similar study conducted among college students ($n = 193$), the presence of a dog caused significant moderation of heart rate ($p < 0.05$), but not blood pressure responses (Friedmann *et al.*, 1988). Grossberg *et al.* (1988) found no buffering effect of the presence of a familiar dog on responses of college students ($n = 16$) to cognitive stressors.

The physiological stress responses of women to a stressful task, mental arithmetic, was undertaken in (a) the presence of a close friend, and (b) the presence of the subject's pet dog (Allen *et al.*, 1991). In the home setting, cardiovascular responses were significantly greater when the friend was present than when only the experimenter was present and when the pet dog was present. Based on these data, the presence of the dog moderated the stress responses more than the presence of a supportive friend. The authors concluded that the non-judgmental aspect of the support afforded by the pet was responsible for decreasing the stress response. This is consistent with other research indicating greater stress responses in the presence of higher-status individuals compared with lower-status individuals (Long *et al.*, 1982).

A more recent study of the effects of the presence of an animal on women's cardiovascular responses to a number of everyday stressors in the normal home environment led to different results (Rajack, 1997). Three different stressors, reading aloud, running up and down stairs, and the sound of an alarm clock, were used in this study. Blood pressures were recorded immediately after each task while heart rate was recorded continuously. There were no differences in the cardiovascular responses of dog owners with their dogs present ($n = 30$) and non-owners ($n = 30$) to

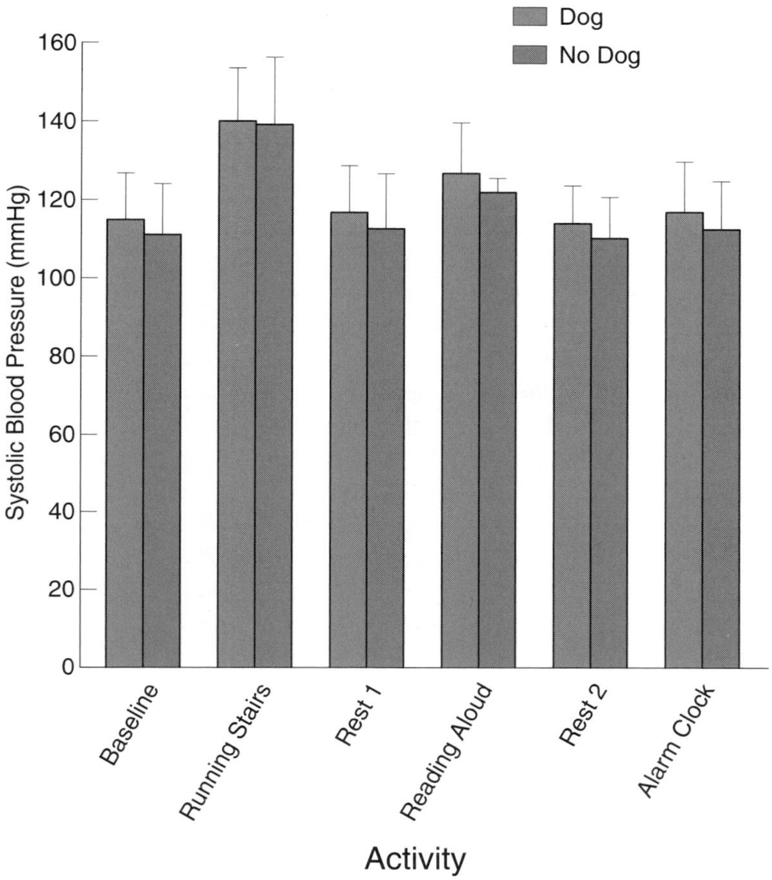

Fig. 8.2. Average systolic blood pressure for dog owners ($n = 30$) with their own dogs present and for dog non-owners ($n = 30$) in their own homes. Blood pressures were recorded while resting at the beginning of the experimental session (baseline) and before (rest) and immediately after three tasks: reading aloud, running up stairs (running stairs), and hearing a sudden alarm clock (alarm clock). Data from Rajack (1997).

running up and down the stairs or reading aloud. Non-owners tended to have a somewhat greater heart rate response to hearing a sudden alarm clock sound ($p = 0.08$; Fig. 8.2). Perhaps the characteristics of the experiment and the many stimuli were so distracting or stressful that the presence of the dog had little effect.

Different individuals are not expected to benefit equally from the presence of pets (Vormbrock & Grossberg, 1988). One study addressed the impact of attitudes towards animals on their de-arousing effects. Blood pressure responses to reading aloud with a dog present were significantly lower for those with a more positive attitude towards dogs than for those with a more negative attitude (Friedmann, Locker & Lockwood, 1993).

On the basis of the research summarized above, the presence of an animal has the potential to influence stress responses but does not do so uniformly. The failure of Rajack's (1997) study to demonstrate any stress-buffering effect of pets may be anomalous, or may be a demonstration of the specificity of the effects of the presence of animals.

The moderating effect of the presence of an animal on stress responses may be influenced by a range of factors. These include the type and familiarity of the setting, the type of stressor, perceptions of the animal, or one's relationship with it. The effects of the type of setting are highlighted in the contrast between the studies of Allen *et al.* (1991) and Grossberg *et al.* (1988). In the former study, the women's stress responses in a home setting were ameliorated when their dogs were present; whereas in the latter study, in a laboratory setting, the subjects' cardiovascular responses did not differ according to whether their dogs were present. The stresses associated with either the setting itself or the nature of the task may negate any stress-moderating effects of the presence of the animal.

Positive perceptions of dogs promote dogs' effectiveness in reducing people's stress responses (Friedmann *et al.*, 1993), and pet owners may benefit more than non-owners in specific situations. However, dog ownership is not apparently necessary for individuals to derive stress-moderating benefits from the presence of a friendly dog. The generalizability of these findings to other animals requires further study. Since attitudes towards particular types of pet are related to a person's choice of pets (Serpell, 1981), particular effort will be required to separate the contributions of attitudes from the more direct effects of the animal.

Good study design is of primary importance. The studies that showed decreased stress responses in the presence of an animal all included cross-over designs (Friedmann *et al.*, 1983, 1986; Allen *et al.*, 1991). That is, the studies included evaluation of the responses of each subject to a stressor both with and without an animal present. In contrast, the studies that did not show an effect did not include cross-over designs (Grossberg *et al.*, 1988; Rajack, 1997). In the latter, each subject was exposed to either the stressor with an animal present or the stressor without an animal present, but not both. Cross-over designs are more powerful than non-cross-over designs because they take into account within-subject variability. This is particularly important for dependent variables such as blood pressure and heart rate which vary tremendously from minute to minute, and from person to person.

Effects of interacting with animals

A smaller body of work has addressed the physiological effects of physically interacting with animals. The variety of ways of interacting with different animals, and the difficulty of standardizing interactions and responses, inhibit research in this area. It is particularly difficult to evaluate the relative contributions to physiological status of the physical movement and exertion involved during interactions. During vigorous interaction, the blood pressure and heart rate depressing effects of the animal may be more than counteracted by the effects of the exertion. Each of the following studies involved instructing participants to interact with animals.

Most early studies observed interactions between people and their pets in order to learn more about the ways people interact with them and with other friendly animals (for example Katcher *et al.*, 1983b). The two most frequent types of interactions with pets were touching and talking to the animal.

Wilson (1991) studied the direct physiological effects of interacting with a pet (talking to and touching it) compared with those of verbal interaction between people. The effects of the activities on state anxiety levels paralleled the effects on cardiovascular indices (Wilson, 1991). While interaction with the friendly animal was less arousing than an interpersonal task, the relative contributions of touching and talking to a dog were not evaluated.

Blood pressures of dog owners ($n = 35$) recruited from a veterinary clinic waiting room were measured while they rested without their pets in a private consultation room, interacted with their pets, or read aloud without their pets in the same room (Katcher, 1981). None of the cardiovascular indices increased while interacting with a pet, but they did increase significantly while reading aloud ($p < 0.05$). Similar effects on blood pressure and heart rate were found when a group of self-selected undergraduate students ($n = 92$) were asked to read aloud, read quietly, or interact with a friendly but unfamiliar dog (Wilson, 1987).

An additional study demonstrated the direct cardiovascular effect of touching a dog. College students' ($n = 60$) blood pressures and heart rates were measured in randomly ordered (six minute) conditions including: resting silently with the dog absent, touching and fondling the dog without talking, calling and talking to the dog from across the room without touching it, talking to and touching the dog, having a casual conversation without the dog present, and having a casual conversation with the dog present (Vormbrock & Grossberg, 1988). As in the previous studies, blood pressures were significantly higher ($p < 0.05$) during the conversations than during the other conditions. Blood pressures also were significantly lower ($p < 0.05$) while touching and fondling the dog without talking than while touching and talking to the dog, or while talking to the dog without touching it. There was no difference in the cardiovascular response patterns between the 30 participants who said they liked dogs and 30 who felt neutral about dogs. None of the participants indicated that they disliked dogs.

The direct physiological consequences of touching animals other than dogs were addressed in two reports. In one case study, a subject's blood pressure during six minutes of touching a pet snake was lower than in the periods of relaxing and looking at the snake that preceded it (Eddy, 1996). The second study examined the physiological responses of a caretaker and nine research assistants to touching chimpanzees (Eddy, 1995). Blood pressures and heart rates were higher when participants touched/tickled the chimps through a barrier than when they rested or observed the chimps from the other side of a protective barrier (see Fig. 8.1). This occurred despite the participants' reported fondness for, and lack of fear of, the animals.

Whether direct effects of petting one's own pet and someone else's pet differ was investigated by Baun *et al.* (1984). Blood pressures decreased significantly from the first to the final assessment when dog owners ($n = 24$) petted their own dogs but not when they petted the unfamiliar animal. Since blood pressures continued to fall in the last period, the duration of the calming effect of petting is unknown. When participants began interacting with their own pets there was an initial 'greeting response', which included the highest blood pressures. If these initial measures are taken into account, differences in responses to the participant's own and others' dogs may not be significant.

Only one study has investigated the stress-moderating effect of touching an animal (Straatman *et al.*, 1997). After a baseline rest period, an unfamiliar, small dog was placed in the laps of the men in the experimental group ($n = 17$) and nothing was placed in the laps of the men in the control group ($n = 19$). The dog remained on the participant's lap during both a seven-minute speech preparation task and a four-minute speech test which followed. There were no significant differences in blood pressure and heart rate responses to the stressors between the experimental and control groups. Thus, having a dog on the lap did not reduce arousal associated with the tasks presented in this study.

It is not known precisely which characteristics of the interactions, the animals, or the people, determine the magnitude of the direct cardiovascular effects of interacting with animals. The non-judgmental aspect of interacting with an animal compared with the demands of interacting with other people is frequently cited as a possible reason for the difference in physiological responses between human–human and human–animal non-threatening interactions (Katcher, 1981; Friedmann *et al.*, 1983; Friedmann & Thomas, 1985; Allen *et al.*, 1991).

The characteristics of human speech directed towards animals may also be a contributory factor (Hirsch-Pasek & Trieman, 1981; Mitchell & Edmundson, 1997). Speaking to an animal is more similar to speaking to a young child than to speaking to another adult (Hirsch-Pasek & Trieman, 1981). The characteristic slow, high-pitched patterns of speech with frequent repetitions and nonsense words may be responsible for the differences in cardiovascular responses. This theory could easily be tested by comparing cardiovascular responses during normal and 'motherese' types of speech.

DISCUSSION

Several epidemiological studies indicate that pet ownership is associated with better health status (Friedmann *et al.*, 1980; Serpell, 1991; Anderson *et al.*, 1992; Friedmann & Thomas, 1995). A number of experimental and quasi-experimental studies confirm that three categories of human–animal association provide physiological benefits to individuals: people explicitly looking at or observing animals or pictures of animals (Lockwood, 1983; Katcher *et al.*, 1983a; Eddy, 1995, 1996); people being in the presence of animals but not interacting with them (Sebkova, 1977; Friedmann *et al.*, 1983, 1986, 1993; Grossberg *et al.*, 1988; Allen *et al.*, 1991); and people touching or interacting with animals (Katcher, 1981; Wilson, 1987, 1991; Vormbrock & Grossberg, 1988; Allen *et al.*, 1991).

More general questions which arise from some of this research include:

1. Do the direct or stress-moderating effects of different types of animals on people's health differ?
2. What are the cultural, experiential and attitudinal bases for these differences?
3. Are the short-term direct or stress-moderating physiological

responses to animals the basis for the differences in health status found in long-term epidemiological studies?

4. Are there differences on the basis of sex or other demographic characteristics?
5. Are animals more effective at moderating the effects of certain types of stressors than others?

Until recently, all species kept as pets were expected to have similar physiological effects on their owners (i.e. Beck & Katcher, 1989). The epidemiological studies reported here (e.g. Serpell, 1991; Friedmann & Thomas, 1995; Rajack, 1997) demonstrate that different types of pets may have different impacts on people's health and its underlying physiology. Thus, comparison of the long-term and short-term effects of dogs and other species considered in the context of a person's animal relationship history may provide important new information.

The studies conducted to date focus primarily on the effects of pets, and of dogs in particular, on an individual's physiology. The effects of specific types of pets on a person's physiology are probably based on that individual's previous direct and indirect experiences, as well as on their beliefs, desires and fears about specific species. We cannot expect specific animal types to evoke uniform physiological responses from different individuals (Friedmann *et al.*, 1993). How the meaning of specific types and even breeds of animals to individuals is responsible for differences in their responses to animals is an area ripe for research. Within one culture, responses also might be expected to differ according to perceptions of the danger inherent in the specific animal-related situation. The interpretation of an animal as safe or unsafe is expected to depend on early learning and/or personal experiences, even for animals that are defined as dangerous. Additional research directly addressing both (1) cross-cultural differences and individual differences within cultural groups, in perceptions of, and responses to, a variety of animals, and (2) the existence and characteristics of innate reactions to animals, would facilitate understanding in this area.

Reviewing the studies from the perspective of cardiovascular health and physiology reveals a different issue. The physiological benefits of animals demonstrated in the epidemiological studies occur largely for men. Over 78% of the participants were men in the large-scale epidemiological studies of heart disease patients (Friedmann *et al.*, 1993; Friedmann & Thomas, 1995; Rajack, 1997). When women were examined separately from men, pet ownership was not related to women's survival (Friedmann & Thomas, 1995; Rajack, 1997). Furthermore, in the CAST study (Friedmann & Thomas, 1995) the interrelationship between type of pet owned, social support and survival differed for men and women. In Anderson *et al.*'s (1992) study of risk factors for coronary heart disease, women benefited from pet ownership only in the menopausal and post-menopausal years. The apparent differences in the predictors of heart disease survival for men and women are consistent with findings in other studies (Herlitz *et al.*, 1995; Mendelson & Hendel, 1995; Iezzoni *et al.*, 1997). In contrast to the epidemiological studies, pre-menopausal women dominate the vast majority of the studies examining responses to animal stimuli (Sebkova, 1977; Friedmann *et al.*, 1983, 1993; Baun *et al.*, 1984; Wilson, 1987, 1991; Allen *et al.*, 1991; Rajack, 1997). The one study including only men (Straatman *et al.*, 1997) did not find that touching an animal moderated the stress response. However, since no other study attempted to investigate this type of effect, differences between men and women in this regard require further investigation. It is equally possible that the support provided by animals impacts on human physiology directly, or as a stress-buffering agent, or both, and in different ways in men and women. There is evidence from studies of other contributors to health that psychosocial variables affect the health of men and women differently. For example, after controlling for physiological risk factors, a lack of social activity and a lack of emotional support were predictors of myocardial infarction among men, while only a lack of social activities was significant for women (Welin, Rosengren & Wilhelmsen, 1996). The field would benefit from attention to addressing possible male–female differences.

In the research investigating the short-term effects of animals, two types of potential health benefits were investigated: direct effects on physiological indicators, and stress-moderating or buffering effects. The experimental and quasi-experimental studies provide evidence that either observing or being in the presence of animals can, in some circumstances, lead to both direct and stress-moderating effects. There is also evidence that interacting with a pet, not necessarily one's own pet, leads to direct de-arousing effects (Katcher, 1981; Baun *et al.*, 1984; Wilson, 1987, 1991) but not to a stress-moderating effect (Straatman *et al.*, 1997). Examination of the studies conducted so far reveals that systematic investigation is required to identify the categories of effects that are likely to occur with each category of human–animal interaction. Studies conducted to date have not addressed these issues in a systematic comparative manner. Too many variables differ between studies to draw meaningful conclusions.

CONCLUSION

The research addressing the effects of animals on human health and indicators of physiological arousal conducted to date provides intriguing evidence that animals can be beneficial, particularly for cardiovascular health. Epidemiological studies indicate that owning pets is associated with better one-year survival of patients after myocardial infarctions, fewer health complaints, and lower cardiovascular risk factors. There is evidence that owning or acquiring dogs may be more beneficial than having cats only. The complex interrelationships among cat ownership, dog ownership, other types of social support, life events, other aspects of psychological and physical health, and gender-related differences in mortality and morbidity must be addressed before firm conclusions are drawn. Variations in research design between studies that have examined the short-term de-arousing effects of animal contact render their findings difficult to interpret and compare.

REFERENCES

Allen, K.M., Blascovich, J., Tomaka, J. & Kelsey, R.M. (1991). Presence of human friends and pet dogs as moderators of autonomic responses to stress in women. *Journal of Personality and Social Psychology*, **61**, 582–9.

Anderson, W., Reid, P. & Jennings, G.L. (1992). Pet ownership and risk factors for cardiovascular disease. *Medical Journal of Australia*, **157**, 298–301.

Baun, M.M., Bergstrom, N., Langston, N.F. & Thoma, L. (1984). Physiological effects of human/companion animal bonding. *Nursing Research*, **33**, 126–9.

Beck, A.M. & Katcher, A.H. (1989). Bird–human interaction. *Journal of the Association of Avian Veterinarians*, **3**, 152–3.

Centers for Disease Control and Prevention (1997). Dog-bite related fatalities – United States, 1995–1996. *Morbidity and Mortality Weekly Report*, **46**, 463–7.

Eddy, T.J. (1995). Human cardiac responses to familiar young chimpanzees. *Anthrozoös*, **4**, 235–43.

Eddy, T.J. (1996). RM and Beaux: reductions in cardiac activity in response to a pet snake. *Journal of Nervous and Mental Disease*, **184**, 573–5.

Friedmann, E. (1995). The role of pets in enhancing human well being: physiological effects. In *The Waltham Book of Human–Animal Interaction*, ed. I. Robinson, pp. 33–54. Tarrytown, NY: Elsevier Science, Inc.

Friedmann, E., Katcher, A.H., Lynch, J.J. & Thomas, S.A. (1980). Animal companions and one-year survival of patients after discharge from a coronary unit. *Public Health Reports*, **95**, 307–12.

Friedmann, E., Katcher, A.H., Thomas, S.A., Lynch, J.J. & Messent, P.R. (1983). Social interaction and blood pressure: influence of animal companions. *Journal of Nervous and Mental Disease*, **171**, 461–5.

Friedmann, E., Locker, B.Z. & Lockwood, R. (1993). Perceptions of animals and cardiovascular responses during verbalization with an animal present. *Anthrozoös*, **6**, 115–34.

Friedmann, E., Locker, B.Z. & Thomas, S.A. (1986). Effect of the presence of a pet on cardiovascular response during communication in coronary prone individuals. Presented at Delta Society International Conference, 'Living Together: People, Animals and the Environment', Boston, MA, August.

Friedmann, E. & Thomas, S.A. (1985). Health benefits of pets for families. Special issue: pets and the family. *Marriage and Family Review*, **8**, 191–203.

Friedmann, E. & Thomas, S.A. (1995). Pet ownership, social support, and one-year survival after acute myocardial infarction in the Cardiac Arrhythmia Suppression Trial (CAST). *American Journal of Cardiology*, **76**, 1213–17.

Grossberg, J.M., Alf, E.F. Jr & Vormbrock, J.K. (1988). Does pet dog presence reduce human cardiovascular responses to stress? *Anthrozoös*, **2**, 38–44.

Herlitz, J. Karlson, B.W., Wilkund, I. & Bergston, A. (1995). Prognosis and gender differences in chest pain patients discharged from an ED. *American Journal of Emergency Medicine*, **13**, 127–32.

Hirsch-Pasek, K. & Trieman, R. (1981). Doggerel: Motherese in a new context. *Journal of Child Language*, **9**, 229–37.

Iezzoni, L.I., Ash, A.S., Shwartz, M. & Mackiernan, Y.D. (1997). Differences in procedure use, in hospital mortality, and illness severity by gender for acute myocardial infarction patients: are answers affected by data source and severity measure? *Medical Care*, **35**, 158–71.

Jenkins, C.D. (1976). Recent evidence supporting psychologic and social risk factors for coronary disease. *New England Journal of Medicine*, **294**, 1033–8.

Katcher, A.H. (1981). Interactions between people and their pets: form and function. In *Interrelationships Between People and Pets*, ed. B. Fogle, pp. 41–67. Springfield, IL: Charles C Thomas.

Katcher, A.H., Friedmann, E., Beck, A.M. & Lynch, J.J. (1983a). Talking, looking, and blood pressure: physiological consequences of interaction with the living environment. In *New Perspectives on Our Lives with Companion Animals*, ed. A.H. Katcher & A.M. Beck, pp. 351–9. Philadelphia: University of Pennsylvania Press.

Katcher, A.H., Friedmann, E., Goodman, M. & Goodman, L. (1983b). Men, women and dogs. *California Veterinarian*, **37**, 14–17.

Lierman, T.L. (ed.) (1987). The role of animals in research. In *Building a Healthy America*, pp. 55–9. New York: Mary Ann Liebert, Inc.

Lockwood, R. (1983). The influence of animals on social perception. In *New Perspectives on Our Lives with Animal Companions*, ed. A.H. Katcher & A.M. Beck, pp. 64–71. Philadelphia: University of Pennsylvania Press.

Long, J.M., Lynch, J.J., Machiran, N.M., Thomas, S.A. & Malinow, K.L. (1982). The effect of status on blood pressure during verbal communication. *Journal of Behavioral Medicine*, **5**, 165–72.

Maslow, A. (1970). *Motivation and Personality*, 2nd edn. New York: Harper and Row.

Mendelson, M.A. & Hendel, R.C. (1995). Myocardial infarction in women. *Cardiology*, **86**, 272–85.

Mitchell, R.W. & Edmundson, E. (1997). What people say to dogs when they are playing with them. Presented at the International Society for Anthrozoology Scientific Sessions, Boston, July 24th–25th, 1997.

Öhman, A., Fredrikson, M. & Hugdahl, K. (1978). Orienting and defensive responding in the electrodermal system: palmar–dorsal differences and recovery-rate during conditioning to potentially phobic stimuli. *Psychophysiology*, **15**, 93–101.

Plaut, M., Zimmerman, E. & Goldstein, R. (1996). Health hazards to humans associated with domestic pets. *Annual Review of Public Health*, **17**, 221–45.

Rajack, L.S. (1997). Pets and human health: the influence of pets on cardiovascular and other aspects of owners' health. Doctoral Dissertation, University of Cambridge, UK.

Rossbach, K.A. & Wilson, J.P. (1992). Does a dog's presence make a person appear more likeable? *Anthrozoös*, **4**, 40–51.

Schantz, P.M. (1990). Preventing potential health hazards incidental to the use of pets in therapy. *Anthrozoös*, **4**, 14–23.

Sebkova, J. (1977). Anxiety levels as affected by the presence of a dog. Unpublished thesis, Department of Psychology, University of Lancaster, UK.

Serpell, J.A. (1981). Childhood pets and their influence on adults' attitudes. *Psychological Reports*, **49**, 651–4.

Serpell, J.A. (1991) Beneficial effects of pet ownership on some aspects of human health and behaviour. *Journal of the Royal Society of Medicine*, **84**, 717–20.

Serpell, J.A. & Jackson, E. (1994). Life events and methodological problems in studies of the health benefits of pet ownership. Paper presented at the Annual Meeting of the International Society for Anthrozoology (ISAZ), New York, October 13th, 1994.

Straatman, I., Hanson, E.K.S., Endenburg, N. & Mol, J.A. (1997). The influence of a dog on male students during a stressor. *Anthrozoös*, **10**, 191–7.

Vormbrock, J.K. & Grossberg, J.M. (1988). Cardiovascular effects of human–pet dog interactions. *Journal of Behavioral Medicine*, **11**, 509–17.

Welin, C.L., Rosengren, A. & Wilhemsen, L.W. (1996). Social relationships and myocardial infarction: a case control study. *Journal of Cardiovascular Risk*, **3**, 183–90.

Wilson, C. (1987). Physiological responses of college students to a pet. *Journal of Nervous and Mental Disease*, **175**, 606–12.

Wilson, C. (1991). The pet as an anxiolytic intervention. *Journal of Nervous and Mental Disease*, **179**, 482–9.

ANTHONY L. PODBERSCEK AND SAMUEL D. GOSLING

9

Personality research on pets and their owners: conceptual issues and review

INTRODUCTION

Personality research on companion animals and their owners has grown steadily over the past few decades. A major focus, inspired by reports on the health benefits of pet ownership (see Friedmann, Thomas & Eddy, Chapter 8), has been on whether the personality traits of pet owners are different to those of non-owners. In addition, there has been a rapid rise in interest in how owners' personalities may influence companion animal behaviour and behaviour problems. More recently, attention has turned to the structure of personality in companion animals and whether it is similar to that of humans.

All these areas are important to furthering our understanding of the human–pet relationship but, to date, no common currency of scientific findings has been produced. This chapter aims to bring the research together for the first time, to evaluate it and suggest ways forward for future study. In addition, we will highlight a number of conceptual issues. To this end, we have divided the chapter into three major parts reflecting three questions: (a) how can we make cross-species comparisons of personality? (b) can human personality influence companion animal personality? and (c) what can we determine about the personalities of pet owners? For the purposes of this review, we define *personality traits* as enduring, relatively stable, characteristics used to explain and describe an individual's behaviour across time and situations (John & Gosling, 2000).

HOW CAN WE MAKE CROSS-SPECIES COMPARISONS OF PERSONALITY?

It is important to clarify at the outset that personality comparisons are more typically made within species (e.g. between two dogs) rather than across species (e.g. between a dog and a cat). Later, we will address issues that may arise from confusing these two types of comparison. Nevertheless, it is worth bearing in mind that although cross-species comparisons of personality are sensible, they are rather rare because most uses of personality are limited to the human domain.

There are two levels at which personality comparisons can be made across species. The first level focuses on the structure underlying personality; for example, are the main personality dimensions associated with horses the same dimensions as those associated with ferrets? The second level focuses on cross-species comparisons of personality traits or dimensions; for example, are rabbits bolder than hedgehogs? In the next two sections we will discuss conceptual issues and empirical research associated with both levels of comparison.

Comparisons of personality structure

Personality structure refers to the interrelation of personality traits: some traits may be conceptually and empirically related, whereas others may be independent of each other. Personality structure is usually detected by reducing a large number of narrow traits into a smaller number of broader dimensions comprised of related traits (Digman, 1990, 1997; Goldberg, 1990, 1992; Goldberg & Digman, 1994). For example, using techniques such as exploratory factor analysis, we may learn that anxious individuals also tend to be jealous, and we may categorize both of these traits under a broad dimension of 'emotional stability' (Goldberg, 1992).

Essentially, cross-species comparisons of personality structure entail comparing the number and content of the major dimensions underlying the personality of one species with those underlying another species (Gosling & John, 1999). To our knowledge, only one study has made direct comparisons of personality structure in companion animals: Gosling and John (1998) quantitatively compared the structure underlying descriptions of dogs, cats and humans. They found a common structure underlying personality descriptions of both dogs and cats that differed from the structure underlying personality descriptions of humans. Specifically, they found that four dimensions characterized descriptions of dogs and cats, which they named 'energy', 'affiliation', 'emotional reactivity', and 'competence'. The first three dimensions showed some similarity to three of the Big Five human personality factors (John, 1990) of 'extraversion', 'agreeableness', and 'emotional stability'. The fourth dimension resembled a blend of the Big Five's 'conscientiousness' and 'openness to experience' factors.

Comparative psychologists can use such structural analyses to relate cross-species similarities and differences in personality structure to cross-species similarities and differences in biology, environment and phylogenetic history (Gosling, in press). Anthrozoologists can use structural analyses to determine whether it is legitimate to compare species in terms of broad personality dimensions. Ideally, one would compare species at the broader dimensional level to provide more reliable estimates than can be provided by single traits. However, as yet, there is no evidence suggesting that the same structure underlies the personality of all the species we examine. Therefore, the cross-species comparisons presented in the next section will be made at the trait, not dimension, level.

Comparisons of personality traits

Personality is based on the concept of individual differences (Pervin & John, 1997). For example, if individuals did not differ in the degree to which they were bold rather than shy, then boldness would not be considered a personality trait. Thus, to say an individual is bold is a shorthand way of saying an individual is bold in comparison to other individuals. When talking about human personality, one can usually get away with this shorthand version because it is clear that the individuals involved in this implicit comparison are the same species (humans). However, when talking about non-human animals, extra specificity is often required. To illustrate the point, consider the following example. Imagine that we tell you there is a black mamba in the next room and you ask us whether it is aggressive. The answer we provide will depend on whether we adopt a within-species framework or a cross-species framework. If we adopt a within-species framework, we may respond 'No, it is very unaggressive. It has only attacked two people in the last hour which is well below the norm for this species of snake'. If, on the other hand, we adopt a cross-species framework, we may respond 'Yes, of course it's aggressive. It's a black mamba, a highly aggressive species of snake'. Therefore, the framework adopted will determine whether we say 'Yes' or 'No' to your question about the black mamba's level of aggression.

In research on companion animals it is often impossible to tell whether a participant in one's research has answered a question using a within-species or cross-species framework. Clearly, it makes no sense to combine the answers of participants adopting within-species frameworks with participants adopting cross-species frameworks. It is likely that a number of factors influence which framework has been adopted. A respondent who is asked to describe the personality of two of her cats is probably more likely to be implicitly comparing them; that is, adopting a within-species framework. On the other hand, a respondent who is asked to describe the personality of her dog and of her cat may be more likely to focus on the cross-species differences. To further complicate the issue, interbreed differences can act somewhat like cross-species differences. For example, if you own a Scottish terrier and rate it as high on excitability, are you thinking the dog is excitable compared with dogs in general or compared with other Scottish terriers?

Ideally, research on companion animal personality will attempt to control for the implicit framework adopted by research participants by specifying whether ratings should be based on cross-species or within-species comparisons. Even this strategy does not alleviate the problem entirely because there is a practical limit on the specificity of the instructions. For example, should participants also be asked to compare the target animal with animals of the same age and sex? Adding specificity may place a burden on the participants, many of whom are volunteers, and compromise the quality of the data. Rather than weighing down participants with too many instructions, we recommend collecting large sample sizes to diminish the effects of the relatively minor rating biases.

A related problem arises when participants have limited exposure to the species (or breed) of animal they are describing. Without exposure to multiple individuals of the same species, it is impossible for participants to distinguish characteristics of the species from characteristics of the individual. For example, if a participant owns a rare species of tropical bird, how is he to know which traits are species specific and which are specific to this individual bird unless he has been exposed to several birds of this species?

The next issue to arise in cross-species comparisons concerns variation in how personality traits are interpreted across species. In some cases, the problem arises because characteristics of different species may emphasize different meanings of already ambiguous traits. For example, when the trait 'touchy' is applied to humans, it is usually taken to mean sensitive, but when loving companion animals are described as touchy, some respondents may think about the degree to which the animal is physically affectionate, as in 'touchy, feely'. Fortunately, this issue can be easily addressed by carefully considering the possible species-specific interpretations and clarifying the questionnaires accordingly. For example, including the word 'sensitive' in parentheses after 'touchy' would prevent misinterpretation of the trait.

A more fundamental problem of interpretation arises for those traits that can be applied literally to one species but only metaphorically to another species. Consider a trait such as 'talkative'. Clearly, a dog does not literally talk so what does it mean when a dog is rated as high on talkativeness? Does it mean that the dog vocalizes a great deal, or that the dog is communicative? Unfortunately, this problem of cross-species interpretation is harder to address than the previous one. Although researchers can attempt to include traits that resolve ambiguity (e.g. by using 'vocal' and 'communicative' instead of 'talkative'), this strategy raises problems of its own. First, it is difficult to anticipate all the traits that will be needed to resolve every ambiguity that might arise when assessing a variety of species. Second, a huge number of additional traits would be needed to clarify all possible ambiguities that may arise in applying personality traits to several different species. Third, species-specific traits that resolve an ambiguity in one species may make little sense when applied to another species. For example, it may be sensible to distinguish vocal macaws from quiet macaws, but the distinction makes less sense for goldfish. A middle ground needs to be found in which a basic set of standard descriptors is supplemented by important species-specific descriptors (Gosling, 1998a).

The final issue to arise in cross-species comparisons concerns the meaning of the ratings themselves. For example, when a participant rates his dog as high on timidity, are we learning about a characteristic of the dog, about the participant making the rating, or about the interaction between the participant and the dog? The ecological approach to perception suggests that such ratings tell us about all three components (Gosling, 1998b). Part of the timidity score reflects real behaviours of the dog that distinguish it from other dogs; part of the score reflects the perceptual and conceptual abilities of the participant that determine what dog behaviours he is able to detect and how he categorizes these behaviours; finally, the score also reflects aspects of the relationship between the participant and his dog. To illustrate this third component, consider how personality ratings made by a cat might differ from those made by a human; we would expect cats to provide a very different personality profile, perhaps one focused more on ferocity than on timidity. Therefore, personality ratings should not be thought of as reflections of objective characteristics of the individual in question but should instead be considered as reflections of how the participant conceptualizes the individual being rated. This does not mean that personality ratings are meaningless, but simply that researchers may need to reconsider what their data reflect.

With the above caveats in mind, we can now examine cross-species differences in the personality of companion animals. We present data from a recent study that assessed six species of companion animals (dog,

$n=1209$; cat, $n=535$; ferret, $n=127$; rabbit, $n=29$; horse, $n=10$; hedgehogs, $n=13$) on the same set of 50 personality traits (Gosling & Bonnenburg, 1998). To provide a conceptual framework for these data we have selected 20 of these traits, four traits from each of the human Big Five personality dimensions (John, 1990), and included self-ratings by human participants ($n=1906$). We have excluded items that are difficult to apply to animals. The cross-species comparisons are presented in Figs. 9.1–9.5 for traits from the human dimensions of extraversion, agreeableness, conscientiousness, emotional stability, and openness to experience (or intellect), respectively.

Although the sample sizes for rabbits, horses and hedgehogs are smaller than optimal, and ratings are only shown for 20 traits, the figures provide a window on the personality profiles of six species of companion animals.[1] Some striking findings are worth mentioning. First, there is considerable cross-species variation in the trait ratings, suggesting that the raters were adopting, at least in part, a cross-species framework when they made their personality judgements. Second, the degree of cross-species variation depends on the trait rated; the means of all seven species (including humans) yield a relatively undifferentiated pattern for 'systematic', whereas there is substantial cross-species variation for traits such as 'jealous', 'temperamental', and 'uncreative'. Third, many of the ratings are consistent with the expected species-typical characteristics; for example, hedgehogs and rabbits were rated as the quietest animals and ferrets were rated as bold and relatively unco-operative. Fourth, with a few exceptions, dogs, cats and, to a lesser extent, horses show similar overall personality profiles.

The ease with which these six species could be compared underscores the usefulness of using the same personality descriptors across studies and species (see Gosling & Bonnenburg, 1998).[2] To facilitate future integration of research on animal personality, we hope that companion animal researchers will follow the lead of human personality research and develop a standard taxonomy of animal personality descriptors.

We now turn to the issue of influences on animal personality. In particular, we examine how one important part of a companion animal's social environment, its owner, may influence its personality.

CAN HUMAN PERSONALITY INFLUENCE COMPANION ANIMAL PERSONALITY?

Whether an owner's personality can influence the personality traits, and hence behaviour, of companion animals is often debated. Prior

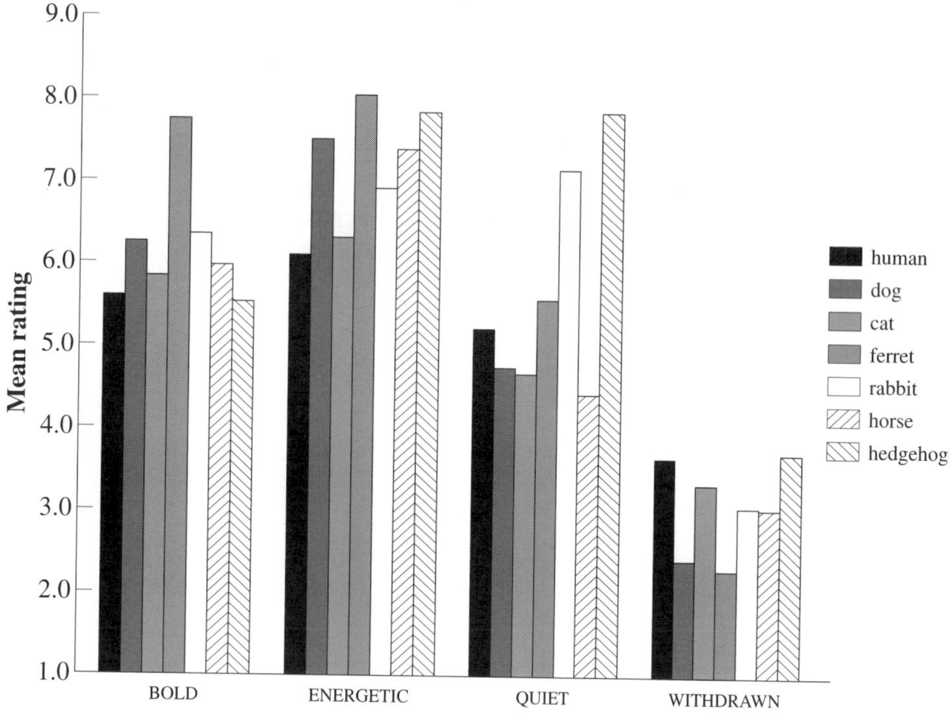

Fig. 9.1. Cross-species comparison of extraversion traits. Ratings made on a nine-point scale (1 = extremely uncharacteristic; 9 = extremely characteristic).

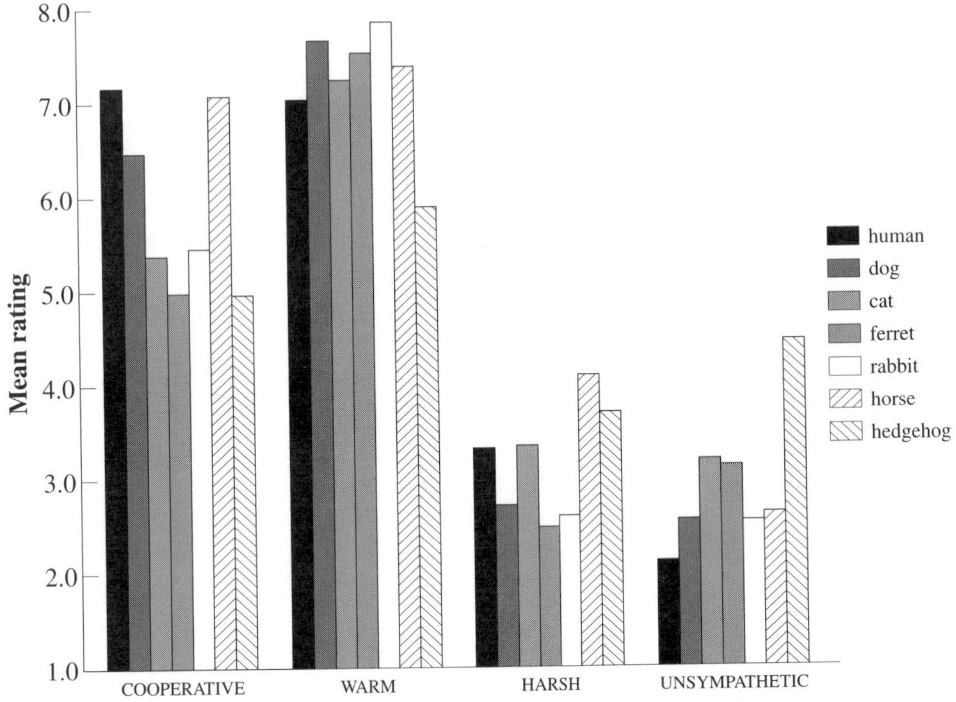

Fig. 9.2. Cross-species comparison of agreeableness traits. Ratings made on a nine-point scale (1 = extremely uncharacteristic; 9 = extremely characteristic).

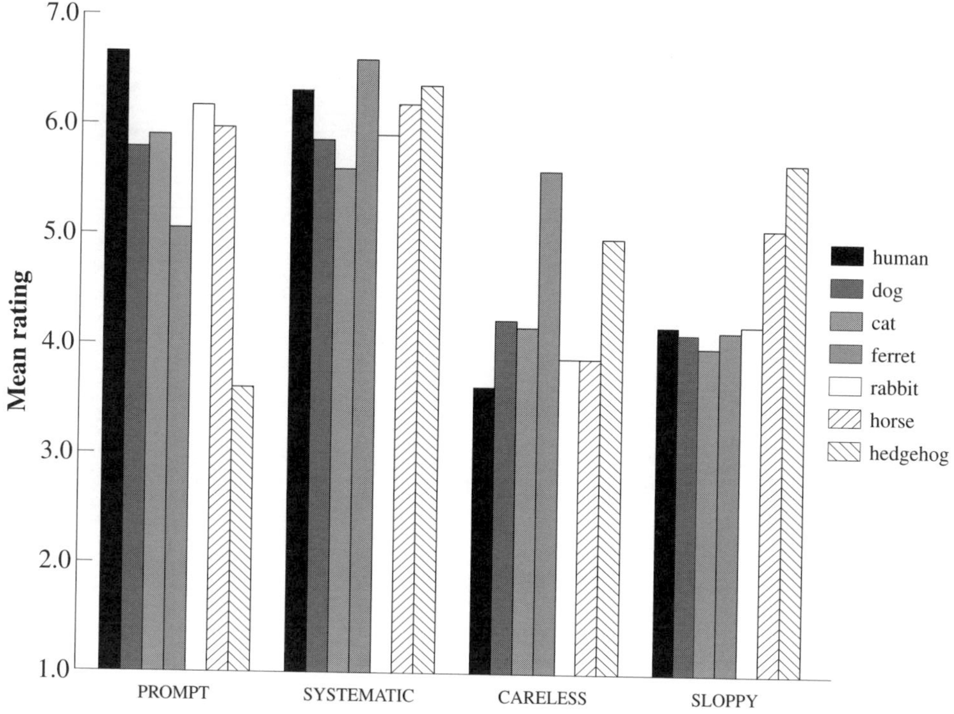

Fig. 9.3. Cross-species comparison of conscientiousness traits. Ratings made on a nine-point scale (1 = extremely uncharacteristic; 9 = extremely characteristic).

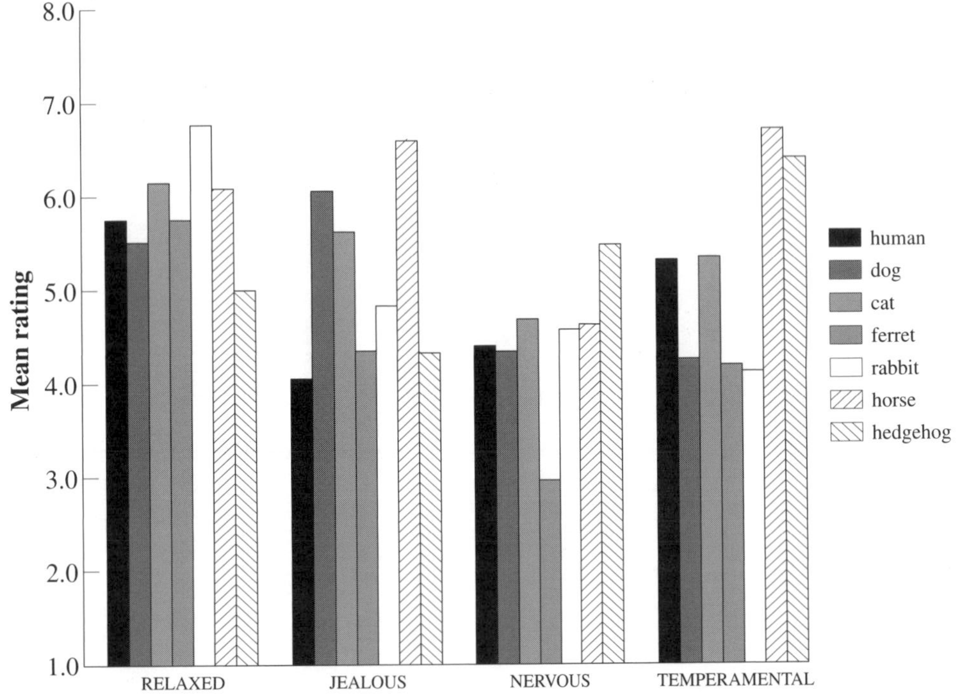

Fig. 9.4. Cross-species comparison of emotional stability traits. Ratings made on a nine-point scale (1 = extremely uncharacteristic; 9 = extremely characteristic).

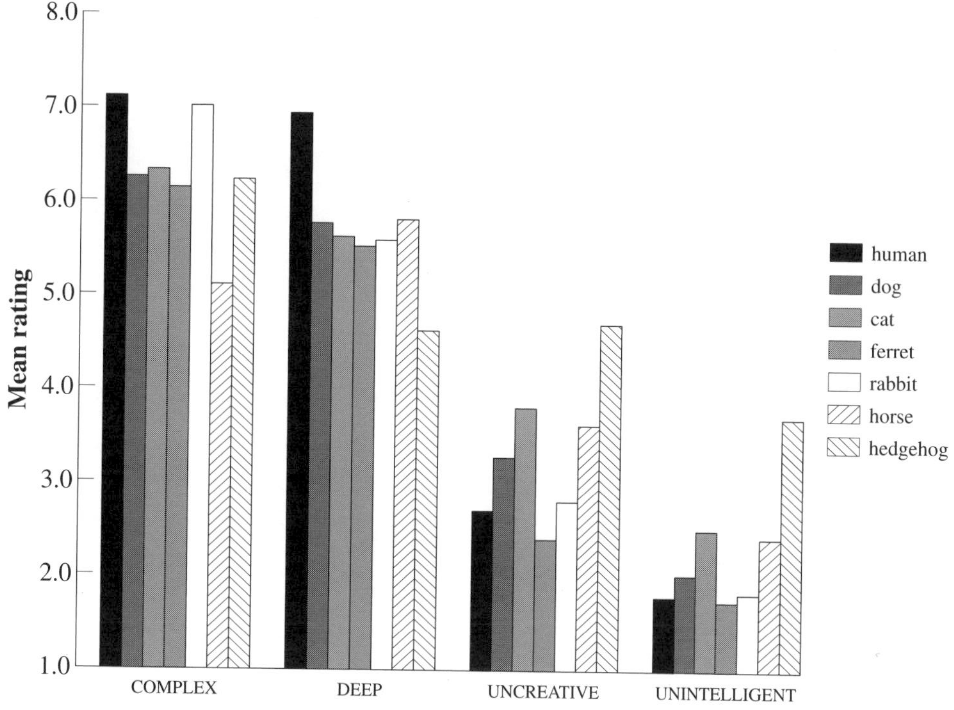

Fig. 9.5. Cross-species comparison of openness traits. Ratings made on a nine-point scale (1 = extremely uncharacteristic; 9 = extremely characteristic).

to the late 1970s it was widely assumed in dog-training circles that most canine behaviour problems were the product of owners failing to exercise sufficient authority, discipline or control over their pets (see, for example, Woodhouse, 1978). This perception also gave rise to the widespread view that the owners of problem dogs are more likely, on average, to be shy, anxious, neurotic and/or submissive (O'Farrell, 1995). However, the contrasting view is also prevalent, as is the opinion that 'blaming' owners for their pets' behaviour problems is both misleading and counterproductive to successful treatment (Mugford, 1995).

As in the field of human developmental psychology, where the sterile 'Nature v. Nurture' debate eventually resulted in a sensible compromise position, it is probable that these dichotomous views of the ontogeny of companion animal behaviour problems will eventually converge on the middle ground; that is, companion animal personality is the outcome of a continuous interaction between the animal's genetic inheritance and the physical and social environment in which it develops (Bateson, 1981; Serpell, 1987). Since owners are significant figures in the social environment of most pet animals, it is plausible to suggest that certain owner characteristics, such as personality, may exert an influence on the animal's personality development. The true extent of this influence, however, has been the subject of surprisingly little empirical research.

An exception is O'Farrell (1987, 1995), who studied 50 dog owners attending a small animal clinic for various veterinary treatments and found a greater prevalence of certain behaviour problems among dogs belonging to owners who obtained high scores on the Neuroticism Scale of the Eysenck Personality Inventory (Eysenck & Eysenck, 1964). Three of these problems, 'sexual mounting of people or inanimate objects', 'destructive when left alone', and 'pestering for attention', would appear to be suggestive of over-intense bonds with the owner, while the fourth, 'biting people', is ambiguous because the author did not indicate who was typically bitten or under what circumstances. Despite this ambiguity, O'Farrell (1995) labels these problems 'displacement activities' and attributes their prevalence to high arousal and conflict states induced by over-anxious and neurotic owners.

Podberscek and Serpell (1997) also found evidence for a link between owner and pet personalities. They studied 128 owners of English cocker spaniels previously classified as 'low' in aggressiveness and 157 owners of English cocker spaniels classified as 'high' in aggressiveness. All owners completed the Sixteen Personality Factor Questionnaire (16PF; Cattell, Eber & Tatsuoka, 1970). Analyses of the data revealed that the owners of high-aggression dogs were significantly more likely to be tense, emotionally less stable, shy and undisciplined than owners of low-aggression dogs.

At least three distinct explanations can be offered for the observed associations between owner personality traits and the animals' behaviour in the above studies. First, and as with previous studies that have relied on subjective assessments of animal behaviour, it is possible that the findings reflect differences in owner perceptions of their pets rather than real differences in the animals' behaviour (see Stevenson-Hinde, Stillwell-Barnes & Zunz, 1980; Feaver, Mendl & Bateson, 1986; Jagoe & Serpell, 1996). In the Podberscek & Serpell (1997) study, owners who obtained high scores on the 'tense' and 'emotionally less stable' personality factors may, for example, have been more sensitive to expressions of aggressive behaviour by the dog, and may therefore tend to give their dogs higher than average scores on the various aggression scales used in the study. The researchers attempted to reduce this kind of subjective bias as far as possible by asking for frequencies of behaviour in specific contexts rather than more general qualitative evaluations. However, in the absence of independent behavioural observations and/or tests of dogs belonging to owners of different personality types, it is impossible entirely to exclude this interpretation of the data.

Second, it could be argued that the differences in owner personality traits are an effect of canine behaviour rather than a cause. For example, perhaps owners of aggressive English cocker spaniels in Podberscek and Serpell's (1997) study tended to become more 'tense' and less 'emotionally stable' through their interactions with the pet. This, however, appears unlikely for two reasons:

(a) All the respondents were adults; indeed, most were aged above 34 years, and therefore we can assume that their personality traits were relatively stable (Pervin, 1970; Ross, 1987; Hampson, 1988; Costa & McCrae, 1994). Change in adult personality is unusual unless major life cycle experiences force people to change (Cavanaugh, 1995: 81).

(b) Pet ownership is very common in Britain, with over 50% of households owning one or more domestic pets (*PFMA Profile*, 1998). Most people have had pets at some stage in their lives and pet owners tend to keep owning the species with which they grew up (Kidd & Kidd, 1980, 1989; Serpell, 1981; Poresky *et al.*, 1988). Therefore, it is very likely that the respondents in this study were experienced dog, if not cocker spaniel, owners. The acquisition of (in most cases) a puppy that subsequently becomes aggressive would therefore be unlikely to have a significant negative impact on the owner's personality. To rule this possibility out, however, would require personality testing of owners before and after the acquisition or disposal of aggressive dogs.

Third, and perhaps most plausibly, it is possible that the findings offer support for the traditional 'dog-trainer' view that anxious, tense and neurotic owners sometimes *cause* their pets to become more aggressive or badly behaved (e.g. Woodhouse, 1978). Even if this interpretation is correct, however, the mechanism of cause and effect is far from clear. O'Farrell (1987, 1995) concluded that the behaviour problems (including 'biting people') of dogs belonging to neurotic owners were the product of owner-induced conflict states and over-arousal. Podberscek and Serpell's (1997) data, however, are more consistent with the idea that these dogs respond to their owners' anxiety, neuroses and/or shyness by becoming more aggressively assertive across a wide range of situations. It will require further research to determine whether the apparent effects of the owners' personalities discussed here were the result of the particular ways in which owners interacted with their pets or due to some other factors.

That owners may influence the personality of their companion animals brings us to a more general question: is the ownership of particular types of pet, or of pets in general, associated with identifiable personality types or dimensions? This is the subject of the final section of this chapter, which is divided into two subsections: (a) pet owner versus non-owner studies, and (b) owners of different species of pet.

WHAT CAN WE DETERMINE ABOUT THE PERSONALITIES OF PET OWNERS?

Pet owner versus non-owner studies

Although relatively little research has compared the personality of pet owners with the personality of non-owners, the work that has been done is diverse. We have summarized these studies in Table 9.1. To provide our review with an organizing framework, we have roughly categorized the traits examined in terms of the Big Five factor model of personality – extraversion, agreeableness, conscientiousness, emotional stability, and openness to experience (or intellect) (John, 1990) – and an extra category, 'other', for those variables which could not be incorporated into this model. The table shows that most of the variables studied

Table 9.1. *Summary of the main studies that compared pet owners and non-owners on personality variables*

Personality variable	Finding	Study	How variable was measured
Extraversion			
Sociable	Pet owners > non-owners	Joubert (1987)	Social Interest Scale
Extraversion	Pet owners = non-owners	Cameron & Mattson (1972); Johnson & Rule (1991)	Eysenck Personality Inventory
Dominance	Pet owners = non-owners	Perrine & Osbourne (1998)	Author-developed rating scale
Sensation-seeking	Pet owners = non-owners	Friedmann *et al.* (1984)	Sensation-seeking Index
Independence	Pet owners = non-owners	Perrine & Osbourne (1998)	Author-developed rating scale
	Pet owners < non-owners	Guttmann (1981)	Author-developed semantic differential scales
Avoid loneliness or being alone	Pet owners > non-owners	Guttmann (1981)	Author-developed semantic differential scales
Agreeableness			
Do not like other people	Pet owners > non-owners	Cameron *et al.* (1966); Cameron & Mattson (1972)	Author-developed rating scale
	Pet owners = non-owners	Perrine & Osbourne (1998)	Used same scale as above
Feeling of being liked	Pet owners > non-owners	Cameron *et al.* (1966); Cameron & Mattson (1972)	Author-developed rating scale
	Pet owners = non-owners	Perrine & Osbourne (1998)	Used same scale as above
Like pets more than people	Pet owners > non-owners	Cameron & Mattson (1972)	Author-developed rating scale
Helpful	Pet owners > non-owners	Kidd & Feldman (1981)	Adjective Check List

Table 9.1. (*cont.*)

Personality variable	Finding	Study	How variable was measured
Social sensitivity, interpersonal trust	Pet owners > non-owners	Hyde *et al.* (1983)	Tennessee Self-concept Scale, Hogan's Empathy Scale, Rotter's Interpersonal Trust Scale
Conscientiousness			
Dependable	Pet owners > non-owners	Kidd & Feldman (1981)	Adjective Check List
Value a clean, tidy home, avoid lasting obligations	Pet owners < non-owners	Guttmann (1981)	Author-developed semantic differential scales
Emotional stability			
Neuroticism	Pet owners = non-owners	Cameron & Mattson (1972); Johnson & Rule (1991)	Eysenck Personality Inventory
	Pet owners < non-owners	Paden-Levy (1985)	
Anxiety, depression	Pet owners = non owners	Friedmann *et al.* (1984)	State–Trait Anxiety Inventory, Cognitive–Somatic Anxiety Scale, Jenkins Activity Scale
Alienation	Pet owners < non-owners	Paden-Levy (1985)	Alienation Scale
Ego strength	Pet owners < non-owners (urban)	Cameron & Mattson (1972)	Barron Ego Strength Scale
Self-acceptance	Pet owners = non-owners	Martinez & Kidd (1980)	Self-acceptance Scale (California Psychological Inventory)

Social self-esteem, general self-esteem	Pet owners = non-owners	Johnson & Rule (1991)	Self-esteem Scale, Texas Social Behavior Inventory, Social Desirability Scale
Self-sufficient, optimistic, self-confident	Pet owners > non-owners	Kidd & Feldman (1981)	Adjective Check List
Other			
Masculinity, femininity, athleticism	Pet owners = non-owners	Perrine & Osbourne (1998)	Bem Sex-Role Inventory, Author-developed rating scale
Type A (coronary prone) behaviour	Pet owners = non-owners	Friedmann *et al.* (1984)	Activity scale
	Pet owners > non-owners	McNicholas & Collis (1998)	15-item Type A Personality Scale

Note:

The personality variables studied have been grouped according to their association with the dimensions of the Big Five factor model: extraversion, agreeableness, conscientiousness, emotional stability, and openness to experience (or intellect). Where the variables did not fit any factor of the model, they were placed in the category 'other.' Note: no variables fitted the 'openness to experience' dimension.

to date are associated with emotional stability, agreeableness and extraversion, whereas virtually no research has been done on the conscientiousness and openness to experience dimensions.

Within each of the dimensions there is much disagreement as to whether or not pet owners are significantly different from non-owners. Because of differences in methodology, personality variables tested and the psychometric tests used (see Table 9.1, column 4), only a few of these studies can be compared directly. These particular studies, though, do not always produce a clear, consensual picture. Two early studies suggested that pet owners: did not like people as much as non-owners; did not feel as liked by others; liked their pets more than they liked people; and that urban pet owners tended to have weak egos (Cameron *et al.*, 1966; Cameron & Mattson, 1972). However, a more recent study, which incorporated some of the questions that Cameron and Mattson (1972) developed for their study, failed to find any differences between current pet owners and non-owners in terms of how much they liked people or how much people liked them (Perrine & Osbourne, 1998).

Interestingly, agreement has been found in two separate studies using the same personality test (Eysenck Personality Inventory) for neuroticism and extraversion. Here, it was found that pet owners were not different from non-owners (Cameron & Mattson, 1972; Johnson & Rule, 1991). However, using a slightly different methodology, Paden-Levy (1985) found that pet owners were less neurotic than non-owners.

A few studies have attempted to examine the same personality variable but have used different psychometric tests. This has also produced some contradictory results. Guttmann (1981) found that non-owners tended to be more independent than pet owners, whereas Perrine and Osbourne (1998) did not find any differences. In a detailed study of current pet owners, former pet owners and those who had never owned pets, Friedmann *et al.* (1984) failed to find any significant differences in Type A (or coronary prone) behaviour. However, McNicholas and Collis (1998) found that pet owners were more likely to have Type A personalities than non-owners.

Not all researchers have been happy with simply separating subjects into pet owners and non-owners. Some have suggested that pet owners should be defined in terms of the strength of the relationship they have with their pets. For example, Brown, Shaw and Kirkland (1972) found that students who had moderate levels of affection for dogs were significantly more affectionately disposed towards people than those with low levels of affection. However, the results did not provide a clear picture, as students with high levels of affection for dogs were less affectionate towards people than those with moderate levels of affection for dogs.

A related study (Lee, 1976, cited in Serpell, 1996) showed that people who interacted a lot with their dogs, as opposed to those who did not, expressed much more desire than non-pet owners to affiliate with people. This was viewed as evidence that dogs did not substitute for human relationships but that dogs may be owned by people who are unable to satisfy their affiliative needs.

In contrast to these studies, Johnson and Rule (1991) did not find any differences between pet owners and non-owners in terms of self-esteem, social self-esteem, neuroticism or extraversion, regardless of the level of attachment the owner had for the pet.

All that can be determined from the array of research studies presented here is that there is no strong evidence for differences between pet owners and non-owners. A problem with almost all the research to date is that it is cross-sectional and does not take into account previous ownership of pets. The categories of 'pet owner' and 'non-owner' are therefore not meaningfully mutually exclusive: non-owners are not necessarily those who have never owned pets. Indeed, the majority of people have owned pets at some stage in their lives, usually childhood (from 88% to 94%; Kidd & Kidd, 1980; Paul & Serpell, 1993), and of those currently without pets, most (53%; Anonymous, 1995) would like to have one. Only one in 20 current non-owners were found not to like pets (McHarg *et al.*, 1995). Therefore, it is not really surprising when differences are not found in studies of pet owners and non-owners. In addition, as adult personality traits are regarded as stable over time and situations (Hampson, 1988), it would be more reasonable to propose that pets help shape a person's personality in childhood. Therefore, measuring adult personality traits without knowledge of the developmental history of the person is not likely to provide us with any useful information.

Future research needs not only to follow the personality development of children (with and without pets) into adulthood (longitudinal studies) but, when conducting cross-sectional studies of adults, collect data on pet ownership history and the number of pets considered important to the person. Paul and Serpell (1993) showed the value of retrospective research when studying the importance of pets on the development of humane attitudes to animals.

Owners of different species of pet

Although there is no clear evidence for differences in the personalities of pet owners compared with non-owners, it is reasonable to suggest that people possessing certain personality traits may be more attracted to or more likely to own a particular species of pet. Indeed, there is some evidence for this.

Kidd, Kelley and Kidd (1984) found some interesting differences in the personalities of owners of horses, birds, snakes and turtles. Male owners of horses were found to be aggressive and dominant, whereas females owners were easy-going and non-aggressive. Bird owners were socially outgoing and expressive, snake owners were relaxed, unconventional and novelty-seeking, and turtle owners were hard-working, reliable and upwardly mobile (Kidd *et al.*, 1984). More recently, it has been shown that there are personality differences between the owners of dogs belonging to different breed groups (Katz *et al.*, 1994). These researchers found that toy breed owners were the most nurturing and least dominant of all the dog owners. Herding breed owners were the most aggressive and orderly, sporting breed owners were the least orderly, non-sporting breed owners were the least nurturing, working breed owners were the most dominant, hound owners were the friendliest, and terrier owners were the least aggressive but the most dependent on others for emotional support. Coren (1998) examined which types of dogs were best suited to which types of owners. He found that shy women were most compatible with clever dogs such as poodles, while men lacking in trust were more suited to independent dogs like Shar-Peis. Authoritarian owners were found to prefer to exert their dominance over assertive dog breeds such as terriers rather than less assertive breeds. Using a similar logic, Russel (1956) reasoned that since dogs are more obedient and subservient than cats, dogs would be preferred by authoritarian types. Consistent with this argument, he showed that people with authoritarian personalities were more likely to admire dogs than cats.

Contrary results also exist. For example, differences between dog and cat owners were not found on the personality variables of extraversion, neuroticism, general self-esteem, and social self-esteem (Johnson & Rule, 1991), self-acceptance (Martinez & Kidd, 1980), and masculinity, femininity, independence, athleticism and dominance (Perrine & Osbourne, 1998).

That studies have not shown any significant differences in the personality traits of dog and cat owners is perhaps not surprising considering dogs and cats are the most common pets owned in Western society (Fogle, 1994), with people often owning both species or having at least owned each at some stage in their lives (McHarg *et al.*, 1995). Kidd *et al.*'s (1984) findings suggest that differences are more likely to be found when comparing the personality traits of owners or enthusiasts of quite different animal species. This is consistent with Gosling and Bonnenburg's

(1998) study of owners of six different species of pet. They did not find any significant differences between dog and cat owners, but the data suggested personality differences between owners of ferrets, rabbits, horses and hedgehogs.

So where does this body of research leave us? Certainly, there is no clear evidence that pet owners and non-owners, and owners of particular types of pets, are different in terms of their personalities. Comparing studies and making sense of the sometimes contradictory findings are difficult because of the variety of methodologies and psychometric tests employed (see Table 9.1). In addition, as almost all of the studies to date have been cross-sectional, it is impossible to determine the importance of early ownership of pets on adult personality traits. Therefore, where differences have been found, it is difficult to say whether they are causes or effects of pet ownership.

To date, the effects of companion animals on the development of human personality have not been explored and therefore their importance in relation to the myriad of environmental factors playing a part in child development remains unclear. For progress to be made in this field we must instigate longitudinal personality studies of the human–animal relationship.

We hope that future studies will develop a battery of standardized (and therefore comparable) personality measures and will employ designs (e.g. longitudinal studies) that will address questions raised by the research reviewed in this chapter.

CONCLUSION

In this chapter we have brought together a broad array of personality research on pets and their owners. We are encouraged by the initial findings about personality differences across species of pet and how pet personality may be related to owner personality. Unfortunately, the much larger body of literature on how pet ownership itself may be associated with certain human personality traits is both confusing and frustrating, as the large variety of methodologies employed make direct comparisons difficult. Much research remains to be done. We have outlined a number of general and more specific suggestions for future research and hope that these will serve as the basis for a more integrative approach to examining personality in pets and their owners.

ACKNOWLEDGEMENTS

Thanks to R. Lee Zasloff for locating some elusive references for us, and to Jennifer Beck, Jill Nicholson, Elizabeth Paul, Michelle Pryor and James Serpell for their insightful suggestions on an earlier draft. The preparation of this chapter was supported, in part, by a research grant from the University of Texas.

NOTES

1. In such data, one should be aware of a pervasive problem with much companion animal personality research – most respondents adore their pets. Thus, for traits with evaluative components, there is a substantial danger of running into issues associated with restriction of range. We therefore recommend taking steps to increase the range of respondents, such as seeking out individuals who are less fond of their pets or asking respondents to describe one of their least favourite pets.
2. To facilitate cross-study integration, Gosling and Bonnenburg (1998) reproduce their rating instrument for use by other researchers.

REFERENCES

Anonymous (1995). *The Power of Pets: a Summary of the Wide-Ranging Benefits of Companion Animal Ownership.* Artarmon, Australia: Australian Companion Animal Council.

Bateson, P.P.G. (1981). Genes, environment and the development of behaviour. In *Animal Behaviour, 3: Genes, Development and Learning,* ed. T.R. Halliday & P.J.B. Slater, pp. 52–81. Oxford: Blackwell.

Brown, L.T., Shaw, T.G. & Kirkland, K.D. (1972). Affection for people as a function of affection for dogs. *Psychological Reports,* **31**, 957–8.

Cameron, P., Conrad, C., Kirkpatrick, D.D. & Bateen, R.J. (1966). Pet ownership and sex as determinants of stated affect toward others and estimates of others' regard of self. *Psychological Reports,* **19**, 884–6.

Cameron, P. & Mattson, M. (1972). Psychological correlates of pet ownership. *Psychological Reports,* **30**, 286.

Cattell, R.B., Eber, H.W. & Tatsuoka, M.M. (1970). *Handbook for the Sixteen Personality Factor Questionnaire (16PF).* Champaign, IL: IPAT.

Cavanaugh, J.C. (1995). Aging. In *Developmental Psychology*, ed. P.E. Bryant & A.M. Colman, pp. 70–89. London & New York: Longman.

Coren, S. (1998). *Why We Love the Dogs We Do: How to Find the Dog that Matches Your Personality.* New York: Free Press.

Costa, P.T. Jr & McCrae, R.R. (1994). Set like plaster: evidence for the stability of adult personality. In *Can Personality Change?*, ed. T.F. Heatherton & J.L. Weinberger, pp. 21–40. Washington, DC: American Psychological Association.

Digman, J.M. (1990). Personality structure: emergence of the five-factor model. *Annual Review of Psychology,* **41**, 417–40.

Digman, J.M. (1997). Higher-order factors of the Big Five. *Journal of Personality and Social Psychology,* **73**, 1246–56.

Eysenck, H.J. & Eysenck, S.B.J. (1964). *Manual of the Eysenck Personality Inventory.* London: University of London Press.

Feaver, J.A., Mendl, M. & Bateson, P. (1986). A method for rating the individual distinctiveness of domestic cats. *Animal Behaviour,* **34**, 1016–25.

Fogle, B. (1994). Unexpected dog ownership findings from Eastern Europe. *Anthrozoös,* **7**, 270.

Friedmann, E., Katcher, A., Eaton, M. & Berger, B. (1984). Pet ownership and psychological status. In *The Pet Connection: Its Influence on Our Health and Quality of Life,* ed. R.K. Anderson, B.L. Hart & L.A. Hart, pp. 300–8. Minneapolis: CENSHARE, University of Minnesota.

Goldberg, L.R. (1990). An alternative 'description of personality': the Big-Five factor structure. *Journal of Personality and Social Psychology,* **59**, 1216–29.

Goldberg, L.R. (1992). The development of markers for the Big-Five factor structure. *Psychological Assessment,* **4**, 26–42.

Goldberg, L.R. & Digman, J.M. (1994). Revealing the structure in the data: principles of exploratory factor analysis. In *Differentiating Normal and Abnormal Personality,* ed. S. Strack & M. Lorr, pp. 216–42. New York: Springer.

Gosling, S.D. (1998a). Personality dimensions in spotted hyenas (*Crocuta crocuta*). *Journal of Comparative Psychology,* **112**, 107–18.

Gosling, S.D. (1998b). The ecological perspective in anthrozoological research. *Newsletter of the International Society for Anthrozoology,* **15**, 11–14.

Gosling, S.D. (in press). From mice to men: What can we learn about personality from animal research? *Psychological Bulletin*, 127.

Gosling, S.D. & Bonnenburg, A.V. (1998). An integrative approach to personality research in anthrozoology: ratings of six species of pets and their owners. *Anthrozoös,* **11**, 148–56.

Gosling, S.D. & John, O.P. (1998). Personality dimensions in humans, dogs, cats, and hyenas. Paper presented at the American Psychological Society (APS) Symposium 'Bridging the gap between personality and animal researchers', Washington, DC, USA, 24th May, 1998.

Gosling, S.D. & John, O.P. (1999). Personality dimensions in nonhuman animals: a cross-species review: *Current Directions in Psychological Science,* ***8***, 69–75.

Guttmann, G. (1981). The psychological determinants of keeping pets. In *Interrelations Between People and Pets*, ed. B. Fogle, pp. 89–98. Springfield, IL: Charles C Thomas.

Hampson, S.E. (1988). *The Construction of Personality: an Introduction*, 2nd edn. London: Routledge.

Hyde, K.R., Kurdek, L. & Larson, P. (1983). Relationships between pet-ownership and self-esteem, social sensitivity and interpersonal trust. *Psychological Reports,* **52**, 110.

Jagoe, A. & Serpell, J. (1996). Owner characteristics and interactions and the prevalence of canine behaviour problems. *Applied Animal Behaviour Science,* **47**, 31–42.

John, O.P. (1990). The 'Big Five' factor taxonomy: dimensions of personality in the natural language and in questionnaires. In *Handbook of Personality: Theory and Research,* ed. L. A. Pervin, pp. 66–100. New York: Guilford Press.

John, O.P. & Gosling, S.D. (2000). Personality traits. In *Encyclopedia of Psychology,* Vol. 6 ed. A.E. Kazdin, pp. 140–4. Washington, DC: American Psychological Association.

Johnson, S.B. and Rule, W.R. (1991). Personality characteristics and self-esteem in pet owners and non-owners. *International Journal of Psychology,* **26**, 241–52.

Joubert, C.E. (1987). Pet ownership, social interest, and sociability. *Psychological Reports,* **61**, 401–2.

Katz, J.S., Sanders, J.L., Parenté, F.J. & Figler, M.H. (1994). Personality traits, and demographic and lifestyle characteristics as predictors of dog breed choice. Paper presented at the Scientific Sessions of the International Society for Anthrozoology (ISAZ) Meeting, New York, USA, 13th October, 1994.

Kidd, A.H. & Feldman, B. (1981). Pet ownership and self-perception of older people. *Psychological Reports,* **48**, 867–75.

Kidd, A.H., Kelley, H.T. & Kidd, R.M. (1984). Personality characteristics of horse, turtle, snake, and bird owners. In *The Pet Connection: its Influence on Our Health and Quality of Life,* ed. R.K. Anderson, B.L. Hart & L.A. Hart, pp. 200–6. Minneapolis: CENSHARE, University of Minnesota.

Kidd, A.H. & Kidd, R.M. (1980). Personality characteristics and preferences in pet ownership. *Psychological Reports,* **46**, 939–49.

Kidd, A.H. & Kidd, R.M. (1989). Factors in adults' attitudes towards pets. *Psychological Reports,* **65**, 903–10.

Lee, R. (1976). The pet dog: interactive correlates of a man–animal relationship. Unpublished report, Department of Psychology, University of Hull.

Martinez, R.L. & Kidd, A.H. (1980). Two personality characteristics in adult pet owners and non-owners. *Psychological Reports,* **47**, 318.

McHarg, M., Baldock, C., Headey, B. & Robinson, A. (1995). *National People and Pets Survey.* Report to the Urban Animal Management Coalition, Australia.

McNicholas, J. & Collis, G.M. (1998). Could Type A (coronary prone) personality explain the association between pet ownership and health. In *Companion Animals in Human Health,* ed. C.C. Wilson & D.C. Turner, pp. 173–85. London: Sage Publications.

Mugford, R.A. (1995). Canine behavioural therapy. In *The Domestic Dog: its Evolution, Behaviour and Interactions with People,* ed. J.A. Serpell, pp. 139–52. Cambridge: Cambridge University Press.

O'Farrell, V. (1987). Owner attitudes and dog behaviour problems. *Journal of Small Animal Practice,* **28**, 1037–45.

O'Farrell, V. (1995). Effects of owner personality and attitudes on dog behaviour. In *The Domestic Dog: its Evolution, Behaviour and Interactions with People,* ed. J.A. Serpell, pp. 153–8. Cambridge: Cambridge University Press.

Paden-Levy, D. (1985). Relationship of extraversion, neuroticism, alienation, and divorce incidence with pet-ownership. *Psychological Reports,* **57**, 868–70.

Paul, E.S. & Serpell, J.A. (1993). Childhood pet keeping and humane attitudes in young adulthood. *Animal Welfare,* **2**, 321–37.

Perrine, R.M. & Osbourne, H.L. (1998). Personality characteristics of dog and cat persons. *Anthrozoös,* **11**, 33–40.

Pervin, L.A. (1970). *Personality: Theory, Assessment and Research.* New York: John Wiley & Sons.

Pervin, L. & John. O.P. (1997). *Personality: Theory and Research,* 7th edn. New York: Wiley.

PFMA Profile (1998). The Pet Food Manufacturers' Association: London.

Podberscek, A.L. & Serpell, J.A. (1997). Aggressive behaviour in English cocker spaniels and the personality of their owners. *The Veterinary Record,* **141**, 73–6.

Poresky, R.H., Hendrix, C., Mosier, J.E. & Samuelson, M.L. (1988). Young children's companion animal bonding and adults' pet attitudes: a retrospective study. *Psychological Reports,* **62**, 419–25.

Ross, A.O. (1987). *Personality: the Scientific Study of Complex Human Behavior.* New York: CBS College Publishing.

Russel, W.M.S. (1956). On misunderstanding animals. *Universities Federation for Animal Welfare Courier,* **12**, 19–35.

Serpell, J.A. (1981). Childhood pets and their influence on adults' attitudes. *Psychological Reports,* **49**, 651–4.

Serpell, J.A. (1987). The influence of inheritance and environment on canine behaviour: myth and fact. *Journal of Small Animal Practice,* **28**, 949–56.

Serpell, J.A. (1996). *In the Company of Animals.* Cambridge: Cambridge University Press.

Stevenson-Hinde, J., Stillwell-Barnes, R. & Zunz, M. (1980). Subjective assessment of rhesus monkeys over four successive years. *Primates,* **21**, 66–82.

Woodhouse, B. (1978). *No Bad Dogs.* Aylesbury, Bucks: Hazell, Watson & Viney.

ELIZABETH S. PAUL

10

Love of pets and love of people

INTRODUCTION

A good proportion of the relatively small amount of anthrozoological research conducted during the 1970s and early 1980s focused on the emotive question of whether or not love of pets is associated with love of people. Some researchers asserted that people who have great affection for animals also have greater warmth of feelings than most for their fellow human beings (Brown, Shaw & Kirkland, 1972; Paden-Levy, 1985). Others argued that the opposite was true; that people turn to pets for affection when their relationships with other people are inadequate or unfulfilled (Cameron *et al.*, 1966; Cameron & Mattson, 1972; Peele & Brodsky, 1974).

As a whole, this discussion probably better resembled an expression of society's divided and ambivalent feelings towards animals than a serious psychological debate (Serpell, 1996). Moreover, such divided opinions were not new to the late twentieth century. Both supporters and detractors of the habit of pet-keeping had for many centuries been disputing the significance of pets for human social relations. Many prominent humanitarians of the eighteenth, nineteenth and early twentieth centuries were also acknowledged animal lovers, and Albert Schweitzer and Mahatma Gandhi explicitly linked individuals' and societies' moral greatness to their regard for the treatment of animals (see Maehle, 1994; Serpell, 1996, for historical reviews of this topic). But pet-keeping has also been viewed with deep suspicion by many, with affection for animals frequently being cited as evidence for witchcraft during the witch hunts of the sixteenth and seventeenth centuries (Serpell & Paul, 1994). Even George Orwell, writing in 1947 (Orwell, 1947, quoted in Serpell, 1996), blamed the rise of pet-keeping on the 'dwindled birthrate'.

As anthrozoological research blossomed during the late 1980s and

1990s, empirical attention given to this issue decreased in prominence, although a parallel line of inquiry, concerning whether cruelty to animals is associated with human-directed violence, has been gaining attention in recent years. My aim in the present chapter is to revive the 'love of people–love of pets' debate, with a view to assessing whether contemporary research has been able to offer any convincing conclusions to this age-old argument. The motive for doing this is not to lend support to either a pro-pet or anti-pet viewpoint. Rather, it is to try to illuminate further both the psychological nature of human–animal relationships, and the nature of people's capacities to feel love, affection or compassion for others, regardless of species or relatedness.

Perhaps the first point to note before embarking on a review of recent work in this area is simply the difficulty of defining precisely the question being asked. The idea of feeling 'love' for someone or something, particularly if the target of this affection is an entire species or kingdom of life on the planet, is probably an impossible notion to define. So what I have looked for in reviewing literature in this field is any study which has approached this general topic by considering one or more of a variety of target feelings, including general affection and compassion (empathy, sympathy) for animals, and similar, positive dispositions towards human beings.

Despite the long-standing dispute between pro-pet and anti-pet factions regarding the true direction of the relationship between love of people and love of pets, popular public opinion in modern, western cultures seems generally to support the view that there is a positive association in the degree of sentiment felt by people for humans and for animals. In her novel, *Devices and Desires*, the crime writer P.D. James (1989) invents a murderer who stalks the streets at night dressed as a woman and walking a dog. In reality neither female nor a pet lover, the killer is thus able to lull his victims into a false sense of security. A number of studies have provided evidence that the assumption upon which this storyline is based is largely accurate. Messent (1983) found that people walking in a park in London were significantly more likely to be spoken to by passersby if they were accompanied by a dog. Similarly, Robins, Sanders and Cahill (1991) found dogs to facilitate contact, confidence and conversation among their owners in a park on the West Coast of the USA. Studies of wheelchair users in the USA have also demonstrated that social acknowledgements (e.g. friendly glances, smiles and conversations) from passersby are substantially more frequent when a service dog is present (Eddy, Hart & Boltz, 1988; Mader, Hart & Bergin, 1989). Experimental studies in which participants are asked to assess photographs or line drawings of people with or without an accompanying pet have also found that people pictured with a dog tended to be perceived as happier, friendlier, more relaxed and less threatening than those pictured without an animal present (Lockwood, 1983; Rossbach & Wilson, 1992). However, recent research has produced equivocal results (Friedmann & Lockwood, 1991; Budge *et al.*, 1996), suggesting that other factors such as a person's gender and the species of pet involved may produce a more complex pattern of perceptions and expectations than was originally reported. Nevertheless, the general conclusion from all of these studies remains the same: that in societies such as Britain and the USA, where pet keeping is accepted and participated in by the majority, it likewise appears to be regarded by the majority as reflecting positive, or at the very least benign, social personalities in those who indulge in it. Such opinions may simply reflect a human tendency to believe that people who live and behave in similar ways to ourselves are likely to be nicer and more trustworthy than those who do not conform to societal norms. Alternatively, they may genuinely be based on personal observations that pet lovers are indeed more pleasant than those who reject the affections of animals.

LOVE OF PEOPLE AND LOVE OF PETS: ARE THEY CORRELATED?

The majority of work in this area has used an 'individual differences' research approach, in which relatively well-matched groups of participants are compared with one another on the basis of their pet ownership status or expressed feelings for pets. Early work by Cameron and his colleagues (Cameron *et al.*, 1966; Cameron & Mattson, 1972) supported the view that pet owners have a lower level of liking for people than do non-owners, and this finding occurred regardless of whether liking for people was directly contrasted with a liking for pets (Cameron & Mattson, 1972) or not (Cameron *et al.*, 1966). However, when Brown *et al.* (1972) classified a group of students as having low, moderate or high affection for dogs, they found that those with 'moderate affection' scored highest on the 'expressed affection (for people) scale' of the FIRO-B test of personality (Schutz, 1958). In contrast, the 'low affection' students scored lowest. They therefore concluded that general affection for pet dogs mirrored similar sentiments towards people, although they did also suggest that very high levels of animal affection may represent a displacement of affection from people to pets. In a study of 18-year-old students, Paden-Levy (1985) found small but significant negative correlations between the ownership of dogs and/or cats during childhood, and self-reported questionnaire scores of 'neuroticism' (Eysenck Personality Inventory, Eysenck & Eysenck, 1965) and of 'alienation' (Dean, 1961). In a more recent study, Perrine and Osbourne (1998) found that students who regarded themselves as 'dog people' or 'cat people' also rated themselves as liking people more and being liked more by others than did those who did not regard themselves as either dog or cat people. Also studying students, Hyde, Kurdek and Larson (1983) found that pet owners tended to score slightly higher than non-owners on 'interpersonal trust' (using Rotter's Interpersonal Trust Scale, Rotter, 1967). Joubert (1987) found that although 'social interest' (measured using a scale developed by Crandall, 1975) was not associated with pet ownership, student pet owners did report spending a significantly greater amount of time each day socializing with others than did their non-pet owning contemporaries. In a study of older women, however, Ory and Goldberg (1984) found apparently conflicting evidence. The size of the women's friend networks were inversely related to their likelihood of owning a pet, but the number of close confidants a woman reported having was positively associated with pet ownership. A number of other studies of the general public in Australia (Ray, 1982), Canada (St-Yves *et al.*, 1990) and the USA (Martinez & Kidd, 1980) have failed to find any significant associations between pet ownership or affection for pets and feelings for people.

An alternative method for assessing the possible nature of any linkage of felt regard for people and for animals is to consider the differences that exist between specific demographic sections of a society, or between different societies around the world, in the overall levels of these two factors. For example, is it the case that cultures where animals are generally held in low esteem also tend to have populations who have low regard for one another? Similarly, in any one culture, is it the case that different demographic groups (for example different socio-economic groups, urban and rural dwellers, men and women) differ systematically from one another in the affection they feel for people and animals?

Unfortunately, with respect to variation between different societies throughout the world, reliable evidence is currently very difficult to obtain. No formal studies of cross-cultural variation in love of people and love of animals have been conducted. Certainly, there is a popular perception that cultures with little emotional or sentimental interest in animals or their welfare also often have less than exemplary welfare protection for their human populations (Grandin, 1988). But any causal explanation for this seems likely to be more to do with these societies' relative levels of wealth than some necessary link between affection for humans and animals. A number of researchers have proposed a connection between 'living on the land' and populations having more utilitarian attitudes towards animals (e.g. see Thomas, 1983; Ingold, 1994). Such societies, based largely on farming economies, tend also to be poorer than their urban and industrialized counterparts, and probably for this reason are less able to offer economic protection and assistance for their human citizens. At an individual level, however, there is no evidence that people in such countries have fewer concerns about each other's well-being.

Within a particular nation or society, any systematic variation between different demographic groups in their apparent affection for people and for pets should also be able to offer some clues concerning the degree to which such sentiments may be psychologically linked. Unfortunately, although several studies in modern Western countries have found pet ownership levels and feelings for animals to vary according to factors such as family composition (Godwin, 1975; Franti *et al.*, 1980; Melson, 1988; Kidd & Kidd, 1989; Endenburg, Hart & de Vries, 1990), socio-economic status (Godwin, 1975; Franti *et al.*, 1980; Wise & Kushman, 1984; Messent & Horsfield, 1985; Marx *et al.*, 1988) and urban/rural residence (Griffiths & Bremner, 1977; Covert *et al.*, 1985; Marx *et al.*, 1988; Wells & Hepper, 1995), no similar studies have been undertaken to look for parallel variation in attitudes towards or feelings for humans. Males and females are known to differ consistently from one another in a variety of aspects of social functioning (for reviews see Maccoby & Jacklin, 1974; Hall, 1984; Eagly, 1987; Eagly & Wood, 1991; Archer, 1996), although it would be difficult to claim that either sex 'loves people' more than the other. Nevertheless, the observation that females tend to be more affiliative and socially demonstrative than males (Hall, 1984; Moore & Boldero, 1991; Franken, Hill & Kierstead, 1994; Kring & Gordon, 1998) could lead to a prediction that females may also be more likely to seek out animals as companions. An alternative prediction, however, could be that males may be more likely to seek out animal companions because they perceive them as less socially demanding and less competitively threatening than fellow humans. In fact, existing research into the relative involvement of males and females with pets has produced mixed and conflicting results. Some researchers have found that more girls and women keep pets than boys and men (Wise & Kushman, 1984; Paul & Serpell, 1992; Rost & Hartmann, 1994), but many other studies have found no significant differences in the levels of male and female pet ownership (Franti *et al.*, 1980; Covert *et al.*, 1985; Marx *et al.*, 1988; Melson, 1988; Wells & Hepper, 1995). Girls have been found to be more likely to report wanting to keep pets (Paul & Serpell, 1992; Rost & Hartmann, 1994), and girls and women have reported stronger emotional relationships with their pets (Ray, 1982; Rost & Hartmann, 1994), although this latter tendency was not found by Stallones *et al.* (1989). There has also been some research suggesting that males and females differ somewhat in their styles of interaction with their pets (Brown, 1984; Mertens, 1991), although other similar studies have found no such differences (Smith, 1983; Mallon, 1993). So, although the actual quality of the relationship between males and females and their pets may differ in some systematic ways (which, in turn, may parallel differences in their relations with other human beings), further research is needed to determine the exact nature and extent of this variation.

EMPATHY WITH ANIMALS

Arguments concerning whether empathy, compassion and concern for the welfare of humans and animals are associated have gone hand in hand with those regarding whether love of pets and love of people are linked. Particularly galling to animal rights and welfare campaigners is the accusation that in caring so much about animals, they neglect or lack compassionate feelings towards their fellow human beings (Paul, 1995). On the other hand, concerns for both people and animals have been explicitly linked by leading humanitarians and moral philosophers (e.g. see Midgley, 1984; Maehle, 1994; Serpell, 1996). Child and animal protection movements have had closely linked histories (Shultz, 1924), and humane educationalists have persistently assumed that engendering empathy with animals will directly encourage the emergence of empathy and sympathy towards people (e.g. see Finch, 1989). This latter hypothesis was recently tested in a study which measured human-directed empathy in fourth and fifth grade children who had taken part in a 40-hour animal-focused humane education programme (Ascione, 1992; Ascione & Weber, 1996). Compared with a group of same-aged controls, these children showed elevated human-directed empathy levels at the end of the programme and one year later. It was concluded that animal-based humane attitudes enhanced by such education programmes do generalize to human-directed empathy. It should be noted, however, that the magnitude of this effect was slight, especially at the one-year follow-up.

Other studies have considered ‘naturally occurring’ variation in human-oriented and animal-oriented empathy or compassion in a variety of populations. Focusing on young children, Poresky and his colleagues found that empathy with other children was positively associated with empathy with pets (Poresky, 1990), and that while pet ownership *per se* was not related to child-oriented empathy, the reported strength of the child's bond with the pet was (Poresky & Hendrix, 1990; Poresky, 1996). Similar results were also reported by Melson (1991). Using a small sample of American psychology students, Hyde *et al.* (1983) observed a marginally significant positive association between pet ownership and empathy (measured using Hogan's, 1969, Empathy Scale). Paul and Serpell (1993) found that a large group of students at a university in England demonstrated significant correlations between level and intensity of childhood pet relationships, concerns for the welfare of animals, and empathy with humans (measured using a modified version of the Mehrabian & Epstein Scale, 1972). In a questionnaire study of 45 British adults, Wagstaff (1991) found that participants who reported strong empathic reactions to a set of photographs of animals in distress tended to be more sympathetic to animal welfare issues and also more positive and sympathetic towards human beings.

Using a questionnaire survey of Scottish adults, Paul (submitted) tested the implicit assumption of many empathy researchers that human-oriented and animal-oriented empathy represent facets of the same basic construct (Mehrabian & Epstein, 1972; Eisenberg *et al.*, 1992). Participants completed the human-oriented questions of the Mehrabian and Epstein (1972) Questionnaire Measure of Emotional Empathy, and a similarly structured Animal Empathy Scale. A small but highly significant correlation between the two scales was found, and this association was maintained when independent analyses were made of male and female, and pet owning and non-owning participants' responses. This indicated that human-oriented and animal-oriented empathy probably do possess some common determinant. However, each also appeared to be independently affected by a variety of factors relating separately to either humans or animals, with pet ownership being positively associated with empathy with animals but not with humans, and child-rearing being positively associated with empathy with humans but not with animals. In conclusion, this study characterized human-oriented and animal-oriented empathy as linked tendencies whose levels of expression are probably largely determined independently by a variety of domain-specific, environmental factors.

A variety of studies have shown that compared with men, women (at least in the Western countries studied) show more empathy with animals (Hills, 1993), demonstrate greater concerns about their welfare and rights (Herzog, Betchart & Pittmann, 1991; Broida *et al.*, 1993; Rajecki, Rasmussen & Craft, 1993; Driscoll, 1995; Wells & Hepper, 1995, 1997; Pifer,

1996, Mathews & Herzog, 1997), are more likely to enrol in relevant pressure groups and campaigning organizations (Shaw, 1977; Plous, 1991; Jasper & Nelkin, 1992; Galvin & Herzog, 1998), and are more willing to donate money to animal charities (Paul, 1992; RSPCA personal communication). This is paralleled by findings that females are also more likely to report and show more human-oriented empathy than males (Eisenberg & Lennon, 1983; Lennon & Eisenberg, 1987).

Finally, a number of studies have found that concern for animal welfare issues, and support for animal rights in particular, tend to be associated with left-wing political attitudes (Furnham & Pinder, 1990; Broida *et al.*, 1993; Furnham & Heyes, 1993), including advocacy of gun control laws, concern about social violence, and support for minority rights (Nibert, 1994). Such findings have been heralded as supporting the view that people's dispositions towards human and non-human animals are linked (Nibert, 1994).

LOVE OF CHILDREN AND LOVE OF PETS

The similarity between people's relationships with pets and with children has been pointed out by a number of researchers and commentators (Beck & Katcher, 1983; Berryman, Howells & Lloyd-Evans, 1985; Bulcroft & Albert, 1987; Serpell, 1996). Many species, especially baby animals and some adult mammals, share the 'cute' features employed by human young to help elicit love, protection and nurturance from their parents (Lorenz, 1943; Jolly, 1972; Gould, 1979). It seems probable that it is this 'cute response' which bears some part of the responsibility for initiating and perpetuating the phenomenon of pet-keeping (e.g. see Serpell, 1996). Domestic dogs have been selected over the centuries to show an increasingly neotenous appearance (Goodwin, Bradshaw & Wickens, 1997) and even teddy bears have undergone selection for cuteness, gradually becoming more appealing to the humans who purchase them (Hinde & Barden, 1985; Morris, Reddy & Bunting, 1995). It could be hypothesized, therefore, that people who are for some reason more susceptible to physical (or behavioural) characteristics of cuteness would show greater affection and affiliativeness towards both animals (especially neotenous or young ones) and human babies and children. In fact, no-one has actually tested this hypothesis explicitly, although observations that women and girls tend to find 'cute' features more appealing than do men and boys (Ritter, Casey & Langlois, 1991; Morris *et al.*, 1995) is consistent with the recent finding of Fridlund and MacDonald (1998) that women more frequently approached a golden retriever puppy, although this sex difference disappeared as the dog grew older. This is also consistent with the observation of Budge *et al.* (1997) that females tend to be expected to keep smaller, more neotenous pets. The finding by Kidd and Feldmann (1981) that pet owners were significantly more nurturant than non-owners (using the Gough & Heilbrun, 1965, 'Adjective Check List') also fits in with the idea that interest in (at least some kinds of) pets might be linked with more general care-giving motivations.

It is also interesting to note that a woman's reproductive status appears to affect her feelings for pets. In a qualitative study, Bulcroft and Albert (1987) found that a number of the women they interviewed reported a dramatic diminution of affection for and involvement with their pets following the birth of their first child. Similarly, using structured questionnaires, Collis, Bradshaw and Cook (1998) found that childless women rated their relationships with their cats as significantly more companionable and offering more emotional support than did those who were either pregnant or already had children. These findings suggest that there may well be some overlap of emotional systems motivating care-giving behaviour and affection towards both pets and children. Fortunately, at least when the children concerned are our own kin, they generally appear to be more potent elicitors of such feelings and behaviour. Serpell (1996), however, quotes a newspaper article describing a dramatic exception to this, when a woman was reportedly more distressed about her dog being destroyed than about the fact that it had just killed her newborn baby.

CONCLUSION

Over the years that research has been conducted in this area, attempts to answer the basic question of whether love of people is correlated with love of pets has been hampered repeatedly by methodological problems of definition and measurement. 'Love of people' is a difficult notion to define, and has been measured by researchers in many different ways, with considerable consequent variability in psychometric validity and usefulness (e.g. see Cameron *et al.*, 1966; Brown *et al.*, 1972; Joubert, 1987). Moreover, as Ray (1982) has pointed out, perhaps for most humans what is important is who we love, not whether we love people in general. Deciding what is meant by 'pet lovers' has also caused problems, with such people frequently being defined simply as pet owners (e.g. see Cameron *et al.*, 1966; Cameron & Mattson, 1972; Hyde *et al.*, 1983; Ory & Goldberg, 1984; Paden-Levy, 1985), a large and heterogeneous group which is unlikely to be able to be characterized by distinctive personality profiles or social characteristics. After all, a so-called pet owner may not be a 'pet lover' at all, being labelled as such simply because someone else in the household has a pet, or because the pet arrived unbidden in the home as a present or a stray. There are also many different species and breeds of pet, each of which might attract different types of owner (e.g. see Kidd & Kidd, 1980; Kidd, Kelley & Kidd, 1984). Someone might love cats but hate dogs, for example, or *vice versa* (Perrine & Osbourne, 1998). Pets may also have different functions in their relations with the owner, such as companion, surrogate child, status symbol, or even fashion accessory (Council for Science and Society, 1988). Given these problems, it is perhaps not surprising that few consistent trends can be extracted from the studies which have tried to find links between love of people and love of pets. If anything, the mounting evidence from an ever-increasing literature in this field suggests that it is other factors, such as childhood experience of pets (Serpell, 1981; Poresky *et al.*, 1988; Kidd & Kidd, 1989; Paul & Serpell, 1993), parental attitudes towards animals (Gage & Magnuson-Martinson, 1988; Schenk *et al.*, 1994) and the behaviour of the individual pet (Serpell, 1996), rather than personality (see also Podberscek & Gosling, Chapter 9) or social tendencies, which are most likely to determine whether someone will be a pet owner or pet lover. If associations do exist in the nature of people's general affectionate relations with pets and humans, they are perhaps most likely to be found by studying very specific groups of pet owners and non-owners. For example, as suggested by Brown *et al.* (1972), people who take pet-keeping to extremes, perhaps keeping large numbers of animals, or lavishing excessive attention on a favourite dog or cat, may indeed be displacing affection from people to pets. Nevertheless, even if this is found to be true, such cases may well be expressing specific 'pathological' problems, and may not extrapolate to the conclusion that all affectionate feelings for animals go hand in hand with less affectionate feelings for people.

The proposal that feeling empathy for people and animals may be correlated is in many ways an easier hypothesis to test than the broad notion that love of people and love of pets are linked. Empathy is a well-defined (albeit complex) psychological construct, the nature of which has been extensively researched (e.g. see Eisenberg & Strayer, 1987; Davis, 1994, for reviews). As a consequence, studies looking for associations between animal-oriented and human-oriented empathy tend to have been more effectively designed, and have yielded more consistent and convincing findings. The results of a number of studies in this area agree that individuals who are more emotionally empathic with people also tend to be more emotionally empathic with animals (Poresky, 1990;

Wagstaff, 1991). That is, people who have a stronger than normal tendency to experience an emotional reaction when witnessing the emotion of another human being will also be more likely than others to experience an emotional response when witnessing the apparent emotion of an animal. Such tendencies also appear to be associated with a childhood history of involvement with pets (Poresky & Hendrix, 1990; Paul & Serpell, 1993; Poresky, 1996). However, these associations do not appear to be strong enough to conclude that the mechanisms determining empathy for both people and animals are entirely interchangeable, as some empathy researchers of the past have assumed (e.g. see Mehrabian & Epstein, 1972; Eisenberg, 1988; Eisenberg *et al.*, 1992). It is clear that there is still considerable variability within individuals in the degree of empathy experienced for either people or animals, with some people being very empathic with people but not with animals, and others being very empathic with animals but not with people. Hitler and some of his fellow Nazis, for example, appear to have maintained a dramatic mismatch of human-oriented and animal-oriented empathy in their aspousal of vegetarianism and animal welfare on the one hand, yet their patent lack of respect for human life on the other (Arluke & Sax, 1992).

There is evidence from research into human-oriented empathy that although the level of empathic responsiveness a person experiences can be greatly influenced by the developmental environment (Koestner, Franz & Weinberger, 1990; Dunn *et al.*, 1991; Eisenberg *et al.*, 1991, 1992), each individual also appears to possess a fundamental inherited tendency to be more or less empathic (Rushton *et al.*, 1986; Davis, Luce & Kraus, 1994). It could be hypothesized, therefore, that studies that have demonstrated a small but significant positive association between human-oriented and animal-oriented empathy may in fact be registering some common underlying genetic predisposition to be more emotionally responsive to the apparent emotion of another, whoever, or whatever, the other may be. Other, non-genetic influences (e.g. experiences with pets and people during early life) may, in addition to this, play a major role in separately determining the overall magnitude of human-oriented and animal-oriented empathy.

People who are particularly emotionally empathic are known to be more likely than others to show concern for the well-being of fellow citizens, to help someone in distress, or to donate money to worthy causes (Eisenberg, 1986). The various studies mentioned earlier which found compassionate social and political attitudes towards both people and animals to be positively linked (Furnham & Pinder, 1990; Broida *et al.*, 1993; Furnham & Heyes, 1993; Nibert, 1994), therefore, might also be explained by this observed positive correlation between human-oriented and animal-oriented empathy. An alternative explanation, however, also exists. Tetlock (1986) and Tetlock, Armor and Peterson (1994) have proposed that many social and political attitudes show strong correlations between one another because they are all ultimately based on a smaller set of higher 'values', such as freedom and equality. Thus, people who strongly support equality as a fundamental moral value could be expected to be in favour of policies or causes which aim to ensure both equality amongst people (as do many left-wing social proposals) and equality between people and non-human animals (e.g. animal rights; see Hills, 1993).

Despite having suggested above that the studies reviewed here have generally failed to find conclusive evidence for a general link between 'love of people' and 'love of pets', it must be said that the accumulating evidence that, compared with men, women tend to show slightly more concern for (and affiliation with) both people and animals does offer support for a degree of association between some facet of these two tendencies. The explanation for such a link may prove to have its origins in one or more of a number of observed differences between males' and females' social, emotional and cognitive functioning (Maccoby & Jacklin, 1974; Hall, 1984; Eagly, 1987; Feingold, 1994; Kring & Gordon, 1998). Hills (1989) pointed out that one difference that has been found clearly to exist between males and females is the degree to which they tend to be oriented towards and interested in either 'things' or 'persons' (Little, 1968, 1983). 'Person specialists' are more often female (Little, 1976, 1983). They tend to be more interested in people than in things or objects, and tend to construe both people and things using more subjective psychological constructs (e.g. desires, moods, etc.). 'Thing specialists' are more often male (Little, 1976, 1983). They are generally more interested in things or objects, and tend to view both people and things using more objective, physical criteria (e.g. size, colour, etc.). Animals occupy ambiguous psychological terrain by possessing facets of both persons and things (e.g. see Arluke, 1988). Thus, the degree to which someone tends to view an animal as a person rather than as a thing, and thus be willing to relate to it socially and view it as having emotions and intentions, may be determined in large part by the extent to which he or she tends to be a 'person' or a 'thing' specialist. Hills (1989) found that this does, indeed, appear to be the case. In a study of 101 Australian adults, she found that person specialists were more oriented towards and interested in animals, and more likely to perceive them 'personalistically' than were thing specialists.

Differences in the tendency to experience nurturant emotions and to offer care to others are another facet of the observed socio-cognitive differentiation of males and females (Feldman, Nash & Cutrona, 1977; Berman, 1980; Berman & Goodman, 1984; Feingold, 1994), and these varying tendencies may also be having parallel influences on men's and women's affiliativeness with, and affection for, both people and animals. It was pointed out earlier in this chapter that people's behaviours towards, and feelings for, pets seem to share many of the characteristics of parents' relations with children, and that females appear to be more responsive to the nurturance-evoking features of both children and pets (Beck & Katcher, 1983; Berryman *et al.*, 1985; Bulcroft & Albert, 1987; Ritter *et al.*, 1991; Serpell, 1996; Fridlund & MacDonald, 1998). Combined with the finding that pet orientation appears to diminish during the phases of life when people are most involved in caring for their own young offspring (Bulcroft & Albert, 1987; Collis *et al.*, 1998), the notion that at least part of the psychological system involved in pet-keeping overlaps with that functionally adapted for the care of human young seems a plausible one.

Finally, many studies have shown that emotional empathy towards humans is stronger in women than in men (e.g. see Eisenberg & Lennon, 1983; Lennon & Eisenberg, 1987), and this parallels the findings of Hills (1993) and Paul (submitted) that women also tend to be more empathic with animals. In fact, the capacity to empathize is probably a very important component of nurturant skills (Wiesenfeld, Witman & Malatesta, 1984), so these findings are perhaps not surprising. The possible nature of a link between human-oriented and animal-oriented empathy has already been discussed above, and will not be reiterated here. Suffice it to say that people's capacities to empathize with both humans and animals appear to be correlated both within and between the sexes. Whether the sources of these correlations are the same or different, and whether they are largely genetically or environmentally determined, will have to be considered by future studies in this area.

REFERENCES

Archer, J. (1996). Sex differences in social behaviour: are the social role and evolutionary explanations compatible? *American Psychologist*, **51**, 909–17.

Arluke, A.B. (1988). Sacrificial symbolism in animal experimentation: object or pet? *Anthrozoös*, **2**, 98–117.

Arluke, A. & Sax, B. (1992). Understanding Nazi animal protection and the holocaust. *Anthrozoös*, **5**, 6–31.

Ascione, F.R. (1992). Enhancing children's attitudes about the humane treatment of animals: generalisation to human-directed empathy. *Anthrozoös*, **5**, 176–91.

Ascione, F.R. & Weber, C.V. (1996). Children's attitudes about the humane treat-

ment of animals and empathy: one year follow up of a school-based intervention. *Anthrozoös*, **9**, 188–95.
Beck, A.M. & Katcher, A.H. (1983). *Between Pets and People: the Importance of Animal Companionship*. New York: G.P. Putnam.
Berman, P. (1980). Are women more responsive than men to the young? Review of developmental and situational variables. *Psychological Bulletin*, **88**, 668–95.
Berman, P. & Goodman, V. (1984). Age and sex differences in children's responses to babies: effects of adults' caretaking requests and instructions. *Child Development*, **55**, 1071–7.
Berryman, J.C., Howells, K. & Lloyd-Evans, M. (1985). Pet owner attitudes to pets and people: a psychological study. *The Veterinary Record*, **December 21/28**, 659–61.
Broida, J., Tingley, L., Kimball, R. & Miele, J. (1993). Personality differences between pro- and anti-vivisectionists. *Society and Animals*, **1**, 129–44.
Brown, D. (1984). Personality and gender influences on human relationships with horses and dogs. In *The Pet Connection: its Influence on Our Health and Quality of Life*, eds. R.K. Anderson, B.L. Hart & L.A. Hart, pp. 216–24. Minneapolis: CENSHARE, University of Minnesota.
Brown, L.T., Shaw, T. & Kirkland, K.D. (1972). Affection for people as a function of affection for pets. *Psychological Reports*, **31**, 957–8.
Budge, R.C., Spicer, J., Jones, B.R. & StGeorge, R. (1996). The influence of companion animals on owner perception: gender and species effects. *Anthrozoös*, **9**, 10–18.
Budge, R.C., Spicer, J., StGeorge, R. & Jones, B.R. (1997). Compatibility stereotypes of people and pets: a photograph matching study. *Anthrozoös*, **10**, 37–49.
Bulcroft, K. & Albert, A. (1987). Similarities and differences between the roles of pets and children in the American family. Paper presented at the Delta Society Annual Conference, 'Living Together: People, Animals and the Environment' October 1987, Vancouver, BC.
Cameron, P., Conrad, C., Kirkpatrick, D.D. & Bateen, R.J. (1966). Pet ownership and sex as determinants of stated affect towards others and estimates of others' regard for self. *Psychological Reports*, **19**, 884–6.
Cameron, P. & Mattson, M. (1972). Psychological correlates of pet ownership. *Psychological Reports*, **30**, 286.
Collis, D., Bradshaw, J. & Cook, S. (1998). Effects of reproductive status on women's 'attachment' to their cats. Paper presented at 'The Changing Roles of Animals in Society', 8th International Conference on Human–Animal Interactions. Prague, September 10th–12th, 1998.
Council for Science and Society (1988). *Companion Animals in Society*. Oxford: Oxford University Press.
Covert, A.M., Whiren, A.P., Keith, J. & Nelson, C. (1985). Pets, early adolescents and families. *Marriage and Family Review*, **8**, 95–108.
Crandall, J.E. (1975). A scale for social interest. *Journal of Individual Psychology*, **31**, 187–95.
Davis, M.H. (1994). *Empathy: a Social Psychological Approach*. Madison, WI: Brown and Benchmark.
Davis, M.H., Luce, C. & Kraus, S.J. (1994). The heritability characteristics associated with dispositional empathy. *Journal of Personality*, **62**, 369–91.
Dean, D.G. (1961) Alienation: its meaning and measurement. *American Sociological Review*, **26**, 753–8.
Driscoll, J.W. (1995). Attitudes toward animals: species ratings. *Society and Animals*, **3**, 139–50.
Dunn, J., Brown, J., Slomkowski, C., Tesla, C. & Youngblade, L. (1991). Young children's understanding of other people's feelings and beliefs: individual differences and their antecedents. *Child Development*, **62**, 1352–66.

Eagly, A. (1987). *Sex Differences in Social Behaviour: a Social Role Interpretation.* Hillsdale, NJ: Lawrence Erlbaum Associates.

Eagly, A.H. & Wood, W. (1991). Explaining sex differences in social behaviour: a meta analytic perspective. *Personality and Social Psychology Bulletin*, **17**, 306–15.

Eddy, J., Hart, L.A. & Boltz, R.P. (1988). The effects of service dogs on social acknowledgements of disabled people. *The Journal of Psychology*, **122**, 39–45.

Eisenberg, N. (1986). *Altruistic Emotion, Cognition and Behaviour.* Hillsdale, NJ: Lawrence Erlbaum Associates.

Eisenberg, N. (1988). Empathy and sympathy: a brief review of the concepts and empirical literature. *Anthrozoös*, **2**, 15–17.

Eisenberg, N., Fabes, R. A., Carlo, G., *et al.* (1992). The relations of maternal practices and characteristics to children's vicarious emotional responsiveness. *Child Development*, **63**, 583–602.

Eisenberg, N., Fabes, R.A., Schaller, M., Carlo, G. & Miller, P.A. (1991). The relations of parental characteristics and practices to children's vicarious emotional responding. *Child Development*, **62**, 1393–408.

Eisenberg, N. & Lennon, R. (1983). Sex differences in empathy and related capacities. *Psychological Bulletin*, **94**, 100–31.

Eisenberg, N. & Strayer, J. (1987). *Empathy and its Development.* Cambridge: Cambridge University Press.

Endenburg, N., Hart, H. & de Vries, H.W. (1990). Differences between owners and non-owners of companion animals. *Anthrozoös*, **4**, 120–7.

Eysenck, H.J. & Eysenck, S.G.B. (1965). *The Eysenck Personality Inventory*. London: University of London Press.

Feingold, A. (1994). Gender differences in personality: a meta analysis. *Psychological Bulletin*, **116**, 429–56.

Feldman, S. S., Nash, S. C. & Cutrona, C. (1977). The influence of age and sex on responsiveness to babies. *Developmental Psychology*, **13**, 675–6.

Finch, P. (1989). Learning from the past. In *The Status of Animals: Ethics, Education and Welfare*, ed. D. Paterson & P. Palmer, pp. 64–72, Wallingford; Oxon: CAB International.

Franken, R.E., Hill, R. & Kierstead, J. (1994). Sport interest as predicted by the personality measures of competitiveness, mastery, instrumentality, expressivity and sensation seeking. *Personality and Individual Differences*, **17**, 467–76.

Franti, C.E., Kraus, J.F., Borhani, N.O., Johnson, S.L. & Tucker, S.D. (1980). Pet ownership in rural northern California (El Dorado County). *Journal of the American Veterinary Medical Association*, **176**, 143–9.

Fridlund, A.J. & MacDonald, M. (1998). Approaches to Goldie: a field study of human approach responses to canine juvenescence. *Anthrozoös*, **11**, 95–100.

Friedmann, E. & Lockwood, R. (1991). Validation and use of the Animal Thematic Apperception Test (ATAT). *Anthrozoös*, **4**, 174–83.

Furnham, A. & Heyes, C. (1993). Psychology students' beliefs about animals and animal experimentation. *Personality and Individual Differences*, **15**, 1–10.

Furnham, A. & Pinder, A. (1990). Young people's attitudes to experimentation on animals. *The Psychologist*, **October**, 444–8.

Gage, M.G. & Magnuson-Martinson, S. (1988). Intergenerational continuity of attitudes and values about dogs. *Anthrozoös*, **1**, 232–9.

Galvin, S.L. & Herzog, H.A. (1998). Attitudes and dispositional optimism of animal rights demonstrators. *Society and Animals*, **6**, 1–11.

Godwin, R.D. (1975). Trends in the ownership of domestic pets in Great Britain. In *Pet Animals and Society*, ed. R.S. Anderson, pp. 96–102. London: Baillière Tindall.

Goodwin, D., Bradshaw, J.W.S. & Wickens, S.M. (1997). Paedomorphosis affects agonistic visual signals of domestic dogs. *Animal Behaviour*, **53**, 297–304.

Gough, H.G. & Heilbrun, A.B. (1965). *The Adjective Check List Manual*. Palo Alto, CA: Consulting Psychologists Press.
Gould, S.J. (1979). Mickey Mouse meets Konrad Lorenz. *Natural History*, **88**, 30–6.
Grandin, T. (1988). Behaviour of slaughter plant and auction employees toward the animals. *Anthrozoös*, **1**, 205–13.
Griffiths, A.O. & Bremner, A. (1977). Survey of cat and dog ownership in Champaign County, Illinois, 1976. *Journal of the American Veterinary Medical Association*, **170**, 1333–40.
Hall, J.A. (1984). *Non-verbal Sex Differences*. Baltimore: Johns Hopkins University Press.
Herzog, H.A., Betchart, N.S. & Pittmann, R.B. (1991). Gender, sex role orientation and attitudes toward animals. *Anthrozoös*, **4**, 184–91.
Hills, A.M. (1989). The relationship between thing–person orientation and the perception of animals. *Anthrozoös*, **3**, 100–10.
Hills, A.M. (1993). The motivational bases of attitudes toward animals. *Society and Animals*, **1**, 111–28.
Hinde, R.A. & Barden, L.A. (1985). The evolution of the teddy bear. *Animal Behaviour*, **33**, 1371–3.
Hogan, R. (1969). Development of an empathy scale. *Journal of Consulting and Clinical Psychology*, **33**, 307–16.
Hyde, K.R., Kurdek, L. & Larson, P. (1983). Relationships between pet ownership and self-esteem, social sensitivity and interpersonal trust. *Psychological Reports*, **52**, 110.
Ingold, T. (1994). From trust to domination: an alternative history of human–animal relations. In *Animals and Human Society: Changing Perspectives*, ed. A. Manning & J. Serpell, pp. 1–22. London: Routledge.
James, P.D. (1989). *Devices and Desires*. London: Faber and Faber Ltd.
Jasper, J.M. & Nelkin, D. (1992). *The Animal Rights Crusade: the Growth of a Moral Protest*. New York: Free Press.
Jolly, A. (1972). *The Evolution of Primate Behaviour*. London: Robert Hardwicke.
Joubert, C.E. (1987). Pet ownership, social interest and sociability. *Psychological Reports*, **61**, 401–2.
Kidd, A.H. & Feldmann, B.M. (1981). Pet ownership and self perceptions of older people. *Psychological Reports*, **48**, 867–75.
Kidd, A.H., Kelley, H.T. & Kidd, R.M. (1984). Personality characteristics of horse, turtle, snake and bird owners. In *The Pet Connection: its Influence on Our Health and Quality of Life*, ed. R.K. Anderson, B.L. Hart & L.A. Hart, pp. 200–6. Minneapolis: CENSHARE, University of Minnesota.
Kidd, A.H. & Kidd, R.M. (1980). Personality characteristics and preferences in pet ownership. *Psychological Reports*, **46**, 939–49.
Kidd, A.H. & Kidd, R.M. (1989). Factors in adults' attitudes towards pets. *Psychological Reports*, **65**, 903–10.
Koestner, R., Franz, C. & Weinberger, J. (1990). The family origins of empathic concern: a 26-year longitudinal study. *Journal of Personality and Social Psychology*, **58**, 709–17.
Kring, A.M. & Gordon, A.H. (1998). Sex differences in emotion: expression, experience and physiology. *Journal of Personality and Social Psychology*, **74**, 686–703.
Lennon, R. & Eisenberg, N. (1987). Gender and age differences in empathy and sympathy. In *Empathy and its Development*, ed. N. Eisenberg & J. Strayer, pp. 195–217. Cambridge: Cambridge University Press.
Little, B.R. (1968). Psychospecialization: functions of differential interest in persons and things. *Bulletin of the British Psychological Society*, **21**, 113.
Little, B.R. (1976). Specialization and the varieties of environmental experience: empirical studies within the personality paradigm. In *Experiencing the*

Environment, ed. S. Wapner, S.B. Cohen & B. Kaplan, pp. 81–116. New York: Plenum Press.

Little, B.R. (1983). *Manual for the Thing–Person Orientation Scale*, 3rd edn. Ottawa: Carleton University, Department of Psychology.

Lockwood, R. (1983). The influence of animals on social perception. In *New Perspectives on Our Lives with Companion Animals*, ed. A.H. Katcher & A.M. Beck, pp. 64–71. Philadelphia: University of Pennsylvania Press.

Lorenz, K. (1943). Die angeborenen Formen möglicher Erfahrung. *Zeitschrift für Tierpsychologie*, **5**, 235–409.

Maccoby, E.E. & Jacklin, E.E. (1974). *The Psychology of Sex Differences*. Stanford, CA: Stanford University Press.

Mader, B., Hart, L. & Bergin, B. (1989). Social acknowledgements for children with disabilities: effects of service dogs. *Child Development*, **60**, 1529–34.

Maehle, A-H. (1994). Cruelty and kindness to the 'brute creation': stability and change in the ethics of the man–animal relationship, 1600–1850. In *Animals and Human Society: Changing Perspectives*, ed. A. Manning & J. Serpell, pp. 81–105. London: Routledge.

Mallon, G.P. (1993). A study of the interactions between men, women, and dogs at the ASPCA in New York City. *Anthrozoös*, **6**, 43–7.

Martinez, R.L. & Kidd, A.H. (1980). Two personality characteristics in adult pet owners and non-owners. *Psychological Reports*, **47**, 318.

Marx, M.B., Stallones, L., Garrity, T.F. & Johnson, T.P. (1988). Demographics of pet ownership among U.S. adults 21 to 64 years of age. *Anthrozoös*, **2**, 33–7.

Mathews, S. & Herzog, H.A. (1997). Personality and attitudes toward the treatment of animals. *Society and Animals*, **5**, 169–75.

Mehrabian, A. & Epstein, N. (1972). A measure of emotional empathy. *Journal of Personality*, **40**, 525–43.

Melson, G.F. (1988). Availability and involvement with pets by children: determinants and correlates. *Anthrozoös*, **2**, 45–52.

Melson, G.F. (1991). Children's attachment to their pets: links to socio-emotional development. *Children's Environments Quarterly*, **82**, 55–65.

Mertens, C. (1991). Human–cat interactions in the home setting. *Anthrozoös*, **4**, 214–31.

Messent, P.R. (1983). Facilitation of social interaction by companion animals. In *New Perspectives on Our Lives with Companion Animals*, ed. A.H. Katcher & A.M. Beck, pp. 37–46. Philadelphia: University of Pennsylvania Press.

Messent, P.R. & Horsfield, S. (1985). Pet population and the pet–owner bond. In *The Human–Pet Relationship*. Proceedings of the International Symposium on the Occasion of the 80th Birthday of Nobel Prize Winner Prof. DDr Konrad Lorenz, held in October 1983, in Vienna, Austria. Vienna: IEMT.

Midgley, M. (1984). *Animals and Why They Matter*. Athens, GA: University of Georgia Press.

Moore, S. & Boldero, J. (1991). Psychosocial development and friendship functions in adolescence. *Sex Roles*, **25**, 521–36.

Morris, P.H., Reddy, V. & Bunting, R.C. (1995). The survival of the cutest: who's responsible for the evolution of the teddy bear? *Animal Behaviour*, **50**, 1697–1700.

Nibert, D.A. (1994). Animal rights and human social issues. *Society and Animals*, **2**, 115–24.

Orwell, G. (1947). *The English People*. London: Collins.

Ory, M.G. & Goldberg, E.L. (1984). An epidemiological study of pet ownership in the community. In *The Pet Connection: its Influence on Our Health and Quality of Life*, ed. R.K. Anderson, B.L. Hart & L.A. Hart, pp. 320–30. Minneapolis: CENSHARE, University of Minnesota.

Paden-Levy, D. (1985). Relationship of extraversion, neuroticism, alienation and divorce incidence with pet ownership. *Psychological Reports*, **57**, 868–70.

Paul, E.S. (1992). Pets in childhood: individual variation in childhood pet ownership. Unpublished PhD thesis, University of Cambridge.

Paul, E.S. (1995). Us and them: scientists' and animal rights campaigners' views of the animal experimentation debate. *Society and Animals*, **3**, 1–21.

Paul, E.S. & Serpell, J. (1992). Why children keep pets: the influence of child and family characteristics. *Anthrozoös*, **5**, 231–44.

Paul, E.S. & Serpell, J.A. (1993). Childhood pet keeping and humane attitudes in young adulthood. *Animal Welfare*, **2**, 321–37.

Peele, S. & Brodsky, A. (1974). *Love and Addiction*. New York: Taplinger Press.

Perrine, R.M. & Osbourne, H.L. (1998). Personality characteristics of dog and cat persons. *Anthrozoös*, **11**, 33–40.

Pifer, L.K. (1996). Exploring the gender gap in young adults' attitudes about animal research. *Society and Animals*, **4**, 37–52.

Plous, S. (1991). An attitudes survey of animal rights activists. *Psychological Science*, **2**, 194–6.

Poresky, R.H. (1990). The young children's empathy measure: reliability, validity and effects of companion animal bonding. *Psychological Reports*, **66**, 931–6.

Poresky, R.H. (1996). Companion animals and other factors affecting young children's development. *Anthrozoös*, **9**, 159–68.

Poresky, R.H. & Hendrix, C. (1990). Differential effects of pet presence and pet bonding on young children. *Psychological Reports*, **67**, 51–4.

Poresky, R.H., Hendrix, C., Mosier, J.E. & Samuelson, M.L. (1988). *Psychological Reports*, **62**, 419–25.

Rajecki, D.W., Rasmussen, J.L. & Craft, H.D. (1993). Labels and the treatment of animals: archival and experimental cases. *Society and Animals*, **1**, 45–60.

Ray, J.J. (1982). Love of animals and love of people. *Journal of Social Psychology*, **116**, 299–300.

Ritter, J.M., Casey, R.J. & Langlois, J.H. (1991). Adults' responses to infants varying in appearance of age and attractiveness. *Child Development*, **62**, 68–82.

Robins, D.M., Sanders, C.R. & Cahill, S.E. (1991). Dogs and their people: pet facilitated interaction in a public setting. *Journal of Contemporary Ethnography*, **20**, 3–25.

Rossbach, K.A. & Wilson, J.P. (1992). Does a dog's presence make a person more likeable? Two studies. *Anthrozoös*, **5**, 40–51.

Rost, D.H. & Hartmann, A. (1994). Children and their pets. *Anthrozoös*, **7**, 242–54.

Rotter, J.B. (1967). A new scale for the measurement of interpersonal trust. *Journal of Personality*, **35**, 651–65.

Rushton, J.P., Fulker, D.W., Neale, M.C., Nias, D.K.B. & Eysenck, H.J. (1986). Altruism and aggression: the heritability of individual differences. *Journal of Personality and Social Psychology*, **50**, 1192–8.

St-Yves, A., Freeston, M.H., Jacques, C. & Robitaille, C. (1990). Love of animals and interpersonal affectionate behaviour. *Psychological Reports*, **67**, 1067–75.

Schenk, S.A., Templer, D.I., Peters, N.B. & Schmidt, M. (1994). The genesis and correlates of attitudes towards pets. *Anthrozoös*, **7**, 60–8.

Schutz, W.C. (1958). *FIRO: a Three Dimensional Theory of Interpersonal Behaviour*. New York: Holt, Rinehart and Winston.

Serpell, J.A. (1981). Childhood pets and their influence on adults' attitudes. *Psychological Reports*, **49**, 651–4.

Serpell, J.A. (1996). *In the Company of Animals: a Study of Human–Animal Relationships*. Cambridge: Cambridge University Press.

Serpell, J.A. & Paul, E. (1994). Pets and the development of positive attitudes to animals. In *Animals and Human Society: Changing Perspectives*, ed. A. Manning & J. Serpell, pp. 127–44. London: Routledge.

Shaw, W.W. (1977). A survey of hunting opponents. *Wildlife Society Bulletin*, **5**, 19–24.

Shultz, W.J. (1924). *The Humane Movement in the United States, 1910–1922*. New York: Columbia University.

Smith, S.L. (1983). Interactions between pet dog and family members: an ethological study. In *New Perspectives on Our Lives with Companion Animals*, ed. A.H. Katcher & A.M. Beck, pp. 29–36. Philadelphia: University of Pennsylvania Press.

Stallones, L., Johnson, T.P., Garrity, T.F. & Marx, M.B. (1989). Quality of attachment to companion animals among U.S. adults 21 to 64 years of age. *Anthrozoös*, **3**, 171–6.

Tetlock, P.E. (1986). A value pluralism model of ideological reasoning. *Journal of Personality and Social Psychology*, **50**, 819–27.

Tetlock, P.E., Armor, D. & Peterson, R.S. (1994). The slavery debate in antebellum America: cognitive style, value conflict and the limits of compromise. *Journal of Personality and Social Psychology*, **66**, 115–26.

Thomas, K. (1983). *Man and the Natural World: Changing Attitudes in England 1500–1800*. London: Allen Lane.

Wagstaff, G.F. (1991). Attitudes towards animals and human beings. *The Journal of Social Psychology*, **131**, 573–5.

Wells, D.L. & Hepper, P.G. (1995). Attitudes to animal use in children. *Anthrozoös*, **8**, 159–70.

Wells, D.L. & Hepper, P.G. (1997). Pet ownership and adults' views on the use of animals. *Society and Animals*, **5**, 45–63.

Wiesenfeld, A.R., Witman, P.B. & Malatesta, C.Z. (1984). Individual differences among adult women in sensitivity to infants: evidence in support of an empathy concept. *Journal of Personality and Social Psychology*, **46**, 118–24.

Wise, J.K. & Kushman, J.E. (1984). Pet ownership by life group. *Journal of the American Veterinary Medical Association*, **185**, 687–90.

Part III
Pets, families and interactions

RACHAEL M. HARKER, GLYN M. COLLIS AND JUNE McNICHOLAS

11

The influence of current relationships upon pet animal acquisition

INTRODUCTION

Deciding to obtain a pet animal constitutes a lifestyle change with far-reaching effects. The animal is likely to become a feature of a person's life for several years. This leads to the question of what is it about an individual's current life circumstances that promotes an impulse to contemplate pet ownership?

It seems likely that individuals who are in some way dissatisfied with their present life circumstances may experience motivation to alter this situation. This intuitive hypothesis has been formalized by Jungermann (1980) as the role of *possibility of continuity* in personal decision making. This refers to the prediction that an individual is not likely to decide upon a course of action if the maintenance of the status quo is a satisfactory option. However, if the continuation of current circumstances is not desirable, some course of action will be taken. Obtaining a pet may be one such course of action to deal with problematic life circumstances. This argument leads to a further question: what kind of change in an individual's lifestyle does the introduction of a pet imply?

Pets as relational partners

It is plausible that introducing a pet animal into a household constitutes an opportunity to form a new relationship. Many owners of pet animals appear to enter into some form of relationship with their pet, analogous to human–human relationships. Hirschmann (1994) argues that the most common role a pet animal can fulfil is that of a friend or companion to its owner. Pet animals are also frequently described as family members (Voith, 1985), or even as surrogate siblings or children (Sanders, 1990; Hirschmann, 1994).

The fact that some individuals view their pet animals as relational partners is a necessary but not sufficient criterion for classifying the interaction between humans and pets as relationship-like. Before deciding whether human–pet associations can be considered as relationships, the term relationship needs clarification. Finding a single definition of a relationship is a difficult task. Research in the field of social relationships tends to be fragmented. Individual types of relationship (e.g. marital, parental, fraternal, friendship) are usually examined as separate entities. Researchers tend to concentrate upon one particular relationship type which interests them, at the expense of examining other relationship structures (Fiske, 1992; Berscheid, 1995). A consequence of this is the absence of a generic perspective on relationships, applicable to all individual instances.

Despite the absence of a unifying definition of relationships, a number of researchers agree that certain elements are present in most relationship structures. It is possible to view relationships as comprising behavioural, cognitive and affective components (Hinde, 1979; Kelley *et al.*, 1983; Fehr, 1996).

The behavioural component of a relationship is manifest in the way individuals act towards one another. Shared activity, mutual assistance and confiding behaviour provide examples of behaviours expected from relational partners (Davis & Todd, 1985). The ways in which relational partners think of each other comprises the cognitive element of a relationship (Blieszner & Adams, 1992). Relational cognition may focus upon such themes as the reliability of the relationship – how likely it is to persist – and authenticity (Wright, 1984) – the extent to which relational partners respond to each other as unique and irreplaceable individuals. The affective domain of relationships involves feelings directed towards relational partners. In addition, it may include emotions concerning the self arising from a particular relationship or from relationships in general. Emotions focused upon the self (e.g. self-esteem, self-worth) can be influenced by relationships in both positive and negative manners (Helgeson, Shaver & Dyer, 1987).

The manifestation of such elements has been studied in a wide range of interpersonal relationships. If these components of relationships are present in associations between people and their pets, this may justify extending the study of relationships to include person–pet relationships.

Behavioural aspects of relationships

One of the most commonly cited reasons for keeping a pet animal is for the companionship it can provide (Horn & Meer, 1984; Endenburg, Hart & Bouw, 1994; Zasloff, 1995). The general term companionship usually translates into partaking in shared activities with the pet animal, such as walking the dog, playing with the cat (Miller & Lago, 1990). However, such behavioural interactions between humans and pet animals are usually asymmetrically organized. Humans tend to interact with their pets when they feel like it, rather than consistently responding to the animal's demands for attention (Smith, 1983). In addition, Smith observes that once interaction between humans and pets has arisen, the termination of such interactional sequences invariably originates from the human. This suggests that activities like walking the dog and playing with the cat only arise when time can be spared from the human owner's other commitments. Therefore, if one views activity as a global concept – involving instigation, performance and termination – labelling certain forms of human–pet interaction as shared activities may be misleading. The performance of an activity could well be seen as shared, yet its instigation and termination are likely to stem from the human pet owner. On the other hand, arguments regarding whether or not a pet animal objectively engages in shared activities cannot detract from observations that many pet owners subjectively perceive that shared activities do take place between themselves and their animals. This is perhaps sufficient justification for the claim that many pet owners view their animals as partners and companions in a range of activities.

The role of pets in providing mutual assistance is highlighted in the case of assistance dogs for blind/deaf/disabled individuals (Fogle, 1981; Hoffman, 1991; Hart, Zasloff & Benfatto, 1996). However, it is possible to perceive the common household pet as providing some assistance to its owners. For instance, an added bonus to many dog owners is the extra security such a pet can provide to their household (Council for Science and Society, 1988). In addition, pet animals are reported to be of great assistance to 'latchkey children', providing support, comfort and value in coping with after school hours before parents return home (Blue, 1986; Guerney, 1991). Such forms of assistance appear somewhat uni-directional: the pet animal may serve a number of functions for its owners, but does the animal receive any form of assistance in return? Bradshaw (1995) believes that pets can enter into a relationship of *mutualism* with their owners. He argues that domesticated breeds of animal would have difficulty surviving without human assistance. Hence, the pet is obliged to perform certain services for its owner in order to ensure that it will receive food and shelter in return. This suggests the form of assistance between pets and humans is organized around reciprocal exchange principles. Whether this form of exchange is akin to the mutual assistance humans provide for each other is debatable. However, some researchers do believe that certain human–human relationships are marked by such reciprocal exchange (Fiske, 1992).

Confiding behaviour, observed in many close interpersonal relationships, also arises between person and pet, in both adults (Cusack, 1988) and children (Bachman, 1975; Guerney, 1991). Cusack claims that animals may be especially important as confidantes as there is no risk of betrayal, even when expressing most intimate feelings and thoughts. However, pets do not possess the cognitive capabilities to understand such communications. Therefore, speaking of one's innermost feelings in the presence of a pet animal may be analogous to talking to oneself, or to inanimate objects (e.g. house plants, soft toys).

Cognitive aspects of relationships

Owning a pet animal can be seen to ensure at least one highly reliable association in an individual's life (Brickel, 1986). Pets are arguably more consistent in their behaviour than humans. Hence, a relationship with a pet is likely to be more constant and reliable than human–human relationships (Katcher, 1983). Paradoxically, the stability and reliability of human–pet association could be used to argue that the term relationship is not applicable to this form of interaction.

Many theorists believe that a core feature of relationships is their instability (Duck, 1990, 1994; Wood, 1994). From this perspective, relationships are seen as processual: they are not settled, stable entities but are constantly evolving and changing as partners create and re-create intimacy. Most human–pet associations centre on established interaction patterns which do not exhibit much variation. However, many pet owners report that their animals are friends (Stewart *et al.*, 1985), companions (Quigley, Vogel & Anderson, 1983), and family members (Sanders, 1990). Hence, it appears that some individuals do construct a cognitive representation of their pets as a definite presence in their network of relationships. In addition, pet animals can also promote perceptions of authenticity if owners believe that their animals view them as unique, irreplaceable and special individuals. Both children (Bryant, 1990) and adults (Loughlin & Dowrick, 1993) appear to value the uniqueness of their relationship with a pet.

Affective aspects of relationships

Strong feelings of love and affection towards pets are reported by child (Bryant, 1990; Covert *et al.*, 1985) and adult (Loughlin & Dowrick, 1993) pet owners. Pet owners in the Loughlin and Dowrick study placed great value on being able to give love and affection to their pets. For some individuals, pet ownership can provide a sense of worth, and maintain or promote self-esteem. Davis and Juhasz (1985) found that adolescent participants rated having a pet as something which made them feel good and satisfied about themselves. Adults also report that keeping animals adds meaning to their lives and has benefits for self-esteem (Loughlin & Dowrick, 1993).

However, one could argue that the sense of worth emanating from keeping pet animals is akin to a sense of worth derived from alternative hobbies and interests. The pride of a stamp collector or art enthusiast may have implications for the individual's self-esteem, yet it would be highly unlikely that one would speak of that person as having a relationship with his or her 'Penny Blacks' or Van Goghs. Similarly, the affection directed towards pets could arguably resemble a fondness for a particular pastime or leisure pursuit. The language we use to describe relational feelings is pervasive, frequently applied to inanimate objects and activities (e.g. 'I love my car', 'I have a passion for football'). In such cases the nature of emotion is perhaps more akin to fondness and enthusiasm than love. Similarly, when people speak of loving pet animals it is possible that such reported emotion differs from true relational affection, and more closely resembles the type of fondness associated with objects and activities.

However, some evidence suggests that the love people feel for their pets may resemble affection found in human–human relationships. It seems that pet owners believe they not only give but also receive love and affection from their animals (Quigley *et al.*, 1983). The perceived reciprocal nature of such affection differentiates it from expressions of partiality towards objects and actions. Most individuals would find it preposterous to describe their car or house as loving them, or to view their favourite sport as possessing feelings towards them. Reciprocal affection is relationships territory, and pet animals may have a foothold in such terrain.

The preceding arguments imply that it may be possible to view associations between humans and pets as resembling relationships. Therefore, the change in an individual's lifestyle, which acquiring a pet implies, may involve an extension of the individual's network of relationships. This proposal raises a further issue: what motivates individuals to seek out new relationships?

Why do we form relationships?

Robert Weiss' (1974) account of the *Social Provision of Relationships* emerges as a useful perspective to explain why individuals are motivated to form new relationships. In his work as a practising psychiatrist, Weiss specialized in social isolation due to a lack of adequate relationships, as well as the dissolution of relationships through bereavement and divorce/separation. His observations regarding the distress associated with an absence of fulfilling relationships led him to conclude that individuals have certain requirements for well-being that can be met only within relationships. In addition, he outlines a hypothesis that an adequate social life is one that purveys a set of *relational provisions.* Weiss identifies six categories of relational provision:

1. Attachment
2. Social integration
3. Opportunity for nurturance
4. Reassurance of worth
5. A sense of reliable alliance
6. Obtaining of guidance.

Attachment provision stems from relationships in which individuals gain a sense of security and place. Attachment-providing relationships make an individual feel comfortable and at ease. The absence of such a relationship may result in feelings of restlessness and isolation. Attachment relationships are likely to be provided by marriage/partnerships, close friendships, sibling or parental relationships, etc.

Social integration is the product of relationships in which participants hold common concerns. This affords the development of a shared interpretation of experience. In addition, there may be opportunities for an exchange of services within the area of common interest. This form of relationship is also a source of companionship.

Opportunities for nurturance arise from relationships in which an individual is responsible for the well-being of another and hence can develop a sense of being needed. Such responsibility can add meaning to an individual's life and add sustenance to wider goals in a variety of areas.

Reassurance of worth is provided by relationships that affirm an individual's competence and value in a given social role. Colleague and familial relationships can function in this way.

A sense of reliable alliance is primarily satisfied via kin relationships. Within a family, it is likely that assistance will be given whether there is mutual affection or not, and irrespective of whether or not reciprocation was given for past help.

Obtaining of guidance becomes important to individuals in stressful situations. Under such circumstances it is important for individuals to have access to a relationship with a respected and trusted figure – someone who can provide emotional support as well as tangible aid in formulating a plan of action to deal with the stressful event.

Weiss believes that, in general, few individual relationships convey all six social provisions. An implication of this is that individuals must maintain a number of different relationships in order to ensure that all provisions are met, thereby allowing them to establish and sustain a sense of well-being.

Relationships and social networks

Most individuals, in keeping with Weiss' argument, do participate in and maintain a range of relationships. Collectively, these relationships form a social network. The term social network describes the total constellation of an individual's relationships (Knipscheer & Antonucci, 1990). Social networks are fluid rather than crystallized structures:

> Because the social network is not an object but a collection of relationships, network change can occur in either the network's structure or in the nature of its component relationships. (Larner, 1990: 181)

Larner believes network change can have an emotional or pragmatic basis. Personal compatibility and emotional satisfaction can play a significant role as an individual elects to dissolve certain relationships and expand upon others. In addition, network reorganization may be the result of changes that remove an individual from contexts in which they meet and interact with others. Larner, like Weiss, implies that individuals attempt to maintain a range of relationships, and that deficiencies in some areas require attention and ultimately compensation, thus rendering networks dynamic structures.

Pet animals within networks of relationships

The failure of a network to meet all necessary relational provisions may drive the individual to seek additional relationships. It is possible that pet animals provide an opportunity to form a novel relationship within the context of a social network. If so, one might expect the structure and function of the person–pet relationship to be partly determined by relational provision available from other social relationships. Indeed, rhetoric suggesting the versatile role of the pet animal as a function of the human owner's need is not uncommon:

> The pet's value as a developmental asset will fluctuate depending on the individual needs, age, and sex of the owner. Therefore, the owner–pet bond is optimally a flexible affiliation. (Davis & Juhasz, 1985: 90–91)

Yet surprisingly, to date, there is little research that specifically aims to establish the degree to which pet animals fulfil various roles dependent upon the needs of their human owner. A possible way to clarify this situation is to examine the pet in the context of an individual's other social relations. If the structure of the human–pet relationship is determined by the composition of an individual's human–human relationships, there may be reason to perceive the pet's value and role as dependent upon human needs.

Measuring network provision

A number of scales have been devised to measure social network provision. One such instrument is Furman and Buhrmester's (1985) Network of Relationships Inventory (NRI). The sub-scales of the NRI provide indices of overall positive and negative relational provision. Sub-scales concerning positive relational provision incorporate five of Weiss' social provisions: the exception being the provision of attachment.

The study reported here examines NRI ratings from adult participants who were either seeking or not seeking to obtain a pet animal. It was predicted that pet-seeking individuals would have more grounds for dissatisfaction with existing relationships, so that these participants would report either more negative provision or less positive provision than individuals who did not wish to become pet owners. In addition, estimates of the extent to which pet-keeping would entail positive and negative implications for the owner were recorded. Such expectations of pet owning may provide an indication of the degree to which motivation to acquire a pet relates to a desire to supplement existing relational provision.

METHOD

Questionnaire design

The *Pet Ownership Questionnaire* used in this study consisted of three sections providing different forms of information: demographic details of participants (A); expected positive and negative implications of pet owning (B); and the participants' current relational provision (C).

Demographic details

Section A of the questionnaire requested details regarding the age, sex, marital status, household composition, income and accommodation status, as well as the present and/or previous pet-owning experience of participants. This information may allow an assessment of whether such demographic variables influence the decision to acquire a pet animal, and whether more complex factors are independent of these demographic variables. This section also asked participants to indicate whether they were 'seriously considering' acquiring a pet animal for themselves. A positive response to this question resulted in further enquiry as to the type of pet being sought.

Expectations of pet ownership

A total of 24 items made up Section B (the expectation scale) of the questionnaire. These items addressed both positive and negative aspects of pet ownership. There were 19 items concerning positive aspects of pet owning (Table 11.1) and five items regarding negative aspects (Table 11.2). Participants were asked to indicate (on a five-point Likert Scale) how much they would appreciate the possible benefits, or find annoyance with potential drawbacks, of owning a pet animal. The scale ranged from 1 (not at all) to 5 (very much). Hence, higher scores on each sub-scale indicated greater estimated appreciation or annoyance. The 24 items in the expectation scale drew upon an interview study in which visitors to local animal shelters declared the benefits they expected to gain from, as well as possible disadvantages to, becoming a pet owner. The scale was included in a pilot study involving 57 adult participants. A principle components analysis of the pilot data confirmed that the expectation questions formed two distinct, unrelated (Pearson's $r = -0.124$, $p > 0.05$) sub-scales, one measuring positive and the other negative items. Test–retest reliability figures also indicated that the instrument was

Table 11.1. *Positive expectation sub-scale items*

Scale item	Description
Companionship	Perception of company provided by a pet
Friendship	Perception of the pet as a friend
Observation	Pleasure gained from watching the pet
Stroking	Pleasure gained from tactile contact with the pet
Affection	Receipt of love/affection from the pet
Training	Pleasure gained from training the pet
Relaxation	Pet's ability to reduce stress/induce relaxation
Nurture	Opportunities to take care of the pet
Esteem support	Reassurance of worth resulting from keeping a pet
Responsibility	Sense of purpose resulting from having a pet
Play	Pleasure gained from play with the pet
Family extension	Perception of pet as a family member
Gentleness	Pet's ability to encourage gentle behaviour
Consideration	Pet's ability to foster consideration for another's needs
Security	Perception of pet as enhancing household security
Entertainment	Perception of pet as a source of fun
Learning to care	Pet care as facilitation of other nurturant behaviours
Loyalty	Perception of the loyalty of pets towards owners
Exercise	Pet as provider of opportunities to exercise

Table 11.2. *Negative expectation sub-scale items*

Scale item	Description
Cost	Concern over financial implications of pet-keeping
Mess	Annoyance with the mess made by a pet
Noise	Annoyance with the noise made by a pet
Damage	Concern over possible damage caused by a pet
Injury	Concern over possibility of pet scratching/biting

stable over a one-week period (negative sub-scale: Pearson's $r=0.948$, $p<0.0005$; positive sub-scale: Pearson's $r=0.889$, $p<0.0005$).

Network of Relationships Inventory

Section C of the questionnaire asked participants to complete Furman and Buhrmester's (1985) Network of Relationships Inventory, with reference to the most important relationships in their lives. Hence, this section provided an indication of the size of an individual's network of close relationships. The NRI consists of 12 sub-scales, each sub-scale relating to a particular relational provision: satisfaction, companionship, intimacy, affection, reassurance of worth, instrumental aid, reliable alliance, opportunities for nurturance, relative power, conflict, antagonism and punishment. Questions regarding conflict, antagonism and punishment comprise a negative sub-scale. Relative power and satisfaction are believed to be distinct properties, whilst the remaining questions form a sub-scale of supportive relational provision. The sub-scale for each provision comprises three items designed to assess the amount of provision supplied by a particular relationship. A total of 36 items constitutes the NRI. Item scoring occurs on a five-point scale ranging from 1 (no provision) to 5 (very much provision). Hence, higher scores represent greater degrees of provision.

Participants

A total of 432 people aged from 18 to 80 years agreed to participate in the questionnaire study. Of these, 228 returned completed forms. This represents a response rate of 53%.

Procedure

Participant recruitment took place over a three-month period, covering March 1996 to June 1996. Outlets for approaching potential participants included local adult education centres and supermarkets. Four supermarkets and three adult education centres granted access for participant recruitment purposes. Two of the supermarkets used were within large council housing estates and the remaining two formed parts of retail parks. Customers at the former stores tended to live nearby (many walked to the store), whilst the latter attracted customers from greater distances – usually travelling by car. The time of visits to these outlets varied with the aim of contacting a representative cross-section of the public.

Individuals approached at recruitment centres received a brief outline of the research project before being asked to participate. Consenting participants were given a copy of the *Pet Ownership Questionnaire* accompanied by a covering letter, and a freepost envelope for return purposes.

Table 11.3. *Frequency of participants seeking particular types of pet animals*

Type of pet	Frequency of participants seeking each type
Dog	50
Cat	22
Small furry animal[a]	8
Bird	2
Fish	1
Horse	1
Total	84

Note:
[a]The category of small furry animals is an umbrella term used where participants indicated a desire to own such creatures as hamsters, gerbils, rabbits, rats and mice.

RESULTS

Questionnaire distribution and response

Of those individuals returning questionnaires, 84 were considering acquiring a pet animal (pet-seeking group), whilst 144 did not wish to obtain a pet (non-seeking group). The types of pets desired by individuals within the pet-seeking group are illustrated in Table 11.3. It is clear that dogs are by far the most popular variety of pet animal sought in the current sample. Cats represent a clear second but are considerably less frequently mentioned than dogs.

Within the pet-seeking group, 28 participants (33%) already owned at least one pet animal, whilst 56 (67%) did not own any pets. There were 54 participants (38%) from the non-seeking group who currently owned pets and 90 (63%) who did not.

Of the 146 participants who did not own any pets, 143 (98%) reported past experience of living in a household with a pet animal. Thus, 225 out of the 228 people in the sample had current or previous experience of pet ownership.

NRI analysis

The 228 respondents provided data on 1274 relationships (mean = 6.8 relationships/participant, minimum = 1, maximum = 13). The 36 NRI items were completed for each relationship. A total score for each of the 12 NRI sub-scales was calculated by summing the ratings obtained for its three composite items. Each participant's mean score across relationships was then calculated for all sub-scales. These means subsequently served as dependent variables. Mean rather than total scores were analysed in an attempt to eliminate bias resulting from unequal numbers of relationships across participants. Pet-seeking status – that is, the categorization of individuals as pet-seekers or non-seekers – served as the independent variable in a series of analyses.

In determining the effect of pet-seeking status upon NRI ratings, it was acknowledged that several other variables may influence reported levels of relational provision. The number of available relationships (network size) and household composition seem intuitively likely to determine the quantity and quality of relational provision. Additional factors such as age, gender, marital status and socio-economic status (here reflected by income and accommodation status) could similarly contribute to differences in the form of relational provision sought and received. The current pet-owning status of individuals may also impinge upon the structure and function of their relational networks. Due to these considerations, analysis of the effect of pet-seeking status on NRI ratings employed a general linear model that partialled out the influence of age, gender, marital status, income and accommodation status, household composition, network size, and pet-owning status.

Group differences in provision

Although data from each NRI sub-scale underwent individual analysis, results are presented in sections corresponding to Furman & Buhrmester's (1985) grouping of sub-scales according to relational provision: conflict, antagonism and punishment collectively assess negative provision; relative power and satisfaction form distinct properties; and all other sub-scales measure supportive relational provision.

Negative provision

NRI ratings from the three sub-scales corresponding to negative relational provision – conflict, antagonism and punishment – were examined for differences between pet-seeking and non-seeking participants. Figure 11.1 shows that the pet-seeking group reported higher levels of conflict, antagonism and punishment than the non-seeking group.

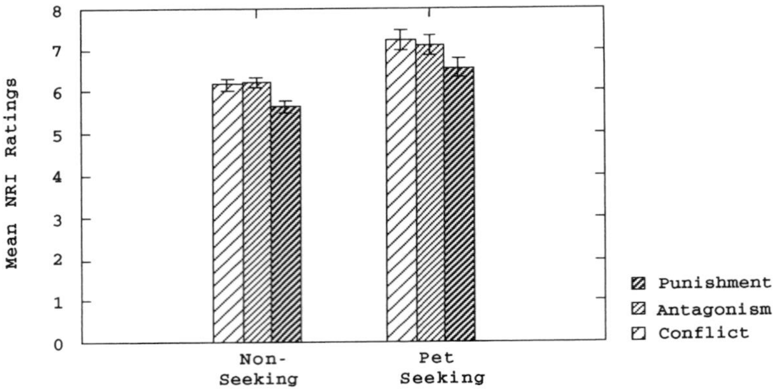

Fig. 11.1. Mean negative sub-scale scores (and standard error) for pet-seeking and non-seeking participants.

Analyses of variance revealed that these differences were all statistically significant:

Conflict: $F_{(1,208)} = 8.506, p < 0.005$
Antagonism: $F_{(1,208)} = 4.309, p < 0.05$
Punishment: $F_{(1,208)} = 5.140, p < 0.05$

Hence, it appears that pet-seeking status significantly affects reported levels of negative relational provision. Pet-seeking individuals provide significantly higher negative ratings than non-seeking participants.

Supportive provision

A difference between the pet-seeking and non-seeking groups was observed concerning NRI ratings obtained for the sub-scale of reliable alliance. Pet-seeking participants reported a mean rating of 12.5, compared to a mean rating of 13.2 for the non-seeking group. An analysis of variance, partialling out the influence of demographic variables, revealed this to be a statistically significant difference ($F_{(1,208)} = 5.782, p < 0.05$).

No further significant differences were observed regarding the other supportive provisions of: companionship ($F_{(1,208)} = 0.041$, $p > 0.05$); nurturance ($F_{(1,207)} = 0.779$, $p > 0.05$); reassurance of worth ($F_{(1,205)} = 0.834$, $p > 0.05$); affection ($F_{(1,204)} = 0.043$, $p > 0.05$); intimacy ($F_{(1,208)} = 0.111$, $p > 0.05$); and instrumental aid ($F_{(1,206)} = 0.686$, $p > 0.05$).

Other provisions

No significant differences were found regarding NRI sub-scales of relative power ($F_{(1,205)} = 1.169$, $p > 0.05$) and satisfaction ($F_{(1,208)} = 0.122$, $p > 0.05$).

Excluding pets from the data base

It emerged that some of the pet-owning participants had included their animals in NRI responses. When data concerning pets were removed and human–human relationships alone were analysed, the same pattern of results emerged. A significant difference was found between the pet-seeking and non-seeking groups regarding sub-scales of conflict ($F_{(1,198)} = 7.856, p < 0.005$), antagonism ($F_{(1,199)} = 9.464, p < 0.005$), punishment ($F_{(1,198)} = 5.863, p < 0.05$) and reliable alliance ($F_{(1,199)} = 9.975, p < 0.005$). No group differences were observed concerning NRI ratings for any other provisions.

Expectations of ownership

Total scores were calculated for both positive and negative expectation sub-scales. Cronbach's alpha coefficients indicated that total scores were representative of individual items for both the negative ($\alpha = 0.872$) and positive ($\alpha = 0.950$) sub-scales. The positive and negative expectation totals did not relate to one another (Pearson's $r = -0.124, p > 0.05$). Hence, total scores for positive and negative expectation sub-scales served as dependent variables in analyses employing pet-seeking status as the independent variable.

As with the analysis of NRI ratings, it was acknowledged that additional variables may influence participants' expectations of pet ownership, current pet-owning status and past experience of pet ownership being obvious examples of variables that may determine expectations of pet ownership. Additional demographic considerations such as age, gender, household composition etc. may also influence positive and negative expectations of pet-keeping. Hence, the analysis of expectation ratings also used a general linear analysis that partialled out the effect of age, gender, marital status, income and accommodation status, household composition, and pet-owning status. Past experience of pet-ownership was not included in the model since this appeared redundant (99% of the sample had experience of pet-keeping).

Figure 11.2 indicates that the pet-seeking group reported higher expectation ratings for positive aspects of pet ownership than the non-seeking group. This difference is statistically significant ($F_{(1, 196)} = 5.424, p < 0.05$). In addition, pet-seeking participants also appeared to place less emphasis upon the negative implications of pet-keeping than did their non-seeking contemporaries. This is highlighted by Fig. 11.3, which indicates that the pet-seeking group provided lower negative sub-scale ratings. Analysis confirmed that this was a statistically significant difference ($F_{(1,198)} = 7.037, p < 0.01$).

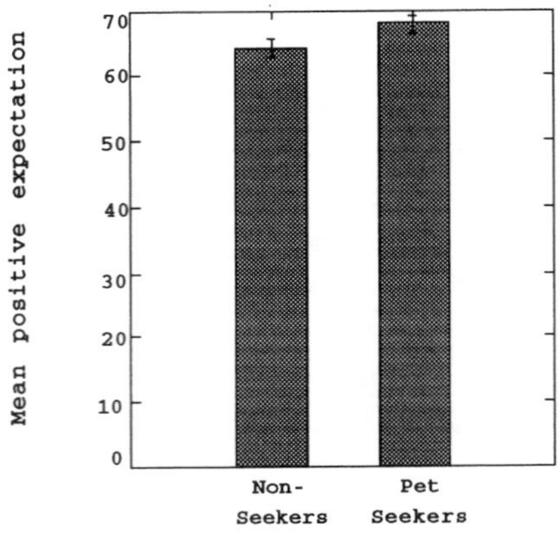

Fig. 11.2. Mean positive expectation scale ratings (and standard error) for pet-seeking and non-seeking participants.

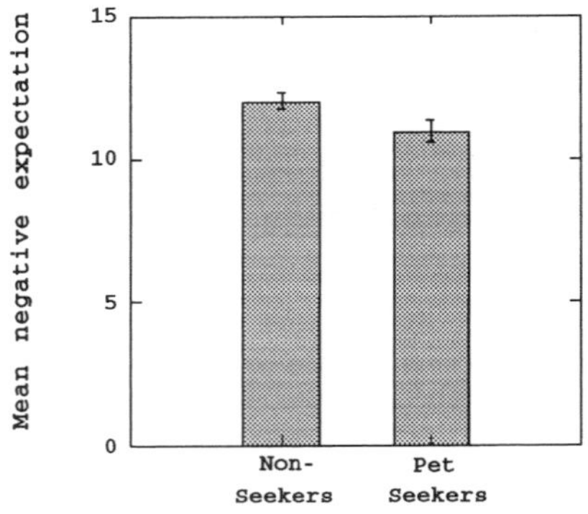

Fig. 11.3. Mean negative expectation ratings (and standard error) for pet-seeking and non-seeking participants.

CONCLUSION

The prediction that differences in NRI ratings would be observed between pet-seeking and non-pet-seeking participants gained partial support from the present study: pet-seeking participants did report significantly higher levels of negative relational provision than non-seekers. In addition, as concerns the supportive provision of reliable alliance, pet-seeking participants reported significantly lower levels of such support when compared to non-seekers. However, other aspects of supportive provision showed no difference between the two groups. Nor were there any between-group differences in levels of reported satisfaction and relative power within relationships.

It seems probable that individuals who report high levels of negative relational provision are less likely to be content with the nature of their current relationships. The higher degree of negative provision may also explain why pet-seeking individuals report their relationships as less stable and reliable than non-seeking individuals. The observation that such individuals are also likely to be considering pet ownership lends support to Jungermann's (1980) view that dissatisfaction with current life circumstances can serve as an impetus for lifestyle change.

It could be argued that the presence of high levels of negative provision amongst pet-seeking participants corresponds to the arguments of Larner (1990) and Weiss (1974): that deficiencies within networks of relationships require attention and new relationships may be formed to compensate for problems in a particular area. If so, it is tempting to suggest that pet animal acquisition is partly motivated by a desire to complement, or compensate for inadequacies with, existing relationships. However, such a contentious statement needs further discussion.

The apparent similarity between ratings of pet-seekers and non-seekers for the satisfaction sub-scale, and all but one of the supportive sub-scales, does not fit with a view of pet-seeking individuals having deficiencies within their relational networks. Higher levels of negative provision in pet-seeking participants do not seem to detract from the supportive elements of their relationships. This may suggest that the motivation to acquire a pet animal has little to do with the expectation that the animal will provide some form of relational provision. It is possible that individuals who experience difficulty with their close relationships would simply appreciate the introduction of any new object of interest, animate or inanimate, to distract their attention from their relationship difficulties.

Nonetheless, the observation that pet-seeking participants produce higher positive expectation ratings than non-seekers is suggestive, as it appears to support the view that individuals expect the animal to fulfil relationship-like functions. The positive sub-scale contains only two items relating to the role of pets as functional equipment or ornaments; the remaining 17 items relate to relationship functions such as companionship, friendship, and opportunities for nurturance. The positive items also request estimates of the expected amount of fun and playful activities associated with the animal. Hence, higher positive ratings imply that the pet is viewed as a potential interactive partner and not as a simple equivalent to an inanimate object of interest. The lower negative expectation ratings provided by pet-seeking participants connote that pet-seekers expect such interaction to take positive forms.

In addition, because almost all pet-seeking participants reported experience of pet ownership, their expectations may be grounded upon previous instances of pet animals fulfilling relational functions. This consideration, coupled with previous results showing a link between relational provision and pet-seeking, lends support to the view that the acquisition of a pet animal can provide an opportunity to form a novel relationship. This reinforces the notion that obtaining a pet animal may relate to a desire to extend a network of relationships.

Nonetheless, further investigation is needed before any definite conclusion can be made about the role of pet animals within their owners' networks of relationships. A follow-up study of participants in the present sample is underway at Warwick University. Individuals who have acquired pet animals are subject to enquiry regarding whether their relationship with the pet serves to fulfil previous expectations. The role of the pet in networks of relationships is also being examined. This should afford an estimate of the extent to which the pet has become incorporated into, and/or increased overall satisfaction with, a relational network. In addition, a concurrent study is examining the relationship networks of established pet owners. In this case participants are asked to include their pets in an assessment of the provision provided by various relationships. This should afford further clarification concerning the nature of person–pet relationships in the context of human–human relationships.

In summary, the findings suggest that current relationships may be an influential factor in the consideration of pet ownership. More specifically, high levels of negative relational provision appear to be associated with the desire to own a pet. One possible explanation for this is that excessive negative provision increases dissatisfaction with current life circumstances and precipitates strategies to deal with this. Pet ownership could be one such strategy. This study revealed that participants seeking pets expect a number of positive relationship-like provisions from the animal. Therefore, there is some evidence compatible with the view that acquiring a pet represents a desire to extend existing networks of relationships.

ACKNOWLEDGEMENTS

Rachael M. Harker is a Graduate Research Assistant funded by the Waltham Centre for Pet Nutrition. June McNicholas is a Waltham Research Fellow.

REFERENCES

Bachman, R.W. (1975). Elementary school children's perception of helpers and their characteristics. *Elementary School Guidance and Counselling*, **10**, 103–9.

Berscheid, E. (1995). Help wanted: a grand theorist of interpersonal relationships, sociologist or anthropologist preferred. *Journal of Social and Personal Relationships*, **12**, 529–34.

Blieszner, R. & Adams, R.G. (1992). *Adult Friendship*. London: Sage.

Blue, G.F. (1986). The value of pets in children's lives. *Childhood Education*, **63**, 84–90.

Bradshaw, J.W.S. (1995). Social interactions between animals and people – a new biological framework. Paper presented at 7th International Conference on Human-Animal Interactions, Geneva, September 6–9, 1995.

Brickel, C.M. (1986). Pet-facilitated therapies: a review of the literature and clinical implementation considerations. *Clinical Gerontologist*, **5**, 309–32.

Bryant, B.K. (1990). The richness of the child–pet relationship: a consideration of both benefits and costs of pets to children. *Anthrozoös*, **3**, 253–61.

Council for Science and Society (1988). *Companion Animals in Society.* Oxford: Oxford University Press.

Covert, A.M., Whiren, A.P., Keith, J. & Nelson, C. (1985). Pets, early adolescents, and families. *Marriage and Family Review*, **8**, 95–108.

Cusack, O. (1988). *Pets and Mental Health.* New York: Haworth Press.

Davis, J.H. & Juhasz, A.M. (1985). The preadolescent/pet bond and psychosocial development. In *Pets and the Family*, ed. M.B. Sussman, pp. 79–94, New York: Haworth Press.

Davis, K.E. & Todd, M.J. (1985). Assessing friendships: prototypes, paradigm case and relationship description. In *Understanding Personal Relationships: an Interdisciplinary Approach*, ed. S.W. Duck & D. Perlman, pp. 17–38. London: Sage.

Duck, S.W. (1990). Relationships as unfinished business: out of the frying pan and into the 1990s. *Journal of Social and Personal Relationships*, **7**, 5–24.

Duck, S.W. (1994). *Meaningful Relationships: Talking, Sense, and Relating.* London: Sage.

Endenburg, N., Hart, H. & Bouw, J. (1994). Motives for acquiring companion animals. *Journal of Economic Psychology*, **15**, 191–206.

Fehr, B. (1996). *Friendship Processes.* London: Sage.

Fiske, A.P. (1992). The four elementary forms of sociality: framework for a unified theory of social relations. *Psychological Review*, **99**, 689–723.

Fogle, B. (1981). *Interrelations Between People and Pets.* Springfield, IL: Charles C Thomas.

Furman, W. & Buhrmester, D. (1985). Children's perceptions of the personal relationships in their social networks. *Developmental Psychology*, **21**, 1016–24.

Guerney, L.F. (1991). A survey of self-supports and social supports of self-care children. *Elementary School Guidance and Counseling*, **25**, 243–54.

Hart, L.A., Zasloff, R.L. & Benfatto, A.M. (1996). The socialising role of hearing dogs. *Applied Animal Behaviour Science*, **47**, 7–15.

Helgeson, V.S., Shaver, P. & Dyer, M. (1987). Prototypes of intimacy and distance in same-sex and opposite sex relationships. *Journal of Social and Personal Relationships*, **1**, 195–223.

Hinde, R.A. (1979). *Towards Understanding Relationships.* London: Academic Press.

Hirschman, E.C. (1994). Consumers and their animal companions. *Journal of Consumer Research*, **20**, 616–32.

Hoffman, R.G. (1991). Companion animals: a therapeutic measure for elderly patients. *Journal of Gerontological Social Work*, **18**, 195–205.

Horn, J. & Meer, J. (1984). The pleasure of their company. *Psychology Today*, **18**, 52–8.

Jungermann, H. (1980). Speculations about decision-theoretic aids for personal decision making. *Acta Psychologica*, **45**, 7–34.

Katcher, A.H. (1983). Man and the living environment: an excursion into cyclical time. In *New Perspectives on Our Lives with Companion Animals*, ed. A.H. Katcher & A.M. Beck, pp. 519–31. Philadelphia: University of Pennsylvania Press.

Kelley, H., Berscheid, E., Christiansen, A., *et al.* (1983). *Close Relationships*. New York: Freeman.

Knipscheer, C.P.M. & Antonucci, T.C. (1990). *Social Network Research: Substantive Issues and Methodological Questions*. Amsterdam: Swets & Zeitlinger.

Larner, M. (1990). Changes in network resources and relationships over time. In *Extending Families: the Social Networks of Parents and their Children*, ed. M. Cochran, M. Larner, D. Riley, L. Gunnarson & C.R. Henderson, pp. 181–204. New York: Cambridge University Press.

Loughlin, C.A. & Dowrick, P.W. (1993). Psychological needs fulfilled by pet birds. *Anthrozoös*, **6**, 166–72.

Miller, M. & Lago, D. (1990). The well-being of older women: the importance of pet and human relations. *Anthrozoös*, **3**, 245–52.

Quigley, J.S., Vogel, L.E. & Anderson, R.K. (1983). A study of perceptions and attitudes towards pet ownership. In *New Perspectives on Our Lives with Companion Animals*, ed. A.H. Katcher & A.M. Beck, pp. 266–75. Philadelphia: University of Pennsylvania Press.

Sanders, C.R. (1990). Excusing tactics: social responses to the public misbehaviour of companion animals. *Anthrozoös*, **4**, 90–2.

Smith, S.L. (1983). Interactions between pet dog and family members: an ethological study. In *New Perspectives on Our Lives with Companion Animals*, ed. A.H. Katcher & A.M. Beck, pp. 29–36. Philadelphia: University of Pennsylvania Press.

Stewart, C.S., Thrush, J.C., Paulus, G.S. & Hafner, P. (1985). The elderly's adjustment to the loss of a companion animal: people–pet dependency. *Death Studies*, **9**, 383–93.

Voith, V.L. (1985). Attachment of people to companion animals. *The Veterinary Clinics of North America. Small Animal Practice*, **15**, 289–96.

Weiss, R.S. (1974). The provisions of social relationships. In *Doing unto Others*, ed. Z. Rubin, pp. 7–19. Englewood Cliff, NJ: Prentice-Hall.

Wood, J.T. (1994). *Relational Communication: Continuity and Change in Personal Relationships*. Belmont, CA: Wadsworth.

Wright, P.H. (1984). Self-referent motivation and the intrinsic quality of friendship. *Journal of Social and Personal Relationships*, **1**, 115–30.

Zasloff, R.L. (1995). Views of pets in the general population. *Psychological Reports*, **76**, 1166.

SHEILA BONAS, JUNE McNICHOLAS AND GLYN M. COLLIS

12

Pets in the network of family relationships: an empirical study

INTRODUCTION

In many western countries, approximately 50% of households contain pets (Marsh, 1994). This high level of pet ownership persists despite many potential costs. In addition to the financial costs of food, veterinary care and other pet products, disadvantages of pet ownership can include: time spent caring for the pet; restrictions on lifestyle; daily hassles resulting from caring for and cleaning up after pets; worry due to destructive or anti-social behaviour of pets; emotional distress, e.g. on the death of a pet; and risks such as bites, allergic reactions and other zoonoses (Plaut, Zimmerman & Goldstein, 1996). Given this long list of potential costs, and that relatively few pets are working animals 'earning their keep' in a practical way, owners presumably perceive substantial benefits from pets to persuade so many to keep them. Pets may have functional roles (Hirschman, 1994) such as impression management (e.g. dogs as fashion accessories, or acquisition of a fierce dog to fit a macho image), or avocation (the pet as a diversion or hobby, e.g. those kept for breeding or competing in shows). However, most accounts of positive aspects of pet ownership focus on pet ownership as a social relationship with advantages arising from relationship-based concepts such as support and attachment (Garrity *et al.*, 1989), and protection against loneliness (Zasloff & Kidd, 1994). There may be other indirect benefits such as those which might result from the additional human contacts made as a result of pet ownership (Messent, 1983; McNicholas *et al.*, 1993). Different types of benefit may combine within a single relationship, for example a pet that is kept for showing may also be valued for companionship, and because it fits the lifestyle image of a particular family. Within a family which shares one pet, each human family member may receive different types and degrees of benefits from the presence of the animal and incur different costs of pet ownership.

Research into pet ownership rarely looks simultaneously at the balance of benefits and disadvantages of pet ownership. Rather, journal articles often seek (and hence find) either positive *or* negative implications. Medical and veterinary publications tend to emphasize zoonoses, whereas journals in the social sciences tend to focus on positive aspects of human–pet relationships and benefits to health. There are exceptions to this general tendency, for example Kidd and Kidd (1994) look at the benefits of pets to the homeless and also the serious problems faced by these people in keeping their animals. This special population can gain benefits of warmth, security and companionship from pet dogs, but the costs can be high if they continue to live on the streets rather than accept accommodation that does not allow pets. Bryant (1990) looked at the costs and benefits of children's relationships with pets. She identified a number of costs arising from child–pet relationships, such as sadness at pet death or illness, distress at not being allowed to care for the pet and worry for its safety. Many of these cost factors are arguably inevitable consequences of the benefits she found. For example, because the children enjoy the companionship and enduring affection given by the animals, they are bound to feel distress at loss or separation from them. Glaser, Angoulo and Rooney (1994) considered the risk of zoonoses versus benefits from relationships with pets for HIV/AIDS sufferers.

The large number of pets in shelters run by animal welfare organizations is testimony that not all human–pet relationships are successful. In the UK in 1995, Wood Green animal shelters took in nearly 13 000 animals; the National Canine Defence League around 10 000. In the USA, Patronek and Rowan (1995) estimated that 7.7% of the dog population was in the care of animal shelters. This gives a population of four million dogs in the USA that have been rejected by their owners for some reason. Animals may be taken in by shelters for reasons other than a failure in the pet relationship, for example owners may have died, moved into residential care, suffered a marriage breakdown, or unemployment. However, when the balance of costs and benefits results in the rejection of pets, there are serious implications for animal welfare.

The study reported in this chapter is primarily concerned with the hypothesis that pet ownership can usefully be conceptualized as a kind of social relationship. Much of the literature on pet ownership has implicitly assumed that human–pet interactions are the basis of social relationships and, further, that it is reasonable to use models of the psychology of human–human relationships to investigate them. Terms such as attachment, companionship and support are frequently used in connection with pet ownership, although this is mostly done unsystematically and uncritically (Collis & McNicholas, 1998).

Given the propensity of humankind to be anthropomorphic with other species, it is unsurprising that pets, especially social animals such as dogs, are sometimes treated 'as if' they are human. Kennedy (1992), claims that we are inclined to anthropomorphism by both nature and nurture. He says (p. 5): 'It (anthropomorphism) is dinned into us from earliest childhood. It has been pre-programmed into our hereditary makeup by natural selection, perhaps because it can be useful for predicting and controlling the behaviour of animals.' Whether or not an animal actually has the mental states attributed to it, anthropomorphism is often a helpful strategy for explaining patterns of events, and hence increasing the ability to predict future events. Collis and McNicholas (1998) observe that it is unlikely that humans have evolved a set of psychological processes specifically to serve relationships with companion animals, therefore if owners perceive pet species as if they are human, and respond accordingly, then they are likely to draw on psychological processes that they use in human–human relationships. To this argument it may be added that, if we had evolved new processes to deal with human–pet relationships, it would be likely that we would have also developed a different vocabulary to distinguish them. We do not, however, have separate terms to describe relationships with pets: they are frequently referred to as man's best friend or as family members; owners say they are attached to their pets, enjoy their companionship, love them and suffer grief at their death.

Thus far, the argument supports the use of models of the psychology of human relationships in looking at human involvement in human–pet relationships. However, what pets bring to the dynamics of a relationship will inevitably differ from what humans bring. A key element that distinguishes human–human from all human–pet relationships is the lack of a shared language. This has important implications for the potential of the relationship, as Hinde (1988: 20) observes: 'The uniquely human attribute of a spoken language is associated with behaviour of a different order of complexity from that found in animals.' This higher order of complexity attributed to human–human relationships would not be available to human–pet relationships. The relative simplicity and lack of language may be part of the appeal for some pet owners, for example pets cannot answer back, pass judgement, be influenced by what others say, or break confidences. Because the pet cannot verbally express an opinion of its owner, the owner may project whatever interpretation he or she wishes on the pet's behaviour, and do so without fear of contradiction.

The lack of a shared language also has importance with regard to the types of psychological models which it may be most fruitful to use in an investigation of human–pet relationships. For example, models such as the one proposed by Duck (1994) are unlikely to be helpful due to the central role of language in enabling relaters to arrive at shared meanings of their relationship. Investigation of the human–pet relationship as a dynamic interaction between two parties will also be restricted by the inability of other species to give an account of their experience of the relationship. Any attribution of *human* motives or emotions to other species could fairly be accused of anthropomorphism. That is not to say that the pet species cannot have some mental experiences analogous to those of humans; however, biological differences between species mean that observations based on *human* experience cannot be assumed to be generalizable to other species. Even when the assumption is made that other species do have mental experiences of some kind, the subjective nature of mental experience may mean that we can never fully grasp how pet species experience their relationships with humans. As Nagel (1974) says, we may guess what it is like to be a bat, but knowing what it is like for a *bat* to be a bat is inaccessible to us. This is not to say that investigations cannot be made into the pet 'end' of the relationship. It is rather to say that these should be based upon observations of behaviour interpreted with reference to the comparative psychology and ethology of the particular species. What the various pet species bring to the human–pet relationships will differ according to the cognitive capabilities and behavioural repertoire of the particular species. In addition, human responses to different pet species are likely to be varied. As Zasloff (1996) observes, 'a dog is not a cat is not a bird'.

Many of the closest human relationships exist within families and it is within the family that pet ownership occurs most frequently. In particular, pet ownership is most prevalent in households with children (Table 12.1). In many cases pets are described in ways that imply that they are considered as significant others. For example, studies which report high percentages of people describing pets as family include: Cain (1983: 87%); Beck and Katcher (1984: 70%); Voith (1985: 99%); Albert and Bulcroft (1988: 87%); Hirschman (1994: 80%). These very high levels of inclusion of pets as family members may have been inflated by the way in which the questions were framed to participants (Fisher, Collis & McNicholas, 1998). However, large numbers of pet owners are willing to include pets in a category of close human relationships. In addition to being described as

Table 12.1. *The prevalence of pet ownership in families*

Size of household	Percentage of households with pets
1	31.0
2	47.8
3	60.6
4	65.1
5+	64.9
Presence of children (age in years)	**Percentage of households with pets**
0–5 only	49.8
6–15 only	73.8
0–5 & 6–15	59.6
No children	43.3

Source: Pedigree Petfoods Pet Ownership Survey (1996).

family members, pets are considered particularly important to children (Levinson, 1972; Bryant, 1982, 1990; Furman, 1989). The family context provides an opportunity to compare human–pet and human–human relationships to examine the extent to which they co-vary between individuals, and to examine how human–pet relationships vary with the family role of the human.

The measure used in this study to investigate the nature of the human–pet relationship is based on Furman's Network of Relationships Inventory (NRI). This has been previously used with children, and to gain information on human–pet relationships (Furman & Burhmester, 1985; Furman, 1989). The NRI is based on Weiss' (1974) proposal that in order to achieve an adequate life organization, individuals require a network of relationships that together provide a number of relational provisions. Individual relationships may be quite specialized, so a person needs to maintain a number of different relationships to provide the range of provisions required, and hence maintain a sense of well-being. Weiss' approach uses a 'needs model' which makes strong assumptions about human motives for participating in social relationships, and about adverse consequences for well-being if some needs are unfulfilled. However, it is possible to interpret data on relational provisions from the NRI in order to assess functions of human–pet relationships without necessarily subscribing to Weiss' model of human needs. Sub-scales within the NRI measure specific relational provisions described by Weiss, such as nurturance, reassurance of worth, and a sense of reliable alliance. The NRI also provides summary measures of overall satisfaction with relationships and the relative power of the two parties in the relationship. The conflict and antagonism sub-scales are of particular interest since these are negative provisions, and help guard against over-emphasis of positive aspects of relationships.

By gathering data from participants on relationships with all immediate family members as well as their relationships with pets, the analysis can examine both the role of pets in terms of relational provisions provided, and the issue of whether the human–pet relationship is used to plug gaps in provisions lacking from the human–human relationships, or simply provides a potential source of extra provisions.

METHOD

Ninety participants from 40 pet-owning households were recruited via pet stores, a veterinary surgery, and an RSPCA shelter. There were two types of household: those with and those without children living at home. All family members in each household aged ten and over were encouraged to participate in the study. This gave rise to six family role types denoted by mothers, fathers, sons, daughters, husbands and wives. The last two refer to couples in households without children living at home.

The NRI sub-scales are summarized in Table 12.2. Each sub-scale was based on three items. Typical items ask how much of a particular relational provision a specific human or pet provides using a five-point scale, e.g. from 1 (not at all) to 5 (very much). For example, a question to determine intimacy or confiding is 'How much do you share your secrets and private feelings with each one?' Participants gave responses for their relationships with every other household member (pets as well as people). A measure of social support was derived by adding the mean scores for the following sub-scales: companionship, instrumental aid, intimacy, nurturance, affection (directed to the participant by the other member of the household), admiration and reliable alliance. A measure of negative interactions was derived from the mean scores of conflict and both antagonism sub-scales.

Participants were also asked to report how much they thought each family member (including themselves) shared in ownership of each household pet. This was measured on a five-point scale: 1 = has no share in

Table 12.2. *NRI sub-scales*

Sub-scale	Description
Companionship	Spending time with others, doing enjoyable things together
Instrumental aid	Others providing help
Intimacy	Confiding in, sharing private thoughts with others
Nurturance	Taking care of, protecting others
Affection (for participant)	Others' love for or care about participant
Affection (for others)[a]	Participant loves or cares about others
Admiration	Respect for participant, approval of participant's actions
Reliable alliance	Participant's belief that the relationship will last
Satisfaction	Participant's satisfaction with the relationship
Relative power	Who makes decisions or is boss in the relationship
Conflict	How much participant and others disagree, or clash
Antagonism (others antagonize)[b]	How much others nag or get on nerves of participant
Antagonism (antagonize others)[b]	How much participant nags or gets on nerves of others

Notes:

[a]Item added by the authors.

[b]The original NRI includes only one set of questions on antagonism, participants being asked, for example, 'How much do you and this person hassle or nag one another?' This study did not assume that the degree of antagonism between family members would always be mutual, e.g. a younger sibling may get on an older sibling's nerves, but not always *vice versa*. As a result, the questions were duplicated, with one set asking participants how much they antagonized others and one set asking how much others antagonized them.

Note: Furman's NRI contains 12 sets of scales, each with three items: companionship, conflict, instrumental aid, satisfaction, antagonism, intimacy, nurturance, affection, punishment, admiration, relative power, reliable alliance. The punishment scale was not used. The wording of some items was slightly altered to make them appropriate for both human and pet relationships.

owning pet; 2 = has a small share in owning pet; 3 = has a moderate share in owning pet; 4 = has a big share in owning pet; 5 = is the only human who owns pet. In order to avoid leading participants to a particular interpretation of pet ownership, no guidance was given to participants on which criteria to refer to in their judgements. Regression analysis was applied to the results to determine which other variables predicted ownership rating.

RESULTS

The 90 participants contributed data on 500 relationship dyads: 256 human–human relationships and 244 human–pet relationships. The number of pets per household ranged from one to nine, with an average of 2.7. The number of other people per household ranged from one to six, with an average of 2.9. Of the human–pet relationships, 105 were with cats and 116 were with dogs. The remaining 23 comprised birds and various small mammals such as hamsters and guinea-pigs. Four relationship types were used in the analysis: human–human; human–dog; human–cat; and human–other. The 'other' category refers to pets other than cats or dogs.

Inter-item reliability of the NRI sub-scales

The reliability of the NRI sub-scales was good, with alpha coefficients for each sub-scale greater than 0.75. This is also supported by the generally good coherence of the sets of sub-scale questions in the principal component analysis (PCA) discussed below.

Components of relationships

Principal components analysis was carried out on the correlation matrix of data from all 39 NRI questions to determine whether they could be interpreted as reflecting a smaller number of underlying dimensions. Decisions on the number of components to retain in PCA were based upon the visual inspection of scree plots of eigenvalues in conjunction with Kaiser's rule that components with eigenvalues less than 1.00 should not be retained. It is now known that the latter criterion on its own results in an overestimation of the number of components (Zwick & Velicer, 1986). Interpretation of the components was facilitated by using varimax rotation.

A PCA on the data from the 244 human–pet dyads revealed two

components, whereas a PCA on the 256 human–human dyads revealed four (see Table 12.3). The first component from the human–pet dyads was interpreted as support; it is consistent with Furman's proposed support component comprising the sub-scales for companionship, instrumental aid, intimacy, nurturance, affection, admiration, and reliable alliance. Questions from these sub-scales were loaded on the first component with loadings greater than 0.50, with the exception of two of the instrumental aid items which loaded 0.49 and 0.38. None of the pets included in the study was a working animal, and all were kept solely as companions. It is therefore surprising that any of the instrumental aid items featured in the support component. The item which did load at greater than 0.5 asked 'How much does the pet help you if you have a problem to sort out?' It is not clear how pets can actually provide practical aid; however, participants reported that they perceived them to do so. The three satisfaction questions also loaded greater than 0.5 on the support component. This is not surprising, given that it is reasonable to assume participants will gain greater satisfaction from relationships offering high support.

The second component comprised all questions in the conflict and antagonism sub-scales, all loading greater than 0.5 except one of the antagonism items (participant antagonizes pet), which loaded at 0.46. Furman grouped these sub-scales (together with a punishment sub-scale which was not used in this study) to form an index of negative interactions. None of the relative power sub-scale questions loaded above 0.50 on either the support or conflict component.

The four components from the human–human dyads were interpreted as: (i) support, (ii) conflict, (iii) relative power, and (iv) intimacy. The support component comprised a similar set of questions to those found from the human–pet relationships data, with the following exceptions: the intimacy sub-scale items comprised a separate component of their own; one companionship item loaded most strongly on the intimacy component; one nurturance item loaded on the relative power component. The conflict component comprised all of the items from the conflict and antagonism sub-scales.

Two main differences may be seen between relationships with humans and relationships with pets. Firstly, relative power features as a component of human relationships, but not of pet relationships. Within the human relationships, two nurturance items loaded most strongly on the relative power component. This component may be due to the influence of parent–child dyads, where power and nurturance both feature strongly in the parent–child relationship. This interpretation deviates from the view that pets have a similar role to children insofar as it implies

Table 12.3. *Components from human–pet and human–human dyads after varimax rotation*

NRI item	Components from human–pet dyads		Components from human–human dyads			
	Support	Conflict	Support	Conflict	Relative power	Intimacy
Affection (i) 1	**0.79**	<0.01	**0.81**	0.27	0.06	0.01
Affection (i) 2	**0.82**	0.08	**0.85**	0.22	0.03	0.15
Affection (i) 3	**0.81**	0.08	**0.77**	0.18	0.06	0.10
Affection (ii)1	**0.69**	0.04	**0.74**	0.29	0.14	0.05
Affection (ii)2	**0.70**	0.02	**0.77**	0.13	0.16	0.08
Affection (ii)3	**0.71**	0.05	**0.59**	0.16	0.26	0.20
Companionship 1	**0.65**	0.23	0.40	0.11	0.21	0.48
Companionship 2	**0.76**	0.16	**0.61**	0.29	0.09	0.34
Companionship 3	**0.72**	−0.11	**0.54**	0.17	<0.01	0.36
Admiration 1	**0.73**	0.04	**0.57**	0.48	0.16	0.29
Admiration 2	**0.61**	0.05	**0.58**	0.19	0.19	0.09
Admiration 3	**0.69**	0.09	**0.53**	0.26	0.13	0.23
Reliable alliance 1	**0.67**	0.21	**0.75**	0.17	−0.04	0.10
Reliable alliance 2	**0.66**	0.34	**0.75**	0.18	0.03	0.13
Reliable alliance 3	**0.63**	0.33	**0.69**	0.18	−0.08	0.20
Satisfaction 1	**0.62**	0.13	**0.65**	0.38	<0.01	0.22
Satisfaction 2	**0.63**	0.34	**0.71**	0.42	−0.15	0.06
Satisfaction 3	**0.74**	0.20	**0.77**	0.28	0.03	0.27
Nurturance 1	**0.53**	0.02	0.39	0.04	0.45	0.25
Nurturance 2	**0.67**	−0.07	0.36	0.05	**0.59**	0.19
Nurturance 3	**0.65**	−0.05	**0.52**	0.01	0.30	0.30
Intimacy 1	**0.51**	−0.05	0.20	0.17	0.15	**0.77**
Intimacy 2	**0.63**	−0.06	0.16	0.20	0.07	**0.84**
Intimacy 3	**0.57**	0.13	0.13	0.12	−0.07	**0.86**
Instrumental aid 1	0.49	−0.15	0.42	0.03	−0.11	0.47
Instrumental aid 2	**0.60**	−0.08	**0.51**	0.11	−0.47	0.39
Instrumental aid 3	0.38	0.03	**0.56**	0.02	−0.39	0.33
Conflict 1	−0.02	**−0.71**	−0.24	**−0.72**	0.12	−0.11
Conflict 2	<0.01	**−0.87**	−0.32	**−0.72**	0.02	−0.09
Conflict 3	−0.03	**−0.79**	−0.24	**−0.73**	0.13	−0.17
Antagonism (iii) 1	−0.08	**−0.81**	−0.40	**−0.70**	0.05	−0.05
Antagonism (iii) 2	−0.04	**−0.74**	−0.34	**−0.65**	0.16	−0.19
Antagonism (iii) 3	0.07	**−0.65**	−0.05	**−0.74**	−0.32	−0.11
Antagonism (iv) 1	−0.13	**−0.53**	−0.34	**−0.67**	−0.23	−0.05
Antagonism (iv) 2	0.16	−0.46	−0.04	**−0.75**	−0.26	0.06
Antagonism (iv) 3	0.23	**−0.54**	−0.16	**−0.64**	0.15	−0.19

Table 12.3. (*cont.*)

NRI item	Components from human–pet dyads		Components from human–human dyads			
	Support	Conflict	Support	Conflict	Relative power	Intimacy
Relative power 1	0.06	0.03	0.17	−0.07	**0.85**	0.01
Relative power 2	0.08	0.32	−0.04	0.04	**0.82**	−0.01
Relative power 3	−0.07	0.08	−0.03	0.03	**0.84**	−0.01
Percentage of total variance explained	30.57	12.94	26.64	14.92	9.64	9.54

Notes:
(i) Affection (for participant from others).
(ii) Affection (of participant for others).
(iii) Antagonism (others antagonize participant).
(iv) Antagonism (participant antagonizes others).
Items with loadings >0.50 in bold type.

that relative power is a salient aspect of parent–child relationships, but not of the human–pet relationships. Secondly, intimacy is included in the general support component for human–pet relationships, but features as a separate dimension in human–human relationships. The intimacy items refer to confiding behaviours such as telling others private thoughts or feelings. These results could be due to participants confiding in human relationships such as with close friends outside the family group, who were not included in the study. Alternatively, it might be that confiding is a more generalized feature of positive relationships with pets but a more specific feature of relationships with people.

The apparent differences in the structure of human–human and human–pet relationships are intriguing and deserve further investigation. Despite the differences, however, the similarity between the two structures overall is also striking. This adds empirical weight to the view that human–pet relationships are similar in nature to human–human relationships and, perhaps more specifically, that the supportive aspects of the two kinds of relationships are broadly similar.

Differences between relationships: support

The index of social support for human–human relationships was higher than for human–pet relationships (see Fig. 12.1). A one-way analysis of variance showed that differences among the four relationship types were significant ($F_{(3,491)} = 63.1, p < 0.001$). *Post-hoc* Tukey tests showed that all pairwise comparisons were significant ($p < 0.05$).

To check whether the differences between relationship types were consistent across the sub-scales that contributed to the overall support index, a two-factor analysis of variance (relationship type × sub-scale) was carried out on sub-scale scores. The interaction was significant ($F_{(18,2946)} = 34.7$, $p < 0.001$), indicating that differences between relationship types were not consistent across sub-scales. The main source of this inconsistency is in the comparison between human–human and human–dog relationships (Fig. 12.2). Scores for instrumental aid, affection and admiration were higher for human–human relationships than for human–dog relationships, whereas scores for companionship, nurturance and reliable alliance were higher for human–dog than for human–human relationships. Support from human–dog relationships was higher than support from human–cat relationships, and the direction of this difference was the same for all seven sub-scales.

As the number of other pets was small, and heterogeneous in the species represented, further analysis of support focused on cats and dogs. Figure 12.3 shows how the support levels from cats and dogs differ by family role. Human–cat relationships broadly follow a trend whereby female family roles report higher levels of support than males. The pattern for human–dog relationships is markedly different, with wives and husbands (from households with no children) reporting the highest levels of support.

A two-way analysis of variance examining the effects of family role and pet type (dog or cat) on level of support from pets showed significant main effects of both variables, and a significant interaction between them: for family role ($F_{(5,206)} = 4.0, p = 0.002$); for pet type (dog/cat) ($F_{(1,206)} = 65.6$, $p < 0.001$); for the interaction of family role × pet type ($F_{(5,201)} = 3.3$, $p = 0.006$). The mean support from dogs was higher than for cats. *Post-hoc* pairwise comparisons on the main effect of family roles showed that the mean level of support reported by sons is significantly lower than for mothers, daughters and wives ($p < 0.05$). Comparison of the family role × pet type combinations indicated that the family roles of husbands and wives reported significantly higher levels of support from dogs compared with cats ($p < 0.05$). For all other family roles, the mean support from dogs was higher than for cats, but the differences failed to reach statistical significance.

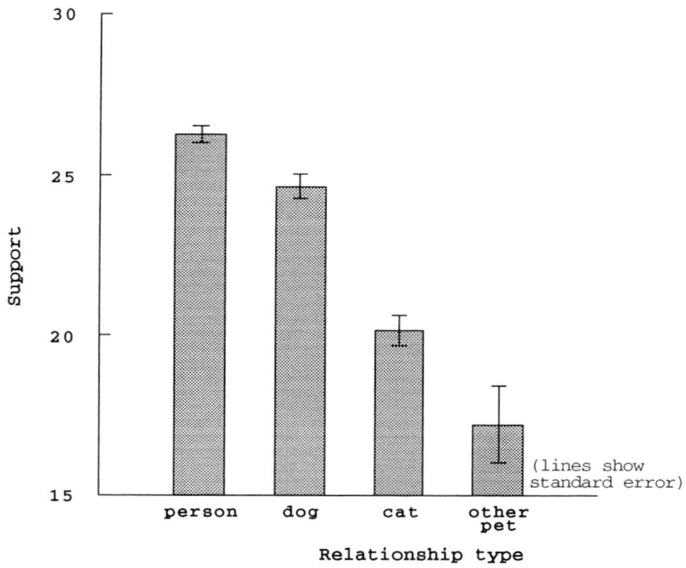

Fig. 12.1. Support by relationship type.

Differences between relationships: negative interactions

The index of negative interactions comprises the conflict and two antagonism sub-scales. Negative interactions were higher for human–human relationships than for human–pet relationship (see Fig. 12.4). A one-way analysis of variance showed that differences among the four relationship types were significant ($F_{(3,492)} = 98.4$, $p < 0.001$). *Post-hoc* Tukey tests showed that all pairwise comparisons between the four relationships types were significant ($p < 0.05$).

The patterns of scores for each of the three sub-scales comprising the negative interaction index were very similar. Despite this, when a two-factor analysis of variance (relationship type × sub-scale) was carried out on mean scores for the conflict and both antagonism sub-scales, a significant interaction was found, ($F_{(3,984)} = 8.8$, $p < 0.001$). This indicates that differences between relationship types were not consistent across sub-scales. There was a difference in the two antagonism sub-scales: in human–human relationships, participants antagonized others more than others antagonized them, whereas in human–pet relationships, pets were perceived to antagonize participants more than participants reported antagonizing pets. However, these effects were small relative to differences between relationship types, which were similar across all three sub-scales (Fig. 12.5).

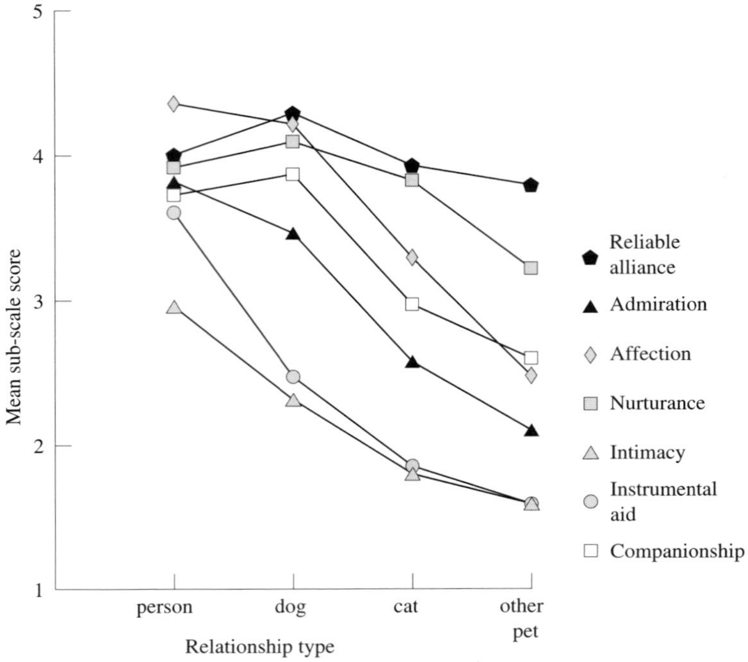

Fig. 12.2. Support sub-scales by relationship type.

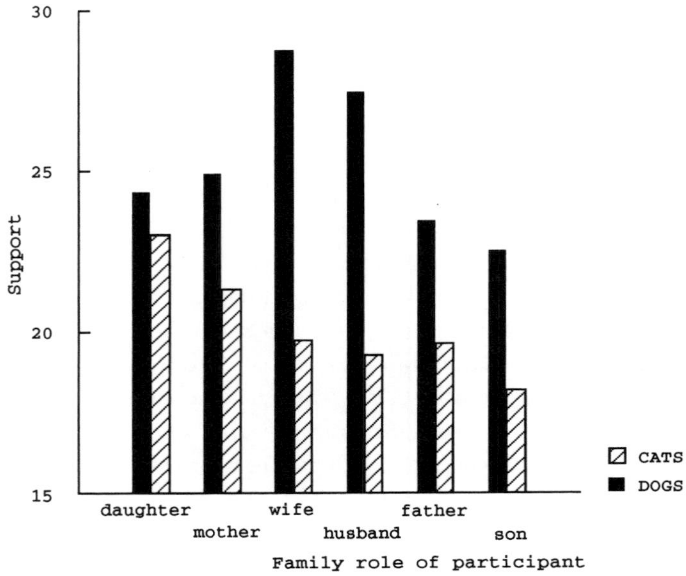

Fig. 12.3. Support from pets by family role and relationship type.

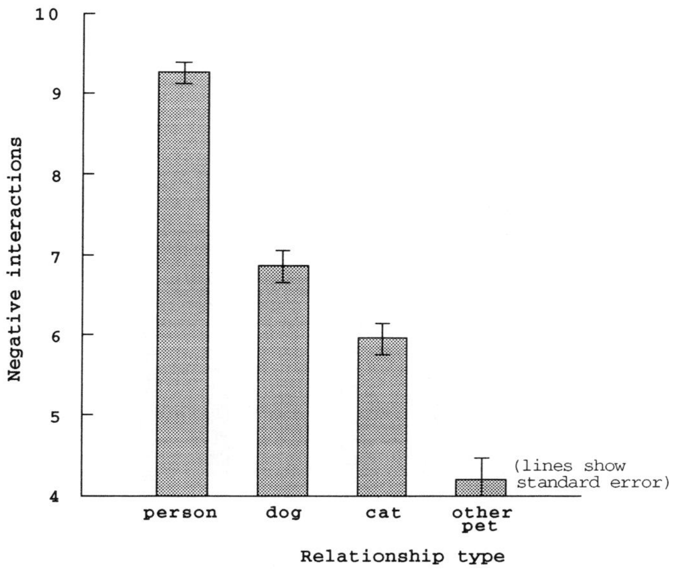

Fig. 12.4. Negative interactions index by relationship type.

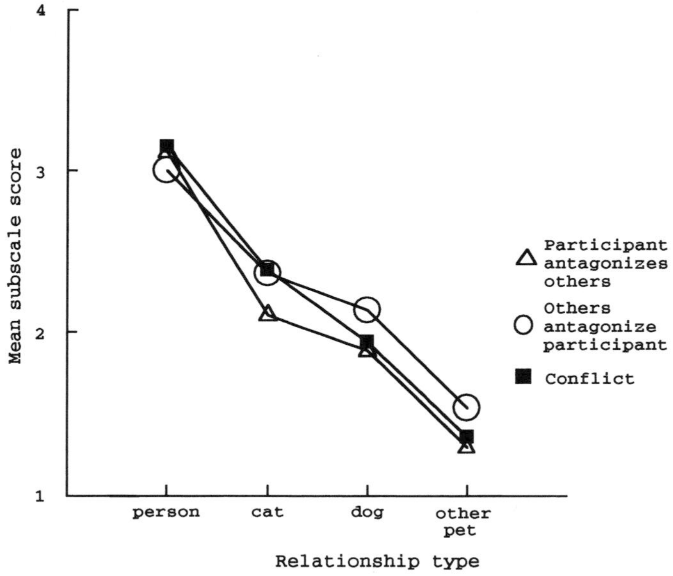

Fig. 12.5. Negative interaction sub-scales by relationship type.

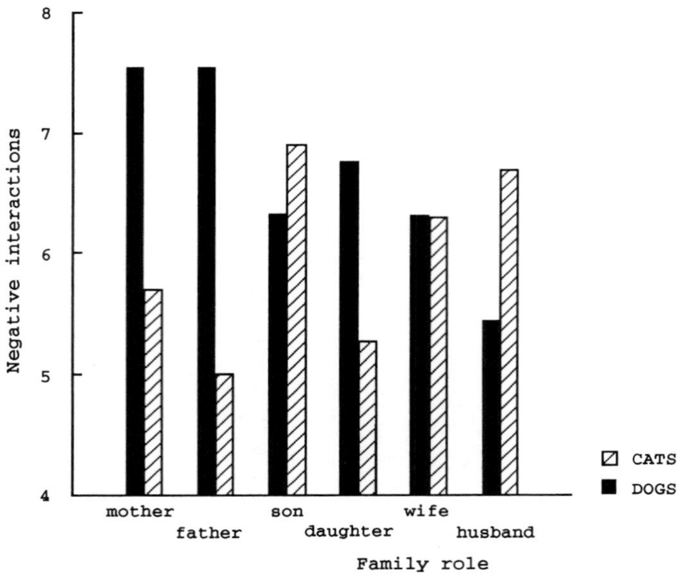

Fig. 12.6. Negative interactions with pets by family role of participant and pet type (cat/dog).

Differences between human–cat and human–dog scores were examined further. A two-way analysis of variance of family role and pet type on negative interactions showed no significant main effect of family role ($F_{(5,207)}=0.6$, $p=0.717$), which is unlike the findings for support. However, there was a significant effect of pet type ($F_{(1,207)}=5.5$, $p=0.020$), with dogs rated higher than cats, and a significant interaction of family role × pet type ($F_{(5,207)}=4.8$, $p<0.001$).

Figure 12.6 shows the ratings of the six family roles for negative interactions with cats and dogs respectively. *Post-hoc* pairwise comparisons were conducted, and showed that fathers and mothers both rated dogs significantly higher on negative interactions than cats ($p<0.05$). Although husbands and sons rated cats higher than dogs, these differences were not significant.

Differences between relationships: other sub-scales

Two sub-scales are not included in either the support or negative interaction indices: satisfaction and relative power. The standardized scores for these are shown in Fig. 12.7.

A one-way analysis of variance on the satisfaction sub-scale showed that differences among the four relationship types were significant

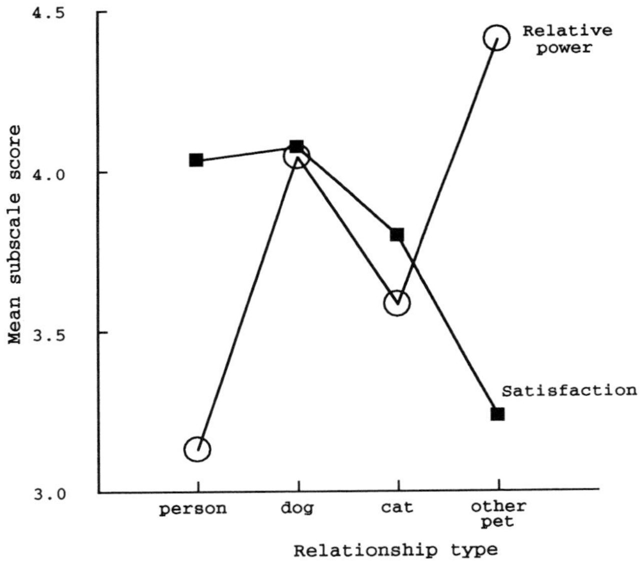

Fig. 12.7. Satisfaction and relative power sub-scales by relationship type.

($F_{(3,492)}$ =9.5, $p<0.001$). Satisfaction was highest with human–dog and human–human relationships. *Post-hoc* Tukey tests showed that all pairwise comparisons were significant ($p<0.05$), except between human–human and human–dog relationships.

Participants reported that they had greater power in relationships with dogs and other pets, and less power in those with cats and people. The difference amongst the four relationship types was found to be significant for the relative power sub-scale by one-way analysis of variance ($F_{(3,492)}$=29.3, $p<0.001$). *Post-hoc* Tukey tests showed that all pairwise comparisons were significant ($p<0.05$), except that between human–dog and human–other pet relationships.

Judgements of ownership share

The data from the ratings of ownership showed that most pets were considered as shared between human family members (see Fig.12.8).

Only ten of the 244 human–pet dyads were allocated to the rating 1 (no share in owning pet) or 5 (the only human who owns pet). These ten dyads with extreme ratings were not typical in that eight of the ten involved small rodents such as hamsters and rabbits rather than the predominant species of cats and dogs. Typical pet species such as cats and dogs were usually considered as shared between the human household

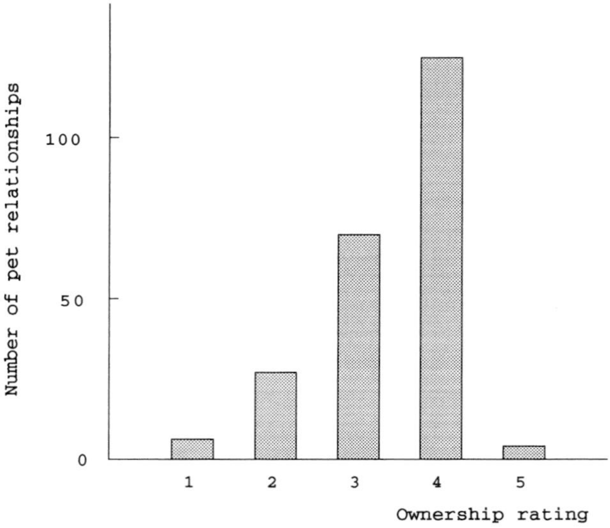

Key:

1 = No share in owning pet 2 = Small share 3 = Moderate share

4 = Big share 5 = Only one owning pet

Fig. 12.8. Number of human–pet dyads reporting each ownership rating.

members. To examine the variables influencing ownership ratings of cats and dogs, regression analysis was computed using family role of participant, pet type (cat/dog), support from pet and negative interactions with pet as predictors (family role of participant and pet type as categorical variables). Results, detailed in Table 12.4, show that family role and the support from human–pet relationships had a significant main effect on ownership ratings.

Differences in ownership rating across family role type are shown in Fig. 12.9. The family role with the highest mean ownership score is husband, followed by mother, wife, daughter, father and son. The level of support from pets increased with increased ownership rating. Neither pet type (dog/cat) nor the negative interactions with pets (sum of mean scores for conflict and antagonism sub-scales) had a significant effect on ownership rating.

Table 12.4. *Regression analysis on participant's self-rating of ownership share of pet*

Variable	*F* ratio	df	Probability
Family role of participant	3.93	5,201	0.002
Pet type (dog/cat)	0.05	1,201	0.827
Support from pet	11.71	1,201	$<$0.001
Negative interactions with pet	0.33	1,201	0.568

Note:
Predictors: family role type and pet type (categorical variables), and indices of support and negative interaction.

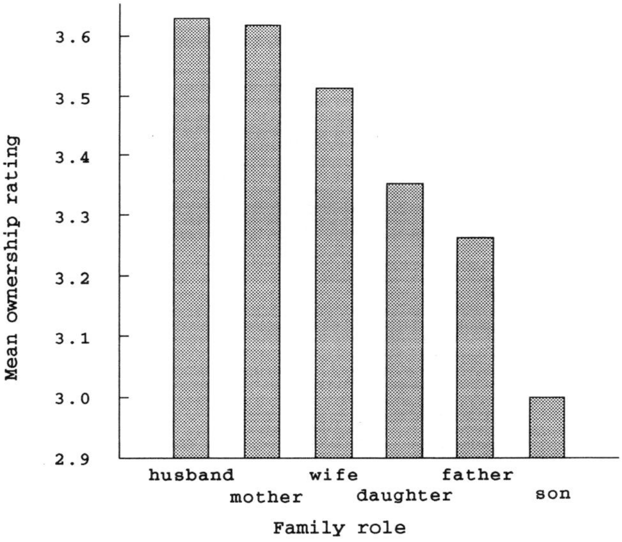

Fig. 12.9. Mean ownership rating by family role.

Do pets compensate for low provisions in human relationships?

If people use provisions from human–pet relationships to compensate for low levels of provisions from other human relationships, a negative correlation between provisions from people and provisions from pets would be expected as the less participants receive from people, the more they might seek from pets.

This general issue can be examined in two ways. First, we can ask whether support and other social provisions experienced by a participant from a typical human relationship are related to what participants experience of social provisions from a typical pet. To do this, we computed two mean scores for each participant for each sub-scale. Averaging across all of a participant's relationships with other people in the family produced a mean score for that participant's human–human relationships, and averaging across his or her relationships with pets produced a mean score for the human–pet relationships. Second, we can ask whether the aggregate social provision experienced by a participant from other people is related to the aggregate level of social provision he or she experiences from pets. To do this, we computed two total scores for each participant for each sub-scale. Totalling across all the participant's relationships with other people in the family produced a total score for human–human relationships, and totalling across his or her relationships with pets produced a total score for human–pet relationships.

Table 12.5 presents the correlations between typical levels of social provision for human–human and human–pet relations, and between aggregate levels of the two sources of provision. The correlations for typical levels of provision were primarily positive, indicating that participants who perceived a higher level of provision from a typical human also tended to perceive higher levels of provision from a typical pet. This indicates that participants who experienced lower than average levels of provision from a typical human in their network also experienced low levels of provision from a pet. This is not what one would expect from the compensation hypothesis. The correlations for aggregate provision are primarily negative, suggesting that, at a descriptive level, low aggregate levels of provision from people are associated with higher aggregate levels from pets. However, it is important to note that this could be explained by a negative correlation found between the number of people in households and the number of pets ($r = -0.23$). It is inevitable that numbers of people and pets will influence aggregate levels of support from people and pets.

The correlations between sub-scale scores could be reflecting patterns of differences between individuals in the experience of social provisions, or patterns of inter-family differences. That is, whole families might well differ from one another in their style and levels of social provision. These issues can be examined by partitioning the overall correlations into within-family and between-family correlations. Within-family correlations have all inter-family differences partialled out so that they reflect individual differences uncontaminated by family effects. As all of the individuals within any family report on the same number of human–human and human–pet relationships, there is a constant linear

Table 12.5. *Correlations of relational provisions from human–pet relationships versus human–human relationships: between-participants analysis*

Relational provision (sub-scale/component)	Correlation of typical levels (*r*)	Correlation of aggregate levels (*r*)
Companionship	**0.23**	−0.21
Instrumental aid	**0.27**	−0.08
Intimacy	0.20	−0.10
Nurturance	**0.24**	−0.20
Affection (1)	0.04	**−0.22**
Affection (2)	0.10	**−0.21**
Admiration	0.08	**−0.22**
Reliable alliance	**0.35**	**−0.22**
Satisfaction	**0.23**	−0.18
Conflict	0.13	−0.03
Antagonism (3)	0.20	−0.05
Antagonism (4)	0.18	−0.09
Relative power	0.10	−0.16
Support index	**0.35**	**−0.23**

Notes:
(1) Others love participant.
(2) Participant loves others.
(3) Others antagonize participant.
(4) Participant antagonizes others.
Significant correlations (at $p < 0.05$) in bold type.

relationship between typical (mean) and aggregate (total) scores, so these two types of variable are not distinguishable in within-family correlational analyses. Between-family correlations reflect only inter-family differences, including the differences in the number of people and pets in different families. Within-family and between-family correlations are shown in Table 12.6.

Within-family correlations of the typical levels of provisions from human–human relationships and human–pet relationships are generally positive and statistically significant. This further supports the earlier finding that subjects reporting low levels of provision from humans also report low levels of provision from pets. This analysis confirms that this pattern operates at the level of individuals rather than families. The levels of provisions reported by individual family members are similar for both human–human and human–pet relationships.

Between-family correlations between typical provisions from pets

Table 12.6. *Correlations between provisions from human–pet relationships versus human–human relationships: within-families and between-families analysis*

Relational provision (Sub-scale/Component)	Within families *r* (df=459)	Between families *r* (df=39)	
	Typical	Typical	Aggregate
Companionship	**0.22**	0.24	−0.25
Aid	**0.25**	0.29	−0.13
Intimacy	**0.22**	0.21	−0.24
Nurturance	**0.23**	0.25	−0.25
Affection (1)	−0.02	0.10	−0.25
Affection (2)	**0.21**	0.01	−0.23
Admiration	0.06	0.10	−0.27
Reliable alliance	**0.33**	**0.37**	−0.30
Satisfaction	**0.26**	0.22	−0.21
Conflict	**0.25**	0.10	−0.06
Antagonism (3)	**0.33**	0.15	−0.10
Antagonism (4)	**0.43**	0.07	0.02
Relative power	**0.12**	0.09	−0.19
Support index	0.21	0.40	−0.26

Notes:
(1) Others love participant.
(2) Participant loves others.
(3) Others antagonize participant.
(4) Participant antagonizes others.
Significant correlations (at $p < 0.05$) in bold type.

and from people were also generally positive and several are of the same order of magnitude as the within-family correlations, so there seem to be family style effects as well as individual effects. Correlations for the aggregate scores were generally negative, which again is probably accounted for by the negative correlation between the number of people and the number of pets in these families. The between-family correlations need to be treated with some care, as only reliable alliance reaches conventional levels of statistical significance ($p < 0.05$). This is because the between-family degrees of freedom are quite small, ($df = 39$), resulting in lower power for the statistical analysis than for the within-family correlations, ($df = 459$). Despite this lack of statistical support for the significance of the apparent trends in correlations, the fact that the supportive sub-scales all follow the same trend suggests that there are some consistent differences between families in the characteristics of their social provision.

The results for each of the sub-scales of companionship, aid, intimacy, nurturance, reliable alliance, satisfaction, and the overall support measure fit the general trends well:

Within families, individuals exhibited a style such that those participants reporting high provisions from people also reported high provisions from pets, and *vice versa*. This may be attributed to the personal style of individuals.

There were family styles, such that individuals in families who reported high mean levels of provision from people also reporteed high levels from pets, and *vice versa*. This was indicated by positive correlations of typical scores between families.

Between families, correlations of *aggregate* scores were negative. As the typical scores of human and pet relationships were positively correlated, the negative correlation of aggregate scores must reflect the negative correlation between the numbers of human and pet relationships in the families.

There are also exceptions to this general trend:

There were no significant correlations between typical scores for human–human and human–pet relationships on sub-scales of affection (from others) and admiration. This suggests that there was no strong individual or family style influencing results for typical levels of these sub-scales. There was, however, a negative correlation of aggregate scores between families, which is likely to be due to the negative correlation between the number of people and the number of pets in families.

The three negative interaction sub-scales showed positive correlations of $r > 0.25$ within families, but not between families (for typical or aggregate data). This suggests that some individual human style is influencing results such that those with high conflict scores for human–human relationships will report high conflict scores for human–pet relationships. There was no corresponding family style influencing the results. Despite the negative correlation between the number of people and the number of pets in families, there was no significant negative correlation between families on the aggregate scores from human–human and human–pet relationships. This may be because the positive correlation at an individual level was so marked.

The sub-scale of affection for others shows a positive correlation ($r=0.21$) between typical scores for human–pet and human–human relationship scores within families, but only a trivially small correlation ($r=0.01$) between families. This suggests that there is an individual style such that participants who report high scores for affection for typical human–human relationships in their family will also report high scores for typical pets. However, there was no corresponding family style influencing the results.

There were no strong correlations between provisions from human–human and human–pet relationships in any of the analyses of the relative power sub-scale.

DISCUSSION

The principal component analysis of NRI data on human–pet relationships revealed two components of the human–pet relationship: support and conflict. This is consistent with the proposed indices of support and negative interactions proposed by Furman (1989). Human relationships between family members gave a more complex pattern, with relative power as an additional component, and intimacy as a separate dimension, independent of the main support component. Despite these differences, the overall componential structure of support in human–human and human–pet dyads was broadly similar. This provides empirical support for the widespread belief that human–pet relationships have similarities to human–human relationships and that the concept of social support can usefully be applied to human–pet relationships, at least at a descriptive level. The concept of social support has its root in the belief that social relationships can protect against ill-health, either by a 'buffering effect' providing protection from the adverse effect stressful life events have on health, or by a 'main effect' giving on-going benefit regardless of particular stress levels (Cohen & Wills, 1985). It does not, however, automatically follow from these findings that support from pets conveys health advantages. That is an empirical issue that needs to be examined in its own right.

Overall, human–human relationships provide significantly more support than human–pet relationships. However, there is evidence that human–pet relationships, particularly those with dogs, provide a source of some elements of support comparable with levels from human relationships. Indeed, on reliable alliance, companionship and nurturance, the mean level of provision from human–dog relationships is higher than that from human–human relationships. Relationships with dogs were rated significantly higher than those with cats, and the cats were in turn rated higher than other types of pet on nearly all measures, both on positive elements of support and negative elements of conflict. This suggests that people in families usually engage in more intense relationships with their pet dogs than with their cats, and that relationships with other pet species are the least intense. When the data were broken down by family role of participant, all family roles reported higher mean levels of support from dogs than from cats, although the difference was statistically significant only for adults without children. The results on negative interactions showed that mothers and fathers reported significantly higher conflict with dogs than with cats. Differences in negative interactions with cats and dogs were not significant for other family roles. These results suggest that cats and dogs may typically interact in different ways with different family role types.

The most frequently owned species, cats and dogs, were usually perceived as shared amongst the human family members, rather than belonging to any individual. The size of the share of ownership was correlated with the family role of the participant, and degree of support from the pet. This suggests that there may be a variety of human–pet relationships within a family sharing a single pet. It therefore seems inappropriate to treat pet ownership as a simple categorical attribute based simply on the presence of a pet in a household.

This issue is particularly important for research investigating associations between pet ownership and advantages for health. There are a number of different models for the causal processes underlying such an association (McNicholas & Collis, 1998). Prominent among these is the hypothesis that human–pet relationships, especially the supportive functions of such relationships, may influence health directly (Collis & McNicholas, 1998). Investigations of this hypothesis need to ensure that human–pet relationships are evaluated in an appropriate manner. If pet ownership is assessed in terms of the presence or absence of a pet in a human's household, as it might well be if a question such as 'do you own a pet?' is posed outside the family context, then these data will not properly represent the nature and variety of human–pet relationships among respondents.

The idea that pet owners use provisions from pets to compensate for shortcomings in other human relationships does not receive support from this study. The correlations between typical levels of provision reported from human–human and human–pet relationships were either non-significant or positive, and suggest that results were influenced by the human style of the individual and/or the family style. The negative correlations between participants' ratings of aggregate provisions from pet relationships and aggregate provisions from human family members reflected the negative correlation between the number of people and the number of pets in families. The finding that families with fewer people have more pets and *vice versa* does not contradict the information presented in the introduction, which showed that larger-sized households are more likely to own pets (Pedigree Petfoods, 1996). Households with more than two members are more likely to have *at least* one pet than those with just one or two people, but of the pet-owning households investigated in this study, those with fewer people had more pets per household. The idea that pet ownership might be a compensatory strategy used by people with low levels of social provision from other humans has often been alluded to in the literature. It receives no support from the data in this study, but this still leaves open the question of why some people have a larger number of pets than others. The family context may well be important: it only needs one family member to decide to acquire several pets for the rest of the household to become multiple pet owners too.

CONCLUSION

This study has succeeded in measuring the provisions of social relationships from humans and from pets in a comparable way, and provides some empirical substance for the hypothesis that what goes on between people and their pets has a lot in common with social relationships between people. The characterization of social relationships in terms of a set of social provisions (Weiss, 1974) is but one of many approaches to the study of human social relationships (Berscheid, 1994; Hinde, 1997), and other facets of relationship theory deserve exploration so that we can better understand the scope and limits of the relationship model of pet ownership.

ACKNOWLEDGEMENTS

Sheila Bonas was supported by a collaborative studentship from the ESRC and the Waltham Centre for Pet Nutrition. June McNicholas is a Waltham Research Fellow.

REFERENCES

Albert, A. & Bulcroft, K. (1988). Families and the life course. *Journal of Marriage and the Family*, **50**, 543–52.

Beck, A.M. & Katcher, A.H. (1984). A new look at pet-facilitated therapy. *Journal of the American Veterinary Medical Association*, **184**, 414–21.

Berscheid, E. (1994). Interpersonal relationships. *Annual Review of Psychology*, **45**, 79–129.

Bryant, B.K. (1982). Sibling relationships in middle childhood. In *Sibling Relationships: Across the Lifespan*, ed. M.E. Lamb & B. Sutton-Smith, pp. 87–121. Hillsdale, NJ: Erlbaum.

Bryant, B. K. (1990). The richness of the child–pet relationship: a consideration of both benefits and costs of pets to children. *Anthrozoös*, **3**, 253–61.

Cain, A.O. (1983). A study of pets in the family system. In *New Perspectives on Our Lives with Companion Animals*, ed. A.H. Katcher & A.M. Beck, pp. 303–17. Philadelphia: University of Pennsylvania Press.

Cohen, S. & Wills, T.A. (1985). Stress, social support and the buffering hypothesis. *Psychological Bulletin*, **98**, 310–57.

Collis, G.M. & McNicholas, J. (1998). A theoretical basis for health benefits of pet ownership: attachment versus psychological support. In *Companion Animals in Human Health*, ed. C.C. Wilson & D.C. Turner, 105–22. London: Sage.

Duck, S. (1994). *Meaningful Relationships: Talking, Sense, and Relating.* Newbury Park, CA: Sage.

Fisher, J., Collis, G.M. & McNicholas, J. (1998). Pets considered as family members. Paper presented at the 8th International Conference on Human–Animal Interactions, Prague, September 10th–12th, 1998.

Furman, W. (1989). The development of children's social networks. In *Children's Social Networks and Social Support*, ed. D. Belle, pp. 151–72. New York: John Wiley & Sons.

Furman, W. & Burhmester, D. (1985). Children's perceptions of the personal relationships in their social networks. *Developmental Psychology*, **21**, 1016–24.

Garrity, T.F., Stallones, L., Marx, M.B. & Johnson, T. (1989). Pet ownership and attachment as supportive factors in the health of the elderly. *Anthrozoös*, **3**, 35–44.

Glaser, C.A., Angoulo, F.J. & Rooney, J.A. (1994). Animal-associated opportunistic infections among humans infected with the human immunodeficiency virus. *Clinical Infectious Diseases*, **18**, 14–24.

Hinde, R.A. (1988). Ethology and social psychology. In *Introduction to Social Psychology*, ed. M. Hewstone, W. Stroebe, J-P. Codol & G.M. Stephenson, pp. 20–38. Oxford: Blackwell.

Hinde, R.A. (1997). *Relationships: a Dialectical Perspective.* Hove, UK: Psychology Press.

Hirschman, E.C. (1994). Consumers and their animal companions. *Journal of Consumer Research*, **20**, 616–32.

Kennedy, J.S. (1992). *The New Anthropomorphism.* Cambridge: Cambridge University Press.

Kidd, A.H. & Kidd, R.M. (1994). Benefits and liabilities of pets for the homeless. *Psychological Reports*, **74**, 715–22.

Levinson, B.M. (1972). *Pets and Human Development.* Springfield, IL: Charles C Thomas.

Marsh, F.O. (1994). International pet food market. Paper presented at the Annual General Meeting of the Federations Europeenne l'Industrie des Aliments pour Animaux Familiers, June, 1994, reported in *Anthrozoös*, 1995, **8**, 55–7.

McNicholas, J. & Collis, G.M. (1998). Could Type A (coronary prone) personality explain the association between pet ownership and health? In *Companion Animals in Human Health*, ed. C.C. Wilson & D.C. Turner, pp. 173–85. London: Sage.

McNicholas, J., Collis, G.M., Morley, I.E. & Lane, D.R. (1993). Social communication through a companion animal: the dog as a social catalyst. In *Proceedings of the International Congress on Applied Ethology, Berlin 1993*, ed. M. Nichelmann, H.K. Wierenga & S. Braun, pp. 368–70. Berlin: Humboldt University.

Messent, P.R. (1983). Social facilitation of contact with other people by pet dogs. In *New Perspectives in Our Lives with Companion Animals*, ed. A.H. Katcher & A.M. Beck, pp. 37–46. Philadelphia: University of Pennsylvania Press.

Nagel, T. (1974). What is it like to be a bat? *Philosophical Review*, **83**, 435–50.

Patronek, G.J. & Rowan, A.N. (1995). Determining dog and cat numbers and population dynamics. *Anthrozoös*, **8**, 199–205.

Pedigree Petfoods (1996). *Pet Ownership Survey*, conducted by GfK. Marketing Ltd., London.

Plaut, M., Zimmerman, E.M. & Goldstein, R.A. (1996). Health hazards to humans associated with domestic pets. *Annual Review of Public Health*, **17**, 221–45.

Voith, V.L. (1985). Attachment of people to companion animals. *Veterinary Clinics of North America: Small Animal Practice*, **15**, 289–95.

Weiss, R. (1974). The provisions of social relationships. In *Doing unto Others*, ed. Z. Rubin, pp. 17–26. Engelwood Cliffs, NJ: Prentice-Hall.

Zasloff, R.L. (1996). Measuring attachment to companion animals: a dog is not a cat is not a bird. *Applied Animal Behaviour Science*, **47**, 43–8.

Zasloff, R.H. & Kidd, A.H. (1994). Loneliness and pet ownership among single women. *Psychological Reports*, **75**, 747–52.

Zwick, W.R. & Velicer, W. F. (1986). Comparison of five rules for determining the number of components to retain. *Psychological Bulletin*, **99**, 432–42.

MARIE-JOSÉ ENDERS-SLEGERS

13

The meaning of companion animals: qualitative analysis of the life histories of elderly cat and dog owners

INTRODUCTION

The human–animal bond has a long history. The first evidence of a probable domestic dog was found in a Palaeolithic grave at Oberkassel in Germany, dated at 14 000 years BP (before present). The first evidence of a special bond between humans and animals was found at the site of Ein Mallaha (Israel), where a grave from 12 000 years BP was found. In it, a human was buried together with a puppy (Clutton-Brock, 1995). From cave paintings, ancient pictures, statues, excavations and literature, we know that animals played diverse and important roles in the lives of our ancestors. They played various religious roles, were 'used' as hunters, guardians, or were just companions.

Today, the roles that domestic animals such as cats and dogs play are just as diverse and depend on the cultural contexts in which they find themselves. In the West they are considered as companion and 'working' animals (e.g. dogs assisting people with disabilities, guarding flocks of sheep, protecting homes from burglary, assisting police and customs officers). In some other cultures, dogs are considered as food, as 'impure' or as 'outcasts' (Serpell, 1995).

The pet as love object, friend, attachment figure, network member

In the last three decades, researchers from different disciplines have focused on the human–animal bond, accentuating interactions and attitudes of humans towards pets, or emphasizing the consequences of this relationship for the physical and/or the mental well-being of the owners. For example, Levinson (1969) argued from within a psycho-analytic framework that pets satisfy vital emotional needs among the elderly by providing a love object. Pets can serve as primitive defence mechanisms, as 'objects of identification', and as 'counterbalances' in situations in which the elderly suffer from the loss of relatives, friends and associates; where they become physically dependent on others; and in the process of accepting oneself as an elderly person. Furthermore, Levinson (1969) stated that pets could pave the way to new friendships and may give the elderly a reason for living. Rynearson (1978) observed that the bond between humans and pets is based on their communality as animals and their mutual need for attachment. He focused on 'pathological attachment relationships' in which relationships between humans and pets served a defensive purpose and disruption of the bond could create enduring psychiatric problems:

> Under normal circumstances they share complementary attachment because of mutual need and response. At times of stress they may temporarily seek out the other for attachment. Under abnormal circumstances of conditioned distrust the human may displace an over-determined need for regressed attachment to the pet. (Rynearson, 1978: 553)

Serpell (1989) discussed whether 'friendship' was an appropriate term for what humans and animals share in a relationship. Although a universal definition of friendship is hard to provide, he argued that human friendships share many features in common with certain social relationships in other species, particularly among non-human primates. One special feature was that, in captivity or under domestication, animals could form cross-species social relationships, which was also the case with companion animals and humans. In most cases, companion animals are socialized to their human owners during early development. Serpell (1989) points out that, as a consequence, the animals tend to behave in an infantile or subordinate way towards their human partners. As such, from the animals' viewpoint, 'friendship' would not be an appropriate description of this asymmetric relationship. From the perspective of the human participant, however, such relationships may qualify as friendships since humans derive certain social and emotional benefits from their relationships with companion animals which are similar to those derived from human friends. Peretti (1990) interviewed elderly people regarding their friendships with other humans and with their dogs. Five variables emerged as important: companionship, emotional bond, usefulness, loyalty and absence of negotiation. In this study, approximately 75% of the subjects mentioned that their dog was their only friend.

Zasloff and Kidd (1994a) explored the various aspects of attachment to feline companions, comparing the benefits of feline and human companionship. Affection and unconditional love from companion cats were mentioned as primary benefits of the human–cat relationship, whereas verbal communication was mentioned as the primary benefit of the human–human relationship. In their discussion, the authors emphasized that both relationships were important for the owners and that cats may serve as a complement to, but not as a replacement for, human contact.

Collis and McNicholas (1998) focused on the suggested association between pet ownership and advantages for human health. They explored extensively the nature and functions of the person–pet relationship. After thorough consideration, Collis and McNicholas (1998) rejected attachment theory, as defined by Bowlby (1969, 1979) and Ainsworth (1989), as a framework for understanding person–pet relationships. They suggested instead that exploring the social support functions of the human–animal bond would be more profitable. Enders-Slegers (1993, 1995) focused on companion animals as social support providers for the elderly, by serving as members of the social network. Reasoning from social support theories, she suggested that having a companion animal may enhance the physical and psychological well-being of the elderly.

Physical, psychological and social well-being

A number of studies have found companion animals to enhance physical well-being (see also Friedmann, Thomas & Eddy, Chapter 8). Friedmann *et al.* (1980) found that pet owners were more likely than non-owners to be alive one year after discharge from a coronary care unit. Baun *et al.* (1984) recorded blood pressures, heart rates and respiratory rates, in sessions in which the participants either petted an unknown or familiar dog, or read quietly. They found that petting a familiar dog significantly lowered both systolic and diastolic blood pressures in the participants. Akiyama, Holzman and Britz (1986) investigated the impact of pet ownership on recently widowed, urban, middle-class women, and found fewer health problems and less drug use among pet owners compared with non-owners. Allen *et al.* (1991) conducted an experiment in which female subjects had to carry out stressful mental arithmetic tasks in the presence of either a pet dog or a close human friend, or alone (control). Autonomic reactivity (skin conductance, blood pressure and pulse rate) was consistently greatest in the 'friend present' condition, and significantly lowest in the 'pet present' condition. Anderson, Reid and Jennings (1992) studied various cardiovascular risk factors in pet owners and non-owners, and concluded that pet ownership reduced risks for cardiovascular diseases. In their study, pet owners had lower systolic blood pressures, serum cholesterol and triglycerides compared with non-owners. Siegel (1990) studied stressful life events, the use of physician services among the elderly, and the modifying role of pet ownership. She found that dog owners in particular were buffered from the impact of stressful life events.

With regard to the effect of pets on psychological well-being, Goldmeier (1986) found that pets reduced feelings of 'lonely dissatisfaction' among the elderly, but only in those living alone. Relationships between loneliness, pet ownership and attachment were also reported by Zasloff and Kidd (1994b) for female students. They found no differences between the loneliness scores of pet owners and non-owners. However, further analysis showed that women living entirely alone were significantly more lonely than those living with pets only, both pets and other people, or other people only (no pets). Garrity *et al.* (1989) found that pet ownership *per se* failed to predict less depression or illness behaviour in a group of elderly people. In a sub-group with particularly severe distress (the bereaved), strong attachment to a pet was significantly associated with less depression, but only when the bereaved person had minimal human support.

The enhancement of social and psychological well-being for the elderly in institutions by companion animals and/or animal-facilitated therapy programmes has been reported by Haggerty Davis (1988) and Manor (1991). Rogers, Hart and Boltz (1993) discussed the role of pet dogs in promoting casual conversations among elderly adults and suggested that the dog is a conversational catalyst for its elderly owner.

Finally, we should not overlook the consequences of the ending of human–animal relationships. The loss of a pet may have serious physical and emotional consequences for the owner. Quackenbush and Glickman (1984), for example, described the provision of social work services in a veterinary hospital to individuals who were distressed about a pet's illness or death, and stressed that disruption of the human–animal bond could have detrimental effects on the health of the owner. Archer and Winchester (1994) found that pet owners experienced intense grief following the death of a pet: this was particularly evident among people living alone. Since pet loss is often not considered a 'significant loss' by family members and friends, the usual sources of social support for recognized grief are not available to a bereaved pet owner (McNicholas & Collis, 1995). Evers (1996) described a clinical case of pathological mourning in a woman following the death of her pet. Positive change only occurred when the therapist applied a strategy based on the mutual acceptance of the emotional significance of her relationship with this animal.

As has just been discussed, the human–animal bond has been found to affect the physical and/or psychological health of pet owners. Processes that may play an important role in this, and may have a direct effect on physical well-being, include daily exercise (walking the dog) and the direct effect of the pet's presence (reduction in blood pressure and heart rate). However, these fail to explain the whole range of therapeutic effects reported. For this reason, researchers have now focused on other explanations, such as whether human–animal relationships offer the same sorts of support as human–human relationships.

A theoretical framework to describe the human–animal bond

The theory of 'social provisions' by Robert Weiss (1974) appears to provide a suitable framework for understanding the apparent links between pet ownership and human health. Although Weiss did not design his conceptual framework to identify social provisions furnished by human–pet relationships, the nature of these relationships as reported by Harker, Collis and McNicholas (Chapter 11), and Bonas, McNicholas and Collis (Chapter 12) justifies considering both relations as comparable.

Weiss (1974) studied human relationships and human resources and assumed that some requirements for psychological well-being could only be met through social relationships. He identified six relational provisions, each ordinarily associated with a particular type of human–human relationship, describing each provision and, in some cases, the affective consequences of its absence:

Attachment: provided by relationships from which participants gain a sense of emotional closeness and security; participants feel comfortable and at home. In its absence, individuals feel lonely and restless.

Social integration: provided by relationships in which participants share a sense of belonging to a group of people who share common interests or concerns. It involves shared interpretation of experience and companionship. An absence of social integration may make life seem dull and the individual may feel socially isolated.

Reassurance of worth: provided by relationships which attest to an individual's competence and skills in social roles.

Reliable alliance: provided by relationships in which one can expect continuing assistance, even in the absence of mutual affection. In its absence, individuals may feel vulnerable and abandoned.

Guidance: provided by trustworthy relationships that, in stressful situations, furnish emotional support, advice and information.

Opportunity for nurturance: provided by relationships in which the participant takes responsibility for the well-being of another person, from which a sense of being needed is developed.

Weiss (1988) divided social relationships into 'primary relationships' and 'secondary relationships'. Primary relationships are close, frequent, face to face, warm, accompanied by commitment, and tend to include one's family and close friends. Primary relationships are divided into relationships of attachment (those whose loss triggers grief) and relationships of community (those whose loss triggers distress and sadness but not severe and persisting grief). Secondary relationships are for the most part instrumental and of little emotional importance: examples would include relations at work, with fellow members of formal organizations, and superficial acquaintances. In general, particular provisions are required from specific relationships . Although some relationships deliver more than one provision, relationships tend to be 'specialized' in their provision. For example, the attachment provision will most often be provided by 'primary relationships', such as partners, close friends or family. As not all relationships provide all provisions, individuals must maintain different relationships to establish the conditions necessary for well-being. Weiss' work on social provisions has influenced research in the field of social support (see Vaux, 1988, for a review). However, the theoretical concepts of social support are not identical and cover a variety of acts or interpersonal transactions in social relationships, and all relate to the relevance and significance of human relationships.

House (1981), for example, distinguished emotional support (behaviour that provides trust and love), informational support (information and advice), instrumental support (material or actual help) and esteem support (information leading the person to believe that he or she is valued). Social support between humans is known to have overall positive effects on well-being and to act as a buffer against the negative effects of stress (Russel & Cutrona, 1991). With age, however, the need for social support increases, while the availability of social support may decline. Elderly people live in a social environment that tends to become smaller and smaller through the death of partners, friends, and age-group members. At the same time, ageing often results in the loss of mobility and financial resources. Both of these factors can influence the physical and psychological health of the elderly.

In this chapter, I report on the meaning elderly people ascribe to their current and previous relationships with pet animals. It is hypothesized, based on the analogy with human–human relations, that the human–companion animal bond enhances the quality of life of the elderly by offering social provisions such as attachment, social integration, reassurance of worth, reliable alliance, guidance and the opportunity for nurturance. Weiss' conceptual framework of social provisions was the starting point for this research. It is hoped the results of this research will provide a better understanding of human–animal relationships and suggest a broader application of social support theories.

METHODS

This research forms part of a longitudinal study of elderly people (70 years of age and above) in which both qualitative and quantitative data were collected. This chapter reports on the qualitative data only.

Recruitment

Participants were recruited in The Netherlands in 1994 with the aid of the 'snowball' method (Snijders, 1992) and via articles in magazines for elderly people and veterinarians. The 'snowball' was started in several areas in The Netherlands, in cities as well as in the country. By this method, people are first invited to participate in the study, and, after the first meeting, are asked to nominate friends or relatives (in the same age-group) who could possibly participate in the study as well. The researcher then contacts these people by telephone or letter and invites them to take part in the research. As this method did not bring in sufficient participants, it was necessary to advertise for more participants through magazine articles.

Participants

In total, 96 participants, mostly female (74%), aged between 70 and 81 years, were interviewed by the author in their homes. Of these, 60 were

Table 13.1. *Demographic variables of the 96 elderly participants in the study*

Demographic	Pet owner $n=60$	Non-owner $n=36$
Living alone	41	22
Living with partner	19	14
Religious	43	29
Not religious	17	7
Male	15	10
Female	45	26
Age 70–75	38	17
Age 76–81	22	19
Living independently	57	27
Living in homes for the aged	3	9
Country dwelling	34	15
Town dwelling	26	21
Dog owner	46[a]	
Cat owner	20[a]	
Education $<$ 12 years	35	21
Education $>$ 12 years	25	15

Note:
[a]Six participants owned cats *and* dogs.

current pet owners and only 14 (15%) had never had pets in their lives. Table 13.1 provides other demographic details of the participants.

Procedure

Interviews began with questions about the person's life history and, where applicable, pet ownership history. This was in order to obtain insights into the motives for, and functions of, pet ownership.

Specific questions were then asked about pet ownership:

What does your companion animal mean to you?
What interactions do you have with your companion animal(s)?
What arrangements do you have for the care of your pet in case you are ill or die?

These three questions will hereafter be referred to as 'meanings', 'interactions', and 'arrangements', respectively. They were analysed in order to determine if social provisions were being derived from pets. Subsequent questions were not specifically aimed at pet ownership. These questions concerned loneliness, life satisfaction, plans for the future, finances, safety, contacts and mobility, and were analysed separately; where they contribute to a better understanding of the human–animal bond, they are discussed. At the end of each interview, a very specific question was asked tentatively: Do you receive emotional support from your pet(s), and if so, how? This question was also analysed separately.

Each interview took approximately two to three hours and was audio-taped.

Analysis

The starting point was a classification satisfying the conditions of content relevancy, logical consistency and a reliable division of all information (Hutjes & van Buuren, 1992), theoretically guided by Weiss' conceptual model of social provisions. The interviews were transcribed verbatim and analysed by the KWALITAN computer program (Peters, Wester & Richardson, 1989). The text of the interview responses was broken down into major themes to which codes were assigned. These codes designated to which social provision the themes could be allocated. Agreement on how to determine themes and categorize them was reached through extensive discussions with peers.

All interviews were analysed. In some interviews several social provisions were found, in others only a few or none were located.

Social provisions in the human–pet relationship were defined as follows:

Attachment: where the respondent reported being attached to the pet, having an emotional bond with the animal, loving the animal, finding it agreeable to have a companion animal to caress or have on the lap, or that the animal made the respondent feel comfortable and 'at home'.

Social integration: if the respondent indicated being part of a group of cat or dog owners (for example walking the dogs together), feeling part of society due to the pet, or that outside contacts were easier to make because of the pet.

Reassurance of worth: if the respondent mentioned that life was worth living due to the companion animal, that the animal made activities meaningful, or that being a pet owner made the respondent feel competent and responsible.

Reliable alliance: if the respondent stated that he or she could always and under any circumstance count on the companion animal. There was a bond of trust; the animal was reliable and predictable.

Guidance, advice and information: where the participant stated feeling guided or informed by companion animal warnings (barking at doorbell, intruders).

Opportunity for nurturance: where the respondent stated feeling happy about being responsible for the care of the pet.

RESULTS

Using the Statistical Package for Social Sciences (SPSS) for Windows, the demographic characteristics of the different groups (owners and non-owners of companion animals; see Table 13.1) were analysed using chi-square for non-parametric distributions, and the Yates correction (continuity correction) (de Heus *et al.*, 1995). No relationships between pet ownership and having a partner, being religious, gender, type of residential dwelling or type of education were found. As pet ownership is not permitted in most of the residential homes for the elderly, not surprisingly, a significant relationship between pet ownership and living circumstances (living in homes for the elderly or living independently) was found.

Role of the pet in the lives of the elderly

Initially, participants were asked about the roles companion animals played in their lives. The reasons for not having a pet any more ranged from: not being allowed to have a pet (in homes for the elderly), no longer wanting obligations, not being able to endure the loss of another pet, or not wanting to leave a pet behind in case of illness or death, lack of money, fear of injury, and that it created extra work. The 14 participants who had never owned pets felt either indifferent to them or did not like them.

The motives for having a pet fell into five main categories: 'love for the pet', 'a need for companionship and/or friendship', 'unable to live without a pet', 'they are an antidote to loneliness', and 'the pet was a gift from the children'. All but one pet owner stated that the pet played an important role in their present lives; most of the time they were very happy having a pet. Sometimes, however, pet ownership was felt as a burden: for example in cases of illness and holidays. In general, the roles

Table 13.2. *Social provisions as identified from the questions 'meanings', 'interactions', and 'arrangements'*

Social provisions as derived from	Meanings $n=82$	Interactions $n=58$	Arrangements $n=40$	Total number of statements
Attachment	79	58	6	632
Social integration	16	5	1	28
Reassurance of worth	29	15	4	75
Reliable alliance	3	0	0	3
Guidance	4	5	1	11
Opportunity for nurturance	46	17	39	185

Note:
n = number of participants who answered the question.

of the pet (friend, companion, playmate, confidant) were not reported as having changed during their lifetime. However, there were periods in the person's life when the pet was viewed as being more important than in other periods. For example, in adolescence, young adulthood (student or worker), when married but without children, or just after retirement, pet ownership was sometimes difficult to establish or of lesser importance in comparison with other interests of the time. The aspect of pet ownership that was said to change during one's lifetime was the responsibility for the care of the pet. In childhood, most respondents left the responsibility and care of the pet to their parents. When they became adults, they took on the role as animal caretakers.

Social provisions from companion animals

Analysis of the three questions on 'meanings', 'interactions' and 'arrangements' produced the following results.

Attachment, emotional closeness

> Attachment: provided by relationships from which participants gain a sense of emotional closeness and security; participants feel comfortable and at home. (Weiss, 1974)

The findings (see Table 13.2) indicated that 'attachment, emotional closeness' was the most important social provision provided from a relationship with a companion animal. Pet owners reported emotions such as 'love, affection, and closeness to the pet' as well as experiences of affection from their pet: they stated that this made them feel comfortable. Pet owners talked about 'attached behaviour' such as physical contact with the pet (stroking the pet, having it on their lap, talking to the pet, sleeping with the pet in the same bed), which made them feel at ease. At the same time they experienced the 'attached' behaviour of their pets: the pet seeking affectionate contact with the owner. Responses indicating 'attachment, emotional closeness' of pet owners included the ideas that the pet owner could not live without the pet, and that the pet was considered a child or family member.

The most common expressions of 'attachment' included emotional dimensions such as: 'He is my (close) companion', 'We love him as our child', 'I love him', 'I am very attached to him', 'He is my dear friend', 'Feel lonely without him'; behavioural expressions of attachment such as: 'I talk all the time to my pet', ' I am fond of stroking her on my lap', ' We always go out together', and more abstract concepts such as: 'I could not do without him', 'He belongs to me, to the family'. These statements were mostly accompanied by demonstrations of love and affection for the animal during the interview (talking to, hugging, kissing, petting and feeding cookies to the cat or dog). The 'attachment' provision was quoted by nearly all participant pet owners and former pet owners (Fig. 13.1). The data suggest a strong emotional bond between the companion animal and the owner.

Opportunity for nurturance

Opportunity for nurturance: provided by relationships in which the participant takes responsibility for the well-being of another person, from which a sense of being needed is developed. (Weiss, 1974)

The second most common social provision was the 'opportunity for nurturance'. The elderly, most of whom lived alone, described being happy to take care of their pet, and feeling needed, responsible and valued. The negative aspects of caregiving were also acknowledged, for example feeling obligated to be at home in time to feed and/or walk the animal.

Typical statements included: 'I like to take care of my pet', 'I am happy to have someone to take care of', 'I wanted to continue caring for somebody', 'Caring gives me the feeling of being useful', 'I feel responsible for his well-being'.

Fig. 13.1. 'Attachment' was the most important social provision derived from pets by the elderly participants in the study.

Reassurance of worth

Reassurance of worth: provided by relationships which attest to an individual's competence and skills in social roles. (Weiss, 1974)

There is a link between this provision and 'opportunity for nurturance'. Being responsible for the well-being of another living being enhances feelings of self-worth and self-esteem. Furthermore, the role as 'caretaker' is preserved, as many other roles disappear at this age. The participants expressed this as follows: 'So distressing when no one needs you; but my dog could not survive without me, he needs me', 'It gives me a reason to live for, it makes me feel good'.

Social integration

> Social integration: provided by relationships in which participants share a sense of belonging to a group of people who share common interests or concerns. (Weiss, 1974)

I realize that the concept of this provision is different in the case of human–animal relationships. It seems unrealistic that in this relationship between species, humans and pets share or have similar interests or concerns. However, pets can function as 'social catalysts', by helping the owner make contact with people and providing security for the owner when out in the community. Some statements were: 'Every day we are walking our dogs together', 'He (the pet) makes me feel part of society', 'It made making contact with other people easy; now we do all kinds of things together'.

Reliable alliance

> Reliable alliance: provided by relationships in which one can expect continuing assistance. (Weiss, 1974)

From the interviews, no resources from the human–animal bond were recognized that clearly fitted this provision. Three pet owners insisted that whatever happened, they could count on their companion animal; the animal would help them through difficult situations. It is doubtful, however, if these statements have the same meaning as Weiss' concept of this provision.

Guidance

> Guidance: provided by trustworthy relationships that, in stressful situations, furnish emotional support, advice and information. (Weiss, 1974)

Although the relationship with the companion animal was on several occasions valued as 'trustworthy', it is not clear to what degree the concepts of 'guidance' in a human–human and in a human–animal relationship are comparable. We scored this provision in 11 cases in which pet owners mentioned feeling 'guided', 'assisted' or 'informed', and being warned of intruders and suspicious visitors. These statements suggested that people were merely talking about a sense of safety derived from the presence of the pet. Another dimension of guidance is 'emotional support in times of stress'. Participants did not talk spontaneously about this issue, so it is not included in Table 13.2. However, when we asked

Table 13.3. *The types and frequencies of responses to the question 'Do you receive emotional support from your pet(s), and if so, how?'*[a]

Statements: emotional support	Number
By their presence	9
By their initiating approaches towards me	7
By their attention to me and noticing my emotions	11
By their sitting on my lap	4
By their demonstrations of affection towards me	2
By providing distractions	2
By providing opportunities to talk and to pet	9
By giving me a reason to go on with my life	5
No experience with the phenomenon	2
No support from the animal at all	4

Note:

[a]44 pet owners answered this question; there were 55 quotations.

about emotional support from the animal in our last question – 'Do you feel supported by your animal in times of stress and sorrow?' – this turned out to be a difficult question to answer for many pet owners. It was noticed that there was embarrassment in admitting that companion animals were emotionally supportive, as if the participants felt ashamed at not having a human source of support. So the question was put in another way: 'Some people report that their pet understands their feelings when they are upset or depressed. The pets behave in a special way to 'comfort' them. Do you recognize this?' The answers are in Table 13.3. [NB. All pet owners used personal pronouns to refer to their pet.]

Out of the 60 pet owners, only 44 answered this question. Sixteen participants did not respond because they considered the question inappropriate or purposeless. Two of these had no experience with the phenomenon of 'animal support' and four denied it could occur.

The responses provided an insight into how animals support the elderly in times of stress and sorrow. It seems that pets help people to cope with new situations by giving 'attention', by 'listening', through 'demonstrations of affection (comforting)', by 'their presence', by acting as a 'distraction', and by being 'meaningful', therefore acting somehow as a member of the social network.

Loneliness, life satisfaction, plans for the future, finances, safety, social contacts and mobility

In addition to social provisions identified in the statements above, analysis of data relating to the issues of loneliness, life satisfaction, plans for the future, finances, hobbies, safety, mobility and religious feelings was carried out. The results of this analysis were an important contribution to a better understanding of the human–animal relationship, and add to the notion of social provisions provided by pets. Talking about loneliness, for example, turned out to be a difficult issue to address. The elderly participants seemed to find it difficult to express feelings of loneliness. Most of them emphasized that they should not complain because many other elderly people were worse off. They stated that they tried to focus on the good things in their lives. However, 19 (out of 96) participants quoted feelings of loneliness. They mentioned living in isolation, without age-group members, without conversation partners, without feelings of being needed. Twenty participants never felt lonely. They stated: 'I never feel alone', 'I like being alone', 'I have a lot to do'. When asked how companion animals influenced feelings of loneliness, individuals stated that having a pet reduced these feelings.

As for safety, we asked about this in general and about feelings of safety connected to the presence of a companion animal. In relation to the latter, two different feelings of safety were mentioned: the feeling of belonging to someone who is devoted to you, 'emotional safety' (provision 'attachment'; $n=6$), and dog owners ($n=36$) mentioned feeling protected from burglary and other aggressive acts from the outside world (provision 'guidance').

The topic 'social contacts' provided information about the contacts that pet owners acquired from walking the dog and looking for the cat. Forty-two participants reported having gained new friends through their companion animals (provision 'social integration'). Another important finding was that companion animals regulated the days of their owners. Two-thirds of the total number of pet owners said that their days were structured by the needs of the animal (food, walk, play, care), the other third had a more flexible daily schedule and stated that the pet had to adjust to them. Pets were said to contribute to life satisfaction, and moreover, the pleasure gained by their companion animals was often mentioned. Observing the actions of a cat or dog made participants smile, have fun, gave them stories to tell to family and friends, and distracted them from physical as well as from psychological distress. Other items that were mentioned were the contact with nature that is maintained, and the physical fitness (mobility) that is stimulated by the companion animal. The topic 'finances' revealed that for some pet owners (those who only have a small pension), pet ownership was a financial burden that was usually borne with pleasure.

DISCUSSION

The results suggest that human–animal relationships and human–human relationships have features in common, and can be analysed in a similar way. Social provisions from Weiss' conceptual framework can be identified in the bond between the elderly and their pets. The most important social provision is 'attachment, emotional closeness', followed by the 'opportunity for nurturance'. Feelings of being useful and being needed indicated that the relationship provision 'reassurance of worth' was also obtained. Furthermore, the pets added indirectly to the provision of 'social integration' (making new friends and acquaintances) and directly to 'the feeling of being guided' in the sense of being 'guarded' and even 'emotionally supported'. The provision 'reliable alliance' seems not to be provided, or at least not for most pet owners. In addition, pets offered pleasure, added to life satisfaction, helped structure daily routines, reduced or prevented feelings of loneliness, provided contact with nature and physical exercise.

Weiss (1988) emphasizes that particular relationships are associated with particular provisions, and although some relationships deliver more than one provision, relationships tend to be specialized. In the case of the human–animal bond, as analysed in this study, it seems that the provisions derive from a primary relationship (such as partners, close friends or family) specialized in providing 'attachment and emotional closeness', 'reassurance of worth' and 'opportunity for nurturance'. Other researchers are supporting this line of thought in describing pets as love objects (Levinson, 1969), attachment figures (Rynearson, 1978; Zasloff & Kidd, 1994a), and friends (Serpell, 1989; Peretti, 1990).

Weiss divides primary relationships into relationships of attachment (those whose loss triggers grief) and relationships of community (those whose loss triggers distress). Focusing on pet loss, further evidence for the human–animal bond being a primary relationship can be found in the studies of Levinson (1969), Rynearson (1978), Archer and Winchester (1994), Quackenbush and Glickman (1984), McNicholas and Collis (1995), and Evers (1996), where grief, pathological mourning and distress after pet loss are described. Grief and distress after pet loss were also mentioned by our participants: statements that an individual could not bear to go through the loss of a pet again (a reason given for not owning a pet) and the stories of pet owners about grief and distress over pet loss in the past. Whether the human–companion animal bond is an attachment relationship or a relationship of community may depend on the individual. This question is beyond the scope of the present study but is one which requires further investigation.

In this study, pets were said to support owners in times of stress and sorrow. It is important to emphasize here that most pet owners in this study lived without human partners. This fact may have influenced the results. However, our findings correspond with the results of the aforementioned studies of Akiyama *et al.* (1986), Garrity *et al.* (1989) and Siegel (1990).

Moreover, the fact that most pet owners reported that pets made them adhere to a strict schedule each day, that they enhanced their mobility, as well as providing relaxation (having fun with the pet, tactile comfort), may add to the physical well-being of the elderly. This would be in line with the research of Friedmann *et al.* (1980), Baun *et al.* (1984), Allen *et al.* (1991), and Anderson *et al.* (1992).

Clearly, some of the data presented here are based on retrospective material (history of pet ownership) which may bias some of the results. Some characteristics of the participants, who survived two world wars and were socialized in a different way from cohorts after the wars, may also limit the generalizability of the results. As was learned from our interviews, this cohort of 'survivors' apparently learned not to complain, to be content with whatever life offered, and not to express emotions unnecessarily. This probably has produced socially desirable answers which may be reflected in topics such as 'loneliness', 'emotional support by the companion animal', 'safety' and so on. More investigations into different age groups should be carried out.

CONCLUSION

It can be seen from the study of the significance of companion animals as perceived by the elderly, using a conceptual framework derived from human relationships, that the relationship with a companion animal offers some social provisions similar to those of interpersonal relationships. This supports an extension of the social support construct, adding companion animals to the human network of social support providers.

ACKNOWLEDGEMENTS

I would like to express my appreciation to Professor Dr M.J.M. van Son, Dr L. Woertman, Dr M. Stroebe and several anonymous reviewers for their comments on a previous draft of this chapter. Thanks also to my students who assisted in typing out the data and in analysing the interviews, as well as to R. Renes for his help with the English language. Finally, I would like to thank the editors of this book, especially Dr A. Podberscek for his patience and generous support.

REFERENCES

Ainsworth, M.D.S. (1989). Attachments beyond infancy. *American Psychologist*, **44**, 709–16

Akiyama, H., Holzman, J.M. & Britz, W.E. (1986). Pet ownership and health status during bereavement. *Omega*, **17**, 187–93.

Allen, K.M., Blascovich, J., Tomaka, J. & Kelsey, R.M. (1991). Presence of human friends and pet dogs as moderators of autonomic responses to stress in women. *Journal of Personality and Social Psychology*, **61**, 582–9.

Anderson, W.P., Reid, C.M. & Jennings, G.L. (1992). Pet ownership and risk factors for cardiovascular disease. *Medical Journal of Australia*, **157**, 298–301.

Archer, J. & Winchester, G. (1994). Bereavement following death of a pet. *British Journal of Psychology*, **85**, 259–71.

Baun, M.M., Bergstrom, N., Langston, N.F. & Thoma, L. (1984). Psychological effects of human/companion animal bonding. *Nursing Research*, **33**, 126–9.

Bowlby, J. (1969). *Attachment*. Harmondsworth: Penguin Books.

Bowlby, J. (1979). *The Making and Breaking of Affectional Bonds*. London: Tavistock Publications.

Clutton-Brock, J. (1995). Origins of the dog: domestication and early history. In *The Domestic Dog: its Evolution, Behaviour, and Interactions with People*, ed. J. Serpell, pp. 7–20. Cambridge: Cambridge University Press.

Collis, G.M. & McNicholas, J. (1998). A theoretical basis for health benefits of pet ownership: attachment versus psychological support. In *Companion Animals in Human Health*, ed. C.C. Wilson & D.C. Turner, pp. 105–22. London: Sage.

de Heus, P., van de Leeden, R. & Gazendam, B. (1995). *Toegepaste Data-analyse*. Utrecht: Lemma.

Enders-Slegers, M-J. (1993). Investigation of the meaning for the elderly of a relationship with a companion animal. In *Science and the Human–Animal Relationship*, ed. E.K. Hicks, pp. 229–35. Amsterdam: SISWO.

Enders-Slegers, M-J. (1995). Do companion animals enhance the quality of life for elderly people? Paper presented at the British Psychological Society (BPS) Symposium on Theoretical and Practical Implications of Person–Pet Relationships, Coventry, UK.

Evers, R. (1996). Berichten uit de dierenhemel; over het overlijden van huisdieren. *Directieve Therapie*, **16**, 64–75.

Friedmann, E., Katcher, A.H., Lynch, J.J. & Thomas, S.A. (1980). Animal companionship and one-year survival of patients after discharge from a coronary care-unit. *Public Health Reports*, **95**, 307–12.

Garrity, T.F., Stallones, L., Marx, M.B. & Johnson, T.P. (1989). Pet ownership and attachment as supportive factors in the health of the elderly. *Anthrozoös*, **3**, 35–44.
Goldmeier, J. (1986). Pets or people: another research note. *The Gerontologist*, **26**, 203–6.
Haggerty Davis, J. (1988). Animal-facilitated therapy in stress mediation. *Holistic Nursing Practice*, **2**, 75–83.
House, J.S. (1981). *Work, Stress and Social Support*. Reading, MA: Addison-Wesley.
Hutjes, J.M. & van Buuren, J.A. (1992). *De Gevalsstudie: Strategie van Kwalitatief Onderzoek*. Meppel, Amsterdam, Heerlen: Boom.
Levinson, B.M. (1969). Pets and old age. *Mental Hygiene*, **53**, 364–8.
Manor, W. (1991). Alzheimer's patients and their caregivers: the role of the human animal bond. *Holistic Nursing Practice*, **5**, 32–7.
McNicholas, J. & Collis, G.M. (1995). The end of a relationship: coping with pet loss. In *The Waltham Book of Human–Animal Interaction: Benefits and Responsibilities of Pet Ownership*, ed. I. Robinson, pp. 127–43. Oxford: Pergamon.
Peretti, P.O. (1990). Elderly–animal friendship bonds. *Social Behaviour and Personality*, **18**, 151–6.
Peters, V., Wester, F. & Richardson, R. (1989). Kwalitatieve analyse in de praktijk en Handleiding bij Kwalitan, versie 2. Nijmegen: Vakgroep Methoden F.S.W. Katholieke Universiteit.
Quackenbush, J.E. & Glickman, L. (1984). Helping people adjust to the death of a pet. *Health and Social Work*, **9**, 42–8.
Rogers, J.W., Hart, L.A. & Boltz, R.P. (1993). The role of pet dogs in casual conversations of elderly adults. *Journal of Social Psychology*, **133**, 256–77.
Russel, D.W. & Cutrona, C.E. (1991). Social support, stress and depressive symptoms among the elderly: test of a process model. *Psychology and Aging*, **6**, 190–201.
Rynearson, E.K. (1978). Humans and pets and attachment. *British Journal of Psychiatry*, **133**, 550–5.
Serpell, J. (1989). Humans, animals, and the limits of friendship. In *The Dialectics of Friendship*, ed. R. Porter & S. Tomaselli, pp. 111–29. London: Routledge.
Serpell, J. (1995). From paragon to pariah: some reflections on human attitudes to dogs. In *The Domestic Dog: its Evolution, Behaviour, and Interactions with People*, ed. J. Serpell, pp. 245–56. Cambridge: Cambridge University Press.
Siegel, J.M. (1990). Stressful life events and use of physician services among the elderly: the moderating role of pet ownership. *Journal of Personality and Social Psychology*, **58**, 1081–6.
Snijders, T.A.B. (1992). Estimation on the basis of snowball samples: how to weight. *Bulletin de Methodologie Sociologique*, **36**, 59–70.
Vaux, A. (1988). *Social Support: Theory, Research, and Intervention*. New York: Praeger.
Weiss, R.S. (1974). The provisions of social relationships. In *Doing unto Others*, ed. Z. Rubin, pp. 17–26. Englewood Cliffs, NJ: Prentice Hall.
Weiss, R.S. (1988). Loss and recovery. *Journal of Social Issues*, **44**, 37–52.
Zasloff, R.L. & Kidd, A.H. (1994a). Attachment to feline companions. *Psychological Reports*, **74**, 747–52.
Zasloff, R.L. & Kidd, A.H. (1994b). Loneliness and pet ownership among single women. *Psychological Reports*, **75**, 747–52.

DENNIS C. TURNER

14

Human–cat interactions: relationships with, and breed differences between, non-pedigree, Persian and Siamese cats

INTRODUCTION

In recent years the domestic cat population has increased, with cats outnumbering dogs in many countries. This increase in popularity is reflected in the large number of studies on their behaviour and interactions with humans, some of which are summarized below.

Background

Karsh (1983, 1984) examined the development of a kitten's first relationship with a human and concluded that the sensitive period of socialization for cats was from two to seven weeks of age. These studies have been substantiated by Turner *et al.* (1986), who also expanded on these findings by including a greater number of influential factors in their study: genes of the father (also examined by McCune, 1995); genetic and modificatory influence of the mother; presence or absence of the mother during encounters with humans; curiosity (level of exploratory behaviour shown) and general indicators of fear in the presence of unknown humans; the effect of stroking the kitten during a sensitive phase of socialization (see also Cook & Bradshaw, 1996); and the effect of feeding the kittens as a facilitator of first close contact (Karsh & Turner, 1988; Turner, 1988, 1995a).

The behaviour of adult cats in their first encounters within a standardized laboratory setting with people unknown to them was studied by Mertens and Turner (1988). Here, interaction patterns between cats and men, women, boys and girls were compared. It was found not only that ethological methods could be used to observe and quantify interspecific interactions, but also that individual personality types amongst the cats were one of the most significant factors influencing their behaviour with people. It was also discovered that there were differences between men, women, boys and girls in the ways in which they interacted with the cats. For example, the adults in the study room vocalized towards the cat earlier and for longer than did the children. Mertens (1991) substantiated these findings with her observations of human–cat interactions in private households.

These studies were followed by an analysis of the influence of family size and the housing conditions of cats on interspecific interactions in home settings. Significant differences between indoor and outdoor cats (e.g. indoor cats are generally more active and spend more time with their owners), single and multiple cat households (e.g. single cats are tolerated better by their owners), and different family sizes (e.g. the smaller the human family, the more social attention the cat gives each member), in both the observed behaviour as well as in the psychological assessments of the cats' personality traits and relationship quality were found (Turner & Stammbach-Geering, 1990; Turner, 1991). The authors were able to show that human–cat relationships are indeed two-way partnerships, with both parties adjusting their behaviour to that of their partners (Turner & Stammbach-Geering, 1990; Turner, 1995b).

The present study

Both the Siamese and Persian are considered to be amongst the oldest purebred cats in the world, as well as being the most extreme in behaviour and character (e.g. Wright & Walters, 1980). Yet, surprisingly, to date, there have not been any ethological studies comparing cat breeds. This chapter presents some of the results of a recent long-term study[1] which compared relationships between people (young vs. older) and Siamese, Persian (Longhair) and non-pedigree cats (Turner, 1995a). Ethological observations of the interactions between owners and their cats, owner-assessment of the character traits of their cats and qualities of their relationships with them are presented.

The rationale for selecting these breeds for observation and comparison with non-pedigree cats was that if behavioural differences between these breeds and their interactive behaviour with humans could not be established, it would be inappropriate to compare other, less extreme lines of cats in future studies.

METHODS

Cat owners were sought in Switzerland through newspaper and magazine articles, and television programmes.

Owner assessments

As in Turner and Stammbach-Geering (1990), the participants were asked to assess their cats' behavioral traits using a series of semantic differential-type rating scales along which owners were asked to mark the position of their 'actual' and their 'ideal' cats. Measurements (mm) were then made of the position of owners' 'actual' and 'ideal' marks relative to each other, and relative to the left-hand end of the scale. The differences between 'actual' and 'ideal' values for each trait were then compared. The traits assessed (33 in total) were the same as those used in Turner and Stammbach-Geering (1990), but also included 'vocalizations' and 'general level of activity'.

Ethological observations

Direct observations of the cats' behaviour and the human–cat interactions were also made in the same private households, but analysed independently, to account for the possibility that subjective ratings by the cat owners might be biased in favour of a particular breed.

Two research assistants visited each cat-owning household on three consecutive days and recorded all interactions observed between the cat (or cats) and the adult (or adults) on portable data recorders. An ethogram was used (see Turner, 1991) which defined 30 behavioural elements that the person, the cat or both could show. Inter-rater reliabilities of the observers were calculated and surpassed the standard 80% for each element analysed later. Analyses of covariance were conducted in which human and cat time present were used as covariables. Factors of interest, in particular cat breed and age of owner (older vs. younger adults), sex of owner, as well as household size (singles vs. couples) and cat housing condition (indoor vs. outdoor access) were the independent variables; all other factors known to influence human–cat interactions were balanced out.

Since previous research (Turner & Stammbach-Geering, 1990) has shown a positive correlation between the number of cats a person had owned (or known) in their lifetime and the degree of 'fussiness' they displayed concerning cat personality traits, the elderly people in this study were expected to be more particular about their cats' character traits than younger adults.

14.1. *Demographic details of the cats and their owners*

	Non-pedigree	Siamese	Persian
Sex of cats			
Male	23	7	16
Female	38	14	19
Housing of cats			
Indoors	17	11	29
Outdoors	44	10	6
Owners			
Singles	15	7	3
Couples	46	14	32
Age of owners			
Young adult	13	14	25
Older (>65 y)	48	7	10

RESULTS

In total, 139 cats and their owners participated in the study (Turner, 1995a). However, only 117 adult cats (> 1 year of age) are discussed in this chapter as this was the number of participants for which full data relating to the variables in question were available. There were 71 female and 46 male owners in this study; mean age 62 years (range: 20 to 85 years). The mean age of the 'younger' adults was 51.8 years, while the mean age of the 'older' adults was 71.4 years. Sixty-one of the participants owned non-pedigree cats, 21 had Siamese cats and 35 had Persian cats (Table 14.1).

Comparisons of trait ratings between non-pedigree, Siamese and Persian cats

Mann-Whitney *U* tests were used to compare the rating differences between the breeds. Table 14.2 lists the significant differences found between the ratings of non-pedigree and Siamese cats by their owners. The owners of Siamese cats rated their animals significantly higher on playfulness, curiosity, friendliness to strangers, proximity (i.e. more often near the person involved), number of vocalizations towards the owner, higher on affection to owner, and significantly lower on laziness.

Table 14.3 shows the results of the comparison of ratings between

Table 14.2. *Comparison of ratings between non-pedigree and Siamese cats*

Trait	Mean rank		
	Non-pedigree	Siamese	*p* value
Playfulness	37.3	53.6	≤ 0.01
Curiosity	37.6	52.9	≤ 0.01
Friendliness to strangers	37.5	53.0	≤ 0.01
Proximity	37.8	52.4	≤ 0.01
Directed vocalizations	38.0	51.6	≤ 0.05
Laziness (inactivity)	44.5	32.7	≤ 0.05
Affection to owner	38.9	49.1	≤ 0.05
Independence	44.0	34.3	≤ 0.1
Predictability	39.2	48.1	≤ 0.1
Enjoyment of physical contact	39.4	47.5	≤ 0.1

Note:

Mann-Whitney *U* tests, corrected for ties. $n = 61$ non-pedigree cats, 21 Siamese cats.

Table 14.3. *Comparison of ratings between non-pedigree and Persian cats*

Trait	Mean rank		
	Non-pedigree	Persian	*p* value
Affection to owner	42.0	59.9	≤ 0.001
Proximity	42.7	58.5	≤ 0.01
Use of cat toilet	44.1	56.2	≤ 0.01
Dietary specialization	43.3	57.5	≤ 0.01
Owner affection	44.0	56.3	≤ 0.05
Urine spraying	52.1	42.2	≤ 0.05
Directed vocalizations	44.7	55.2	≤ 0.05
Cleanliness	45.4	54.0	≤ 0.05
Predictability	44.8	54.9	≤ 0.05
Friendliness to strangers	45.0	54.7	≤ 0.05
Aggressiveness	51.8	42.7	≤ 0.1
Playfulness	51.4	43.5	≤ 0.1
Independence	51.3	43.6	≤ 0.1

Note:

Mann-Whitney *U* tests, corrected for ties. $n = 61$ non-pedigree cats, 35 Persian cats.

Table 14.4. *Comparison of ratings between non-pedigree and pedigree cats*

	Mean rank		
Trait	Non-pedigree	Pedigree	*p* value
Affection to owner	49.8	69.0	≤ 0.001
Proximity	49.5	69.3	≤ 0.001
Friendliness to strangers	51.5	67.2	≤ 0.01
Directed vocalizations	51.7	67.0	≤ 0.01
Dietary specialization	52.4	66.1	≤ 0.05
Use of cat toilet	53.9	64.6	≤ 0.05
Owner affection	53.3	65.2	≤ 0.05
Curiosity	52.9	65.6	≤ 0.05
Predictability	53.0	65.5	≤ 0.05
Urine spraying	63.4	54.3	≤ 0.05
Independence	64.3	53.2	≤ 0.05
Aggressiveness	63.7	53.8	≤ 0.1
Enjoyment of physical contact	54.6	63.8	≤ 0.1
Cleanliness	55.4	62.9	≤ 0.1

Note:
Mann-Whitney *U* tests, corrected for ties. $n = 61$ non-pedigree cats, 56 pedigree cats.

non-pedigree and Persian cats. Persian cat owners rated their animals higher on affection to the owner, proximity, use of the litter box, dietary fussiness, their own affection towards the cat, number of vocalizations directed towards the owner, general cleanliness, predictability and friendliness towards strangers.

The results of comparisons between non-pedigree cats and both purebred breeds pooled together are shown in Table 14.4 Ratings favoured the purebreds on all but one of the significant differences found, namely degree of dietary specialization. Purebreds were fussier eaters but otherwise were better behaved and more interested in their owners than the non-pedigree cats.

When one compares rating differences between the two pedigree breeds directly (Table 14.5), one finds only three significant differences, all of which had been hypothesized based on character descriptions in the popular literature. The Siamese cats were more playful, more active and presented a problem more often when it came to using the litter box than the Persian cats.

A final comparison tested the hypothesis that prior experience of keeping cats influences the degree of 'fussiness' about one's own cat. The amount of difference between the 'actual' and 'ideal' ratings of younger

Table 14.5. *Comparison of ratings between Siamese and Persian cats*

	Mean rank		
Trait	Siamese	Persian	*p* value
Playfulness	36.7	23.6	≤ 0.01
Laziness (inactivity)	22.5	32.1	≤ 0.05
Use of cat toilet	24.8	30.7	≤ 0.05
Curiosity	32.2	26.3	≤ 0.1

Note:
Mann-Whitney *U* tests, corrected for ties. $n = 21$ Siamese cats, 35 Persian cats.

Table 14.6. *Significant rating differences between actual and ideal values comparing younger and older adult owners*

Breed	Trait	Direction of difference between younger (Y) and older (O) adult owners
House cats	Nocturnality	Y > O
	Independence	Y > O
Siamese	No significant differences	
Persian	Enjoyment of physical contact	Y > O
All breeds pooled	Independence	Y > O

versus older adults was examined. Table 14.6 shows that greater differences were found between the actual and ideal ratings of younger adults than between those of retired people. In every case, the older people showed more tolerance or acceptance of the cats' behaviour, and the younger adults desired more conformity to their own human lifestyles. Considering the data from owners of the three breeds separately (Table 14.6), the younger adults wanted cats that were quiet during the night, less independent, and that would enjoy more physical contact. Over all breeds pooled, the older people accepted the independence of their cats better than the younger adults.

Ethological observations

The results of the direct observations of behaviour and interactions in the same private households, tested by analyses of covariance,

Fig. 14.1 Elderly people appear to be more tolerant of their cats and to enjoy 'closer' relationships with them than younger adults. © D.C. Turner, I.E.T.

corroborated those from the subjective ratings of cat traits and relationship qualities by the owners.

When the data from both men and women were considered together (Table 14.7), the cat breed comparisons yielded the following results. More total interaction time was noted in the relationships with Persian (a tendency) and Siamese (significantly) cats than with non-pedigree cats, as well as significantly more interaction time with Siamese than Persian cats. No significant differences were found between the breeds in the number of interactions per hour, but the mean duration of interactions was significantly higher for both pedigree breeds. Time spent in close proximity between cat and human (less than 1 m) was also significantly greater for both purebreds than non-pedigree cats, and

Table 14.7. *Significant effects in the analysis of covariance (men and women considered together)*

	Comparisons						
Dependent variables	P–Np	S–Np	S–P	Indoor–outdoor	Singles–couples	Younger–older	Intercept
Total interaction time (%)	9.4(*)	20.1***	10.8*	n.s.	11.8***	n.s.	33.4
No. interactions per hour	n.s.	n.s.	n.s	0.70**	0.42*	1.04***	2.57
Mean duration of interaction	4.12*	6.52***	n.s.	n.s.	2.16(*)	−3.16*	8.16
Close proximity (< 1 m)	10.5*	18.2***	7.7(*)	n.s.	10.0**	n.s.	25.8
Interaction from a distance	n.s.	n.s.	3.09*	2.47*	1.83(*)	2.77*	7.62
Play time	n.s.	1.32(*)	1.93**	1.54**	1.11*	n.s.	1.66
Petting time	n.s.	2.69***	1.64*	n.s.	1.71***	n.s.	3.10
Speaking to cat per hour	n.s.	n.s.	2.16(*)	2.91**	4.61***	4.52***	10.0

Notes:

P = Persian cats; Np = non-pedigree cats; S = Siamese cats. Levels of significance, *** = ≤ 0.001; ** = ≤ 0.01; * = ≤ 0.05; (*) = ≤ 0.1; n.s. = no significant difference. Intercept = model means, either in percentage of the time present or frequencies per hour, and percentage point or frequency differences between the independent variables compared.

tended to be greater with the Siamese than the Persians. Interaction at a distance (greater than 1 m), time spent playing, and petting time were also significantly greater for the Siamese than for the Persian cats and, for petting time, also significantly greater than for non-pedigree cats. Therefore, the subjective ratings of the cat owners regarding their purebred animals were also reflected in the observed behavioural interactions.

Turner's (1991) results, which indicated a significant effect of housing condition (indoors vs. outdoor access), were substantiated in the present study and examined in greater detail. Owners interacted with indoor cats significantly more often, and for longer when they were further apart (more than 1 m), and spoke more often to their cats (see Table 14.7) than with outdoor cats.

The comparison of people living alone with those living in couples supported a number of further hypotheses (Table 14.7). Single people spent significantly more time interacting with their cats, interacted significantly more often, and tended to interact longer each time with their cats than people with partners. The same trend is found for all other dependent variables, which is not surprising considering they are components of total interaction time.

Regarding the comparison of younger and older people, no difference was found in total interaction time (although a higher value for the older owners might have been expected), but two differences in the structure of those interactions were found. Younger adults interacted significantly more often with their cats, but when older people interacted, they did so for significantly longer. Younger people interacted significantly more from a distance, and spoke significantly more often to the cat than the elderly did (Table 14.7).

Since Mertens and Turner (1988) and Mertens (1991) have shown differences in interactive behaviour between men and women in previous studies, and both men and women lived in most of the households in the present study, their behaviour was compared by Wilcoxon matched-pairs signed-ranks tests (Table 14.8). Women spent significantly more time petting the cats than men did, spent a greater proportion of the time interacting from a distance, and spoke more often to the cat than men did. These results substantiate those of Mertens (1991). The present study, however, was the first one in which Turner's (1991) measure of relationship quality – willingness to comply with the partner's interactional wishes – could be applied. This was defined as a positive reaction to an approach or to a directed vocalization by the partner (i.e. an intent by the partner to interact). The cat's willingness to comply was found to be significantly higher towards women than men.

Table 14.8. *Comparison of women (W) and men (M) on various variables by Wilcoxon Matched-pairs Signed-ranks Test*

Variable	Result	p value
Petting time	W > M	< 0.05
Interaction from a distance (> 1 m)	W > M	< 0.001
Speaking to the cat (number of times per hour)	W > M	≤ 0.001
Cat's willingness to comply with partner's intention to interact	W > M	< 0.01

Given that differences were also found between men and women in the present study, further analyses involved interactional data between women and cats only. The data were adjusted for the cat's and woman's presence time within the house.

From Table 14.9 one notes that the results are similar to those presented in Table 14.7; however, the dependent variable 'vocalizations per hour' reached significance in the second test. The number of directed cat vocalizations per hour was significantly higher in the Siamese cats than in either the non-pedigree or Persian cats. This substantiates the well-known vocal behaviour of the Siamese breed. It would appear that Siamese cats use their vocalizations to initiate contact (considering just the women's interactions, no differences were found between speaking to cats of different breeds) and data on general contact initiation also support this.

Table 14.10 lists the absolute and relative figures for 'intents to interact' by the women and cats in these relationships, broken down by breed of cat. One notes that a higher percentage of the intents to interact come from the Siamese cats, allowing us to expect a higher total interaction time with these cats. These data were not statistically examined because they are cumulative values from all animals in all relationships examined and, therefore, not independent.

CONCLUSION

Not only were differences in the subjective ratings of cat behavioural and relationship traits found between the breeds (most of which favoured the purebred animals), these were by and large substantiated by the direct observational data. Further, ratings for the traits 'predictability' (higher in the purebreds) and 'independence' (lower) suggest that selective breeding (probably convergent selection) has made the original,

Table 14.9. *Significant effects in the analysis of covariance*

	Comparisons						
Dependent variables	P–Np	S–Np	S–P	Indoor–outdoor	Singles–couples	Younger–older	Intercept
Total interaction time (%)	n.s.	20.1***	11.7*	n.s.	12.3**	n.s.	31.7
No. interactions per hour	n.s.	n.s.	n.s	0.51*	0.42(*)	1.11***	2.45
Mean duration of interaction	3.68*	6.42***	2.74(*)	n.s.	2.34*	−2.95*	8.02
Close proximity (< 1 m)	9.3*	18.9***	9.5(*)	n.s.	10.5**	n.s.	24.2
Interaction from a distance	n.s.	n.s.	n.s.	1.67(*)	1.80(*)	3.91***	7.43
Play time	n.s.	1.23(*)	1.90*	1.45**	1.21*	n.s.	1.42
Petting time	n.s.	3.01***	1.92*	n.s.	1.76**	n.s.	2.95
Speaking to cat per hour	n.s.	n.s.	n.s.	1.89(*)	4.81***	5.12***	9.19
Cat vocalizations per hour	n.s.	1.94*	2.38**	n.s.	n.s.	n.s.	3.11

Notes:

P = Persian cats; Np = non-pedigree cats; S = Siamese cats. Levels of significance, *** = ≤ 0.001; ** = ≤ 0.01; * = ≤ 0.05; (*) = ≤ 0.1; n.s. = no significant difference. Intercept = model means, either in percentage of the time present or frequencies per hour, and percentage point or frequency differences between the independent variables compared.

Only interactions with women are considered.

Table 14.10. *Absolute and relative frequencies of intentions to interact by women and cats broken down by cat breed*

	Intents by women	Intents by cats	Total number of intents
Non-pedigree cats	722 (61.9%)	444 (38.1%)	1166
Persian cats	497 (60.3%)	327 (39.7%)	824
Siamese cats	358 (53.2%)	315 (46.8%)	673

unpredictable, independent cat not only more sociable towards humans, but also more predictable and dependent in purebred forms. Cats appear to take on the role of a significant partner in relationships involving people living alone.

Contradictory to the prediction that elderly people would be more particular about their cats' character traits, over all breeds pooled and for non-pedigree cats, they accepted the independence of their cats better than the younger adults. The original hypothesis was based upon the positive correlation found in a previous study between the number of cats a person had owned and the number of traits showing significant differences between actual and ideal ratings. Perhaps the elderly people in the current study had (by chance) owned fewer cats over their lives, but this was, unfortunately, not assessed. Turner and Stammbach-Geering (1990) were able to demonstrate that accepting the cat as it is, is one of the keys to a harmonious human–cat relationship. Elderly people appear not only to be more tolerant of their cats (from the rating data), but also to enjoy 'closer' relationships with them (from the interactional data).

The results presented above also have implications for the selection of animals involved in specific situations such as in psychotherapy or social support programmes for people. A cat that 'assists' the therapist in his or her practice by being there, as a topic for conversation, need not be particularly active. A cat placed in a family as part of an intervention therapy will probably influence the social dynamics of the family differently, depending on its own activity levels, readiness to play, or vocal attributes and attributes of the family members. This is one reason why the research by Bradshaw and his co-workers in Southampton (Bradshaw & Limond, 1996; Cook & Bradshaw, 1996) on the stability of cat personality traits is so important. Different human problems, physical and emotional, will also require animals with different character traits. For example, less active older people and individuals with depressive tendencies might benefit more from individual cats or breeds that are more active in contact initiation.

Many important aspects of human–cat interactions and relationships remain to be discovered and it is hoped that the information presented in this chapter will stimulate others to consider these fascinating animals.

ACKNOWLEDGEMENTS

I thank the following research assistants and all of my graduate students for their help on various aspects of this long-term project: C. Mertens, K. Stammbach, J. Stocker, F. Widmer, and especially Daniel Zbinden. This study was financed by the Waltham Centre for Pet Nutrition in England, after initial work was supported by Effems AG, Zug, the Swiss National Science Foundation and the University of Zurich.

NOTE

1. First published in greater detail by Turner (1995a) in a German language book.

REFERENCES

Bradshaw, J.W.S. & Limond, J.A. (1996). Personal perceptions and affect towards household cats. Paper presented at the ISAZ '96 Conference 'The Animal Contract', Cambridge, 24–26th July, 1996.

Cook, S.E. & Bradshaw, J.W.S. (1996). Reliability and validity of a holding test to measure 'friendliness' in cats. Paper presented at the ISAZ '96 Conference 'The Animal Contract', Cambridge, 24–26th July, 1996.

Karsh, E.B. (1983). The effects of early handling on the development of social bonds between cats and people. In *New Perspectives on Our Lives with Companion Animals*, ed. A.H. Katcher & A.M. Beck, pp. 22–8. Philadelphia: University of Pennsylvania Press.

Karsh, E.B. (1984). Factors influencing the socialization of cats to people. In *The Pet Connection: its Influence on Our Health and Quality of Life*, ed. R.K. Anderson, B.L. Hart & L.A. Hart, pp. 207–15. Minneapolis: CENSHARE, University of Minnesota.

Karsh, E.B. & Turner, D.C. (1988). The human–cat relationship. In *The Domestic Cat: the Biology of its Behaviour*, ed. D.C. Turner & P. Bateson, pp. 159–77. Cambridge: Cambridge University Press.

McCune, S. (1995). The impact of paternity and early socialisation on the development of cats' behaviour to people and novel objects. *Applied Animal Behaviour Science*, **45**, 111–26.

Mertens, C. (1991). Human–cat interactions in the home setting. *Anthrozoös*, **4**, 214–31.

Mertens, C. & Turner, D.C. (1988). Experimental analysis of human–cat interactions during first encounters. *Anthrozoös*, **2**, 83–97.

Turner, D.C. (1988). Cat behaviour and the human/cat relationship. *Animalis familiaris*, **3**, 16–21.
Turner, D.C. (1991). The ethology of the human–cat relationship. *Swiss Archive for Veterinary Medicine* (SAT, in German), **133**, 63–70.
Turner, D.C. (1995a). *Die Mensch–Katze–Beziehung. Ethologische und psychologische Aspekte.* Stuttgart: Gustav Fischer Verlag (later, Enke Verlag).
Turner, D.C. (1995b). The human–cat relationship. In *The Waltham Book of Human–Animal Interaction: Benefits and Responsibilities of Pet Ownership*, ed. I. Robinson, pp. 87–97. Oxford: Pergamon Press.
Turner, D.C., Feaver, J., Mendl, M. & Bateson, P. (1986). Variations in domestic cat behaviour towards humans: a paternal effect. *Animal Behaviour*, **34**, 1890–2.
Turner, D.C. & Stammbach-Geering, K. (1990). Owner assessment and the ethology of human–cat relationships. In *Pets, Benefits and Practice*, ed. I. Burger, pp. 25–30. London: British Veterinary Association Publications.
Wright, M. & Walters, S. (1980). *The Book of the Cat*. London: Pan Books.

Part IV
Welfare and ethics

ARNOLD ARLUKE

15

Secondary victimization in companion animal abuse: the owner's perspective

INTRODUCTION

Researchers have documented the personal meaning of being criminally victimized. Fischer and Wertz (1979) were among the first to observe that criminal victims undergo a transformation in their lives as they are forced to develop new understandings and behaviours in the aftermath of crime. According to Young (1991), crime victims can pass through as many as three stages of adjustment to their trauma. Immediately after a crime it is common for victims to experience an 'acute crisis stage' involving shock and sometimes rage. This acute crisis is often followed by a second stage which Young calls the 'emotional effort to survive', involving grief, guilt and depression. Although the evidence suggests that the majority of victims of most types of crimes recover from the psychological effects of these two stages, Young claims that a substantial minority of crime victims enter a third stage of adjustment during which they experience lasting changes in their thinking, feeling and behaviour (Maguire & Corbett, 1987). This third stage, 'living after death', entails a lingering and heightened sense of vulnerability leading to increased vigilance, social cautiousness, and avoidance of others.

Less is known about the impact of crimes on those close to victims. It is speculated that there is a ripple effect of crime on these 'indirect' (Morgan, 1988) or 'secondary' victims (Knudten *et al.*, 1976), causing them some of the same distress as that experienced by primary victims. One can easily imagine the common thread running across all victimizations – the emotional upheaval that results from the shattering of assumptions we generally hold about ourselves and the world – being empathically shared by those close to the primary victims of crime (Janoff-Bulman, 1985).

Psychologists have provided support for the idea of secondary victimization. 'Vicarious traumatization' or 'secondary catastrophic stress response' has been observed among those close to victims of highly stressful but not necessarily criminal experiences (Figley, 1983). In the process of attending to the victimization of a family member, supportive family members themselves are touched emotionally, albeit indirectly (Figley, 1989), as are disaster workers by the victims they help (e.g. McCammon *et al.*, 1988) and mental health professionals by their patients (e.g. McCann & Pearlman, 1990).

Clinical reports of rape more directly support the notion of secondary victimization in criminal cases. Significant others of rape victims have been found to experience guilt, shame and anger as well as symptoms of post-traumatic stress disorder (Holmstrom & Burgess, 1979). In the only empirical studies of secondary victimization in rape cases, significant others experienced measurable distress after the assault (Veronen, Saunders & Resnick, 1989), although when compared, this distress was equally pronounced in cases of non-sexual assault (Davis, Taylor & Bench, 1995). There is also evidence that significant others experience similar psychological effects in cases of homicide (Amick-McMillan, Kilpatrick & Veronen, 1989) and robbery, assault and burglary (Friedman *et al.*, 1982).

This research asks if secondary victimization is experienced by companion animal owners when their animals are abused and, if it is, does their victimization entail a transformation similar to that observed among primary crime victims? Of course, it could be argued that owners in these cases are primary and not secondary victims in that companion animals are regarded as property under the law. However, such an approach misses the point of the present research: namely, that at least among the companion animal owners whose incidents are discussed below, relationships with their animals were far closer to person than property relationships. As such, companion animals become primary victims in such relationships, from the owners' perspective.

Researchers have not focused on the companion animal owners' experience in abuse cases, instead concentrating exclusively on the abusers of animals. For example, there have been studies of the psychopathology of animal abusers (e.g. Rigdon & Tapia, 1977), the epidemiology of their crimes (Vermeulen & Odendaal, 1993; Arluke & Luke, 1997; Donley & Patronek, 1997), and the link between their actions and violence towards humans (e.g. Kellert & Felthous, 1985; Ascione, 1998; Arluke *et al.*, 1999). This focus on abusers is reasonable because there is a need to understand and prevent their behaviour. Nevertheless, such a focus neglects others who might be affected by the abuse – sometimes intimately and at great emotional cost. More specifically, the human victims of animal abuse – animal owners – are nowhere in the research picture.

To generate insight into the social experience of secondary victimization in animal abuse cases, 18 owners of abused animals were interviewed to elicit their perceptions of the impact of animal abuse on their lives. Half of the owners had companion animals that were abused by individuals outside the household, such as strangers or neighbours (i.e. 'remote cases'), while the other half had companion animals that were abused by domestic partners (i.e. 'intimate cases'). Through convenience sampling, the former cases were obtained from a Massachusetts animal protection organization and the latter from a Massachusetts shelter for battered women. These two groups of victims were compared because they represent different ways that such abuse is manifested and handled. In remote cases, respondents took action and filed complaints of single, unpredictable incidents of companion animal abuse as opposed to intimate cases, where respondents did not take action and the companion animal abuse was chronic and known. Since the nature of and response to animal abuse by owners appeared to vary between these two groups, it was assumed that their experience as victims would be different.

Prior research suggests that crime victims given large-scale surveys or police-like interrogations are less likely to provide accurate information than when they are interviewed in depth about their experiences (Maguire, 1985; MacLeod, 1989). In line with this finding, a semi-structured, open-ended interview was used that allowed respondents to explore their experiences in a supportive and non-judgemental setting. Most interviews lasted approximately 45 minutes, had a high degree of co-operation, and examined the following general topics: discovering the abuse, confronting the abuser, dealing with law enforcement officials, and long-term coping with the abuse. All interviews were tape recorded, transcribed, and analysed for recurrent themes that characterized each group's perspective towards the abuse of companion animals. As described below, three major themes emerged from the analysis that were consistent with Young's stages of adjustment to victimization.

ACUTE CRISIS

Young (1991) found that after their victimization, targets of crime first experience an acute crisis, involving shock, disbelief and sometimes rage because no one expects to become a victim or to learn that a family member has been victimized. Moreover, the crime itself, as well as the injury or death of a loved one, are hard to understand or comprehend, as in the case of parents of kidnapped or killed children (Donnelly, 1982).

Shock and disbelief

Clearly, some degree of shock was also reported by owners in this study. For example, after one owner's companion animals were killed, she recalled: 'It was like – what do they call it – post-traumatic stress? I couldn't laugh, couldn't cry. When it really sunk in what happened, you know, they put me on an anti-depressant medication. I was totally shut down.'

However, it was far more common in remote rather than intimate cases for this initial stage to be delayed because owners were completely unaware that their animals had been intentionally harmed. If there were no witnesses to the incident or a precipitating event, some owners did not know why their animals had been injured or killed, leaving them 'in the dark' as to what happened. As one owner remembered: 'The dog came home and you know it was acting strange. And it just died right on the living room floor. I mean, we thought that he might have been hit by a car or something because he had just a little puncture wound in the back, and we didn t know what had happened until later.'

Sometimes it took days to locate missing animals and discover their abusers, an understandably anxiety-ridden period for owners. Although there was an end to the uncertainty surrounding their companion animal's disappearance, harm or death, the discovery process led some owners to a disturbingly graphic 'scene of the crime' where they saw their dead or severely injured companion animals. In one instance, for example, an owner became suspicious that something had happened to her 'outdoor' dog when it did not return home after two days, and began a neighbourhood search for him the next day. One neighbour who was approached for help was a local dog officer who denied seeing this dog, although he had killed it. The following day, armed with photographs of her companion animal, the owner encountered a young girl who recognized the dog and claimed that she saw the dog officer pick up this animal a few days earlier. The owner then returned to the dog officer's home and inspected his yard, fearing that he might have destroyed her companion animal as a 'stray'. In her words:

> I was really upset now because I had heard stories that Mr S. was killing dogs right away. I knew that he had a big open pit where he threw dead animals. We went through the woods to this pit. There were dead dogs in it and a couple of calves. There was snow on them, so I knew Igor was not in there because he had been home the day after it snowed. There was a big box on a wagon nearby. I went over to it and opened the door. The bottom of this box was covered with dogs. My Igor was one of them. I climbed into the box and took Igor out . . . the box was a home-made gas chamber.

Owners' initial confusion turned to shock and disbelief once they discovered that their animals had been deliberately harmed or killed. Upon discovery, they suddenly faced what seemed like an impossible situation to believe, especially because of the apparent senselessness and randomness of the abuse. Speaking about a neighbour who abused her dog, for instance, one owner said of him: 'He's kind of a degenerate that lives up the road and he gets into a lot of trouble. But we never had any problems with him. Don't know why he did it. He never complained about the dog – he never said anything. It could be that Blackie bothered his rabbits, but he never said anything and we never knew of anything that happened. So I'm not sure why he did it.'

In a few remote cases, discovery was particularly horrifying because owners directly observed the abuse of their companion animals, creating disturbing memories that lingered indefinitely. For example, one woman was routinely washing dishes and casually looked out of her kitchen window. On the other side of her fence, she saw a boy throw a rock at her dog that was leashed in the yard; the animal was struck, causing bleeding and seizures but not death. In several cases, these memories were particularly difficult for owners because they witnessed abuse of their children as well as of their animals. In one case, while sitting in her living room, a woman heard her neighbour say, 'Prince, go get him'. When she looked out the window, she saw his dog biting her kitten. As she yelled for him to stop, he called off his dog but retorted that her kitten had crawled into his yard. He then called for her 11-year-old son to approach the fence and as he did, the neighbour tossed the dead kitten over the fence, telling the young boy, 'Now take care of your kitten!'

In most of the intimate cases, the shock of the first stage of victimization was somewhat muted compared with that experienced by owners in remote cases. The former typically came upon the abuse as one further violent episode in what was often a long history of domestic violence. While clearly disturbed by what they discovered, owners in intimate cases usually saw their partner's abuse as part of a pattern of indiscriminate violence towards objects, people and animals. As one woman observed of her partner: 'He would destroy everything breakable'. She elaborated by saying that her partner was equally likely to throw an appliance out of the window as he was to hit her companion animal or herself. While many of the intimate owners claimed that their partners liked their companion animals or were indifferent to them, several owners knew that their partners had long histories of animal abuse. In one case, for instance, the owner believed that her partner disliked animals. In her words: 'He don't like cats and dogs. He don't like no animals, not even birds. Every time he sees a pigeon in the street, he tries to run it over'. And in another case, the owner recounted her partner's admitted history of animal cruelty:

> He said when he was a kid, you know, they used to blow up frogs and toads and things like that. And he hunts. He says, you don't have animals as pets. You kill them. He shoots squirrels and things like that. He had a great big shepherd that he liked because it was so aggressive, but he had no appreciation of animals having feelings or anything like that.

Rage

Not surprisingly, intense anger is commonly observed in the acute crisis stage because victims feel helpless and out of control. Acting on vague impulses to clear up what happened, punish abusers, or prevent such incidents from occurring again, most owners in remote cases wanted to confront abusers directly and immediately. These efforts failed most of the time; owners often felt frustrated and angry when they confronted alleged abusers because the latter typically denied harming the companion animals or blamed the incident on the animals themselves. As one owner noted about the abuser of her dog: 'The boy totally denied everything . . . blah, blah, blah, and so did his parents'. In one blatant case, an owner said that he ran to his front door after he heard his cat scream, only to see a man behind a bush. Seconds later he saw his cat motionless on the sidewalk in front of this man and his shepherd dog. When confronted by this owner, the man claimed that the cat had bothered his dog, who in turn bit and killed the cat. On closer inspection by the owner and a veterinarian, the cat was found to have three broken legs and internal injuries, but it had no saliva, bite wounds, or puncture marks that would indicate a dog attack. When prosecuted in court, the man was charged with beating the cat, but the case was dismissed by the judge.

Rather than after the abuse, owners in intimate cases often confronted their partners as the abuse was happening, although these confrontations did not stop the abuse from happening again. Like remote owners, they too felt rage at this time, in part because they felt helpless to stop the abuse. As one owner said: 'He would hit my cat. I would jump in and grab my kitten. My baby. That was my baby. That would make me very angry and I used to cry, like, don't do that to my cat cause I love cats'. Interestingly, many of these owners said when they were the targets of abuse they would not confront their partners. One subject said: 'If I saw him abusing the dog I would fight for the dog, whereas I wouldn't for myself, oddly enough'.

Complicating the rage experienced by owners in intimate cases was the frustration of confronting partners who denied the abuse, were forgiven, and then later repeated the abuse. This experience paralleled the cycle common in domestic violence cases, where co-dependencies with partners permit abuse to continue over time in a dance of accusation, denial, anger and forgiveness. In one case, for example, the owner felt trapped in a long-term relationship with an abusive partner who she intermittently blamed and forgave for the harm of her parrot. She recounted this experience with anguish:

> I had come home and my parrot was really sick. She was almost dead. The vet did an x-ray and there was a BB lodged in her wing. And he had BB guns in my house. He had shot her. I was like horrified, you know, nobody hurts my animals. You'd be better off hurting me. I couldn't believe it. I was furious, and then I came back to the house from the vet's office, and he had made me this nice dinner. I asked him, 'Do you have anything to tell me?' 'No.' I said, 'Well, how did she get a BB in her wing? You have BB guns here. You shot my bird.' 'No I didn't.' 'Yes, you did.' And finally, I was just like screaming and he said, 'OK, I'm leaving.' As he was grabbing all his stuff to run out the front door I just grabbed him . . . You know, when he was straight, he was this really nice person. I just kept wanting to help him. He finally admitted it. A few weeks later, I went to work and when I came home the bird's whole face was black and blue, and she was cut. So he hurt her even more. She didn't have that cut before. I was in a state of denial over it. I was afraid for me and I was afraid for my other animals. I wanted to get him out of here, but he kept saying, 'Give me another chance'. It was so hard.

There was further frustration when police were contacted by owners. Owners in intimate cases rarely sought police involvement, a finding consistent with an Humane Society of the United States (HSUS) (1996) survey claiming that 58% of those who personally witnessed animal abuse never reported it to authorities. However, owners in remote cases frequently called the police but were often disappointed with the response they received. Lack of interest in pursuing cases, or even inappropriate jokes or comments, made owners feel that the police were minimizing or invalidating the significance of what had transpired. As one owner summed up: 'We called the police right away, but they didn't seem to do much. I think as far as they were concerned, that was the end of it because it was a dog and not a human.' Although certainly not as extreme as the recriminations, social rejection and ostracism making rape victims reluctant to contact the police (Peterson, 1991), some owners claimed that their experiences would make them hesitant to go to the police should they encounter future cases of animal abuse. In general, owners were more satisfied with official investigation when they called the law enforcement department of the local animal protection society, typically doing this after their disappointing experiences with the police.

If their cases made it to court, almost all owners experienced additional frustration because alleged abusers were found not guilty or received minor punishments. For example, one owner angrily remarked, 'So my cat was killed and he [abuser] gets six months probation and a $36 fine. I was appalled. I felt everything should have been more severe. This sentencing stinks.' Another owner described the sentencing of her companion animal's abuser: 'All he got was probation. I really don't think he was punished enough. I wasn't happy with that at all.' Many owners wanted the abusers to think and feel differently towards animals. As one owner noted: 'I didn't want to kill him. I wanted him to suffer though, so he wouldn't think that animals have no feeling and that was an OK thing to do. I wanted him to know just how bad they felt and how terrified they must have been.' Similarly, in another case where the abuser only had to pay for veterinary bills and perform community service in the local animal shelter, the owner said: 'I think he should have spent a little time in jail, and let him sit there and think about what he did to a poor helpless animal.' Owners were particularly angry when the abusers of their companion animals were not found guilty. As one owner noted: 'I think whenever somebody gets away with something and you're the victim, it's kind of a shock that something like that can happen to you. And of course you can't fix it, but nothing else happens. It's just dropped and it's upsetting. When it hits you, of course it's upsetting.'

EMOTIONAL EFFORT

Young (1991) found that after their initial shock and anger, victims commonly experienced feelings of grief and guilt. This stage of 'emotional effort' involved working through the sadness and losses sustained in their victimization as well as the feelings of responsibility for what happened to victims. Owners in companion animal abuse cases exhibited similar responses.

Grief

When abusers killed companion animals, owners typically experienced a complicated grief reaction. Most owners felt that it was harder to mourn the loss of an abused companion animal than it was to mourn animals that died in more 'natural' ways. In one case, the owner compared the loss of one of his dogs that died after it was accidentally hit by a car to another one of his dogs that was deliberately run over and killed: 'With the greyhound, I believe that was accidental – that night it was pouring and you couldn't see in front of you. But it was devastating with Gunther. This fellow went over Gunther once, he backed up over him a second time, and then went over a third time to make sure it was done.' Continuing, the owner commented: 'It's bad enough when you lose a dog, but it's even more upsetting to lose a dog for no good reason. What do you expect if your animal is murdered? You hurt so much for so long.'

A number of owners recalled, with great emotional intensity, lingering images of how their animals died or were harmed. Compounding these memories were guilty feelings of not being there to care for animals while they were dying. As one owner summed up: 'It bothered me very much. I mean, it's sad to lose a pet through natural causes and this was just such a violent, violent thing. And, you know, I sort of resented the fact that I didn't get my last moments with Moe.' The struggle of some owners to make sense of such traumatic memories of abuse may have prevented them from dealing with issues of loss and resolving their grief.

A form of grief was reported even in cases where animals were not killed but had permanent physical or psychological damage. In these instances, owners grieved the loss of the animal's former being. Each time they observed their abused animals, owners claimed they felt a vague sense of grief. One owner claimed that 'I have to look at my dog every day with the bandaged leg. I change that every four days. Every time I take it off, she chews the leg. She'll not run like she used to. So I am confronted with that every day. I'll never forget it.'

Guilt

Interestingly, in almost all remote cases, grief was not compounded by guilt, even in cases where owners put their companion animals at some risk by allowing them to roam their neighbourhoods without supervision. Despite their possible culpability for the abuse, these owners felt that they could not have anticipated the abuse, so reports of subsequent guilt were rare.

However, many owners in intimate cases admitted feeling guilty because they did not prevent the abuse of their companion animals and because they felt that the abuse was really aimed at them. As one woman noted: 'I wish I had done something before he did it – like leaving.' Another woman sadly reported: 'It just made me sick. You know, I went and looked and there was blood on the place that my cat always slept. And all I could imagine was this poor trusting creature being slammed into the wall. It just made me, like how could I let that happen to my animals?' Perhaps feeding this sense of guilt, battered women commonly believed that their companion animals were abused because of them. That is, they believed that their partners were not angry with their animals *per se*, but with them; the animals were in the wrong place at the wrong time, and served as convenient surrogate targets. As one woman noted of her partner's behaviour: 'He was just going off at me in general and she [dog] was sitting here and, you know, she just got it. Just 'cause she was sitting there.' Similarly, another battered woman noted: 'John was my batterer. One day we got into a real, real big argument, and like the cat was right there on the couch. He just like picked it up and threw it against the wall.' Their partners' animal abuse was sometimes seen as a tool to 'get at' them. One woman, whose partner 'gave away' her dog without her consent and then told her the dog was 'lost', noted: 'He was using the dog to get at me, which he really achieved when he gave the dog away. He knew how much I loved him [the dog].'

Although owners in intimate cases reported guilty feelings for not preventing the abuse of their companion animals, the vast majority felt that they tried to protect their animals more than they did themselves. One woman said that she would not usually yell at her partner when he abused her, but that she would yell at him to stop his abuse of her companion animal. Unfortunately, when she did this, 'It would just change his focus, and then he would direct it [abuse] at me.' Another woman said that she held her companion animal in her arms to protect it from her partner's blows. And in yet another instance, the owner claimed that she took her companion animal with her whenever she left home so that it would not be alone with her partner.

Despite these reported attempts to protect them, companion animals were still often harmed. In several cases of intimate abuse, owners claimed that they felt even more guilty than they would have otherwise because friends blamed them for permitting the abuse. As one owner recounted: 'Well, one friend said to me "It was your fault, you let him into your house". And I haven't talked to her since, and its been two years. I said, "You are not a friend if you can say that to me." That made me furious. I wanted to kill her . . . You know, I was a victim. I wasn't the culprit.'

LIVING AFTER LOSS

According to Young (1991), crime victims usually become survivors. That is, after enduring trauma, people resume their everyday routines, although their experience of life is altered. They may have feelings of vulnerability not felt before as well as more 'positive' changes in their outlook and behaviour. This third stage of victimization was also evident among the owners in the present study.

Vulnerability

Both remote and intimate owners reported feeling vulnerable long after their animals were abused. In the former cases, some owners continued to be worried, months after making their official complaints, that abusers might retaliate against their property, animals, family or themselves. One owner even refused to be interviewed, fearing that the abuser – a prominent community member who allegedly killed her dog 20 years earlier – might learn about this and 'make life miserable' for her. She commented: 'I don't know how he might react in the future, and I have two cats.'

Over time, most owners managed this vulnerability by becoming vigilant about further abuse. They took measures to detect and prevent abuse of their companion animals as well as harm to their children or property. This was certainly true in remote cases of abuse. For instance, after a woman's cat was stoned to death, she never closed her venetian blinds to permit a constant, unobstructed view of children in the neighbourhood and never let her cats go outside if they were 'too friendly'. Another owner said that 'if I hear anything outside, I just run right out.' In another case, the owner said: 'From that point on, we pretty much keep the cats in the house. And I shortly thereafter hired an electrician to put flood lights out front and in the backyard.' Children were also protected by parents who feared that abusers would attack them too. After a woman's dog was badly stoned by a local boy, the owner said: 'For a while I was nervous about my son being out. I was afraid there was going to be some retaliation. I told him to stay away from him [abuser]. "Just don't get near him. If you see him coming, you go the other way, and you come right home".'

This vigilance was also evident in intimate cases of abuse. One owner, for instance, claimed that she became increasingly suspicious of her partner's actions after she discovered that he abused one her dogs. After her second companion animal was injured, she noted: 'One time I spent the night away from where we were living. My dog had been supposedly hit by a car, and I was thinking to myself, "now, could he have done that?".' After separating from abusive partners, intimate owners maintained that they became more suspicious of new partners' behaviour towards their companion animals. As one woman noted: 'I just split up with this guy – a pathological liar – who was insanely jealous of my daughter when she came to visit me for a week. So I can just imagine what he'd do to a pet. You know, I would definitely look at how my next partner would treat an animal.'

In a few cases, the vulnerability was generalized beyond the particular abuser in question. The experience of animal cruelty made some owners feel wary about other people, suspicious and distrustful of their intent, such that owners became more hesitant than usual to allow newcomers into their lives. An owner summarized this sentiment:

> It's made me very cautious. What it's done is to create boundaries. I won't let anyone near me. It's like I don't trust them first, like I used to. I used to just trust people first until they proved that they couldn't be trusted. Now they have to prove that they can be trusted before I let them get close to me or anything, or in my home. I've just changed a lot. There's no unsafe people in my life any more. I worked really hard to not let people get close to me until I really have a feeling that they are trustworthy. Even then, I'm just real suspicious.

Seeing good

As they coped over long periods of time with the abuse, owners in remote cases often sought ways to see something 'good' coming out of these untoward events, perhaps to lessen lingering discomfort about the senselessness of what happened. Most common in remote cases were attempts to prevent further incidents of abuse in their communities, sometimes going so far as to 'reform' the abuser of their animals. For instance, after one owner's dog was killed by a neighbourhood boy, the owner spoke to him several times about the importance of companion animals, insisted that the fine imposed in court be paid directly out of this boy's allowance, and intervened whenever she saw him fighting with other children. Most common were attempts to prevent further incidents of abuse in general. Some owners, for example, were eager to co-operate with humane society efforts to publicize their abuse cases, even though personal information would be revealed in the press. 'I was ready to do anything', noted one owner.

Similarly, owners in intimate cases claimed that the abuse of their animals led to some 'good'. Many of these owners felt that companion animal abuse was a trigger or catalyst for them to end their own abusive situations. A number of respondents described a particular incident of companion animal abuse as the 'last straw' that precipitated the end of an emotionally and/or physically damaging relationship with their partners. Some claimed that they left their own abusive situations only because they could no longer tolerate the abuse of their companion animals. As one woman remarked: 'He would sometimes kick Bushy, and he would yell at him too, but when he finally killed him, that caused the end of our marriage. That was it. I think it would have ended anyway, but I also know that the abuse of my dog was a huge factor. Especially what he did at the end.' Another woman had her husband committed to a psychiatric facility shortly after he deliberately stepped on her dog's leg and broke it, thereby ending several years of animal and human abuse in her household.

The belief that companion animal abuse facilitated the end of their own abusive situations perhaps indebted these owners to their animals and strengthened the bonds between them. They came to see their companion animals as co-travellers through difficult periods in which they both endured and supported each other – in some cases, even experiencing the same type of abuse. One woman noted: 'He had kicked the dog and he had kicked me a couple of times. Now, she's uncomfortable around men, and she's grown very close to me.' Some owners reported a new sense of 'appreciation' of their companion animals after the abuse occurred. As one interviewee noted:

> As a matter of fact, I enjoy them more. Instead of just having them and they are a job, I just totally enjoy them. I just appreciate them, and I'm really happy they are in my life. I don't let people tell me that it's not the right thing to do because people before were saying, 'You need to get rid of some of your animals, and you need to let more people in your life'. And I was doing that. And now it's like, no, this is my life, and it makes me happy, and this is what I do.

Further strengthening their bonds, some owners felt that their companion animals tried to protect them from abusive partners. One woman noted: 'He was nasty to the dog, and he tried to hit him. But the dog was very protective of me. When he was in one of his tempers, the dog kept an eye on me, so he (the dog) bit him. I think he would have been a lot more physical with the dog and, ah, maybe me, but I think he [partner] was afraid of him [dog] after that.' Another woman claimed: 'With him, it was either the cat or me. When he got like this [abusive], her hair would rise up and her fangs would come down. She was really mad and all her claws would pop out. She would like snap back at him.'

CONCLUSION

In general, the short-term and long-term responses of companion animal owners to animal abuse cases parallel the responses of victims of other crimes. Short-term emotional reactions, such as anger, shock and frustration, are common among victims and were clearly experienced by respondents in this study. Although many recover from this initial crisis stage, a substantial minority of victims report lasting changes in their attitudes and behaviour because basic assumptions are attacked about the meaning and predictability of life. This, too, was a recurrent theme in respondents' reports.

However, there were some differences when comparing the experiences of secondary victims in companion animal abuse cases with those of other victims of crimes. In the emotional aftermath of crime, victims are heavily dependent not only on their families but on the larger community (Young, 1991). With many crimes, the public is sympathetic and concerned for the victim, often sharing the victim's anger towards the offenders. Yet, among the respondents studied, there appeared to be little evidence of an outpouring of emotional support from outside immediate families. Future research should clarify whether this apparent indifference is a response to secondary victimization *per se*, because community members may perceive this to be less serious than primary victimization, or is a response to animals being involved in the victimization, because some community members may not regard such situations as serious or worthy of their support.

A second difference stemmed from respondents' reluctance to report the companion animal abuse to authorities and, if reported, their dissatisfaction with local police response. With many crimes, there is more willingness to report offenses to authorities and greater satisfaction with police investigation. In this regard, the experience of companion animal owners with law enforcement authorities is more consistent with the law enforcement experience of rape victims, for whom reports of police insensitivity and subsequent reluctance to report such crimes are common.

The results of this study need to be approached cautiously. All of the respondents in remote cases chose to report animal abuse to the authorities. There is no way to estimate the proportion of people who do not report such crimes and how their experiences as secondary victims might be different from the experiences of those who do. Similarly, all respondents in intimate cases chose to go to shelters, and there is no way to gauge the proportion who do not and how their secondary victimization might be different from that of those who do. A second limitation of this study is that respondents were predominantly white, female and working class. Researchers have documented that styles of coping with victimization vary according to the demographic background of victims (Walklate, 1989). Future research needs to consider this variation by examining owners of abused animals in more varied sociodemographic contexts than those studied here.

Despite these weaknesses, the research discussed in this chapter suggests a new social policy approach to combat companion animal abuse that focuses as much on the human as on the animal side of this issue. While concern for the welfare of companion animals, in and of itself, should be sufficient grounds for animal abuse cases to be taken seriously by law enforcement authorities, this has not been the case, as reflected in the adjudication records of abusers (Arluke & Luke, 1997). Although the reasons for this indifference are unclear (Arluke & Lockwood, 1997), the finding that animal cruelty can have a substantial emotional impact on the people who own these animals may encourage professionals to pay more attention to this issue than they now do. To the extent that professionals are unfamiliar with the human consequences of animal abuse, evidence such as that provided in this chapter must be disseminated to social workers, police, teachers, lawyers, judges and others whose work brings them into direct contact with animal abusers.

There are other policy implications of this research that address the human side of animal abuse. In some intimate cases of abuse, respondents claimed that they delayed going to shelters because, in their absence, domestic partners might further harm companion animals. Humane organizations need to develop 'foster' programmes that would house and care for companion animals until owners can safely return to their domestic situations. Clearly, interventions such as this would not only benefit owners. By widening public and professional awareness and concern to include the human victims in cases of animal abuse, companion animals themselves will surely gain too.

ACKNOWLEDGEMENTS

I wish to thank Carter Luke, Ken Shapiro and Tina Pereira for their valuable comments on earlier drafts of this work. Funding was generously provided by the President's Fund of the Massachusetts Society for Prevention of Cruelty to Animals and the Geraldine R. Dodge Foundation.

REFERENCES

Amick-McMillan, A., Kilpatrick, D. & Veronen, L. (1989). Family survivors of homicide victims: a behavioral analysis. *The Behavior Therapist*, **12**, 75–9.

Arluke, A., Levin, J., Luke, C. & Ascione, F. (1999). The relationship of animal abuse to violence and other forms of antisocial behavior. *Journal of Interpersonal Violence*, **14**, 963–75.

Arluke, A. & Lockwood, R. (1997). Understanding cruelty to animals. *Society and Animals*, **5**, 183–93.

Arluke, A. & Luke, C. (1997). Physical cruelty towards animals in Massachusetts, 1975–96. *Society and Animals*, **5**, 195–204.

Ascione, F. (1998). Battered women's reports of their partners' and their children's cruelty to animals. *Journal of Emotional Abuse*, **1**, 120–33.

Davis, R., Taylor, B. & Bench, S. (1995). Impact of sexual and nonsexual assault on secondary victims. *Violence and Victims*, **10**, 73–84.

Donley, L. & Patronek, G. (1997). The epidemiology of animal cruelty, abuse and neglect in Massachusetts. Unpublished report. North Grafton: Center for Animals and Public Policy, Tufts University School of Veterinary Medicine.

Donnelly, K. (1982). *Recovering from the Loss of a Child*. New York: Macmillan.

Figley, C. (1983). Catastrophes: an overview of family reactions. In *Stress and the Family*, Vol. 2. *Coping with Catastrophe*, ed. C. Figley & H. McCubbin, pp. 3–20. New York: Brunner/Mazel.

Figley, C. (1989). *Helping Traumatized Families*. San Francisco: Jossey-Bass.

Fischer, C. & Wertz, F. (1979). Empirical phenomenological analyses of being criminally victimized. In *Duquesne Studies in Phenomenological Psychology*, Vol. III, ed. A. Giogi, R. Knowle & D. Smith, pp. 135–58. Pittsburgh: Duquesne University Press.

Friedman, K., Bischoff, H., Davis, R. & Person, A. (1982). *Victims and Helpers: Reactions to Crime*. Washington, DC: US Government Printing Office.

Holmstrom, L. & Burgess, A. (1979). Rape: the husband's and boyfriend's initial reactions. *The Family Coordinator*, **28**, 321–6.

Humane Society of the United States (HSUS) (1996). Telephone survey. Denver, CO: Penn and Sehoen, December 21–3.

Janoff-Bulman, R. (1985). Criminal vs. non-criminal victimization: victims' reactions. *Victimology: An International Journal*, **10**, 498–511.

Kellert, S. & Felthous, A. (1985). Childhood cruelty toward animals among criminals and noncriminals. *Human Relations*, **38**, 1113–29.

Knudten, R., Meade, A., Knudten, M. & Doerner, W. (1976). *Victims and Witnesses: the Impact of Crime and Their Experience with the Criminal Justice System*. Washington, DC: US Government Printing Office.

MacLeod, M. (1989). Interviewing victims of crime. In *Crime and Its Victims: International Research and Public Policy Issues*, ed. E. Viano, pp. 93–100. New York: Hemisphere Publishing.

Maguire, M. (1985). Victims' needs and victim services: indications from research. *Victimology: An International Journal*, **10**, 539–59.

Maguire, M. & Corbett, C. (1987). *The Effects of Crime and the Work of Victim Support Schemes*. Aldershot: Gower.

McCammon, S., Durham, T., Allison, E. & Williamson, J. (1988). Emergency workers' cognitive appraisal and coping with traumatic events. *Journal of Traumatic Stress*, **1**, 353–72.

McCann, I. & Pearlman, L. (1990). Vicarious traumatization: a framework for understanding the psychological effects of working with victims. *Journal of Traumatic Stress*, **3**, 131–49.

Morgan, J. (1988). Children as victims. In *Victims of Crime: a New Deal*, ed. M. Maguire & J. Pointing, pp. 74–82. Milton Keynes: Open University Press.
Peterson, S. (1991). Victimology and blaming the victim: the case of rape. In *To Be a Victim: Encounters with Crime and Injustice*, ed. D. Sank & D. Caplan, pp. 171–8. New York: Plenum Press.
Rigdon, J. & Tapia, F. (1977). Children who are cruel to animals – a follow-up study. *Journal of Operational Psychiatry*, **8**, 27–36.
Vermeulen, H. & Odendaal, J. (1993). Proposed typology of companion animal abuse. *Anthrozoös*, **6**, 248–57.
Veronen, L., Saunders, B. & Resnick, H. (1989). Self-reported fears of rape victims: a preliminary investigation. *Behavior Modification*, **4**, 383–96.
Walklate, S. (1989). *Victimology: the Victim and the Criminal Justice Process*. London: Unwin Hyman.
Young, M. (1991). Survivors of crime. In *To Be a Victim: Encounters with Crime and Injustice*, ed. D. Sank & D. Caplan, pp. 27–42. New York: Plenum Press.

JOANNA SWABE

16

Veterinary dilemmas: ambiguity and ambivalence in human–animal interaction

INTRODUCTION

The relationship between humans and domesticated animals is replete with contradictions. It can be characterized by both intimacy and exploitation. Our conduct towards other animals is often seen in very black and white terms: some species we keep and use for food; upon others we lavish affection. In reality, however, the borders are rather more nebulous. While dogs are commonly kept and cherished as pets, they may also be severely maltreated and abandoned by their owners, or employed for sporting or scientific research. Similarly, an animal destined for the dinner plate may receive a great deal of respect, care and affection throughout its lifetime, and will not be treated at all as if it were merely an animate unit of production. We must therefore avoid making sweeping statements about the relationship between ourselves and other animals. One dog owner may dote upon his dachshund, another may beat his. As Arluke and Sanders (1996: 4) have pointed out, 'one of the most glaring consistencies' in our interactions with other animals 'is inconsistency'. It is largely for this reason that we should be cautious when referring to the small domesticated animals in our homes and gardens as 'companion animals'. Frequently, this descriptor will fit the bill, but not always.

The human–animal relationship can thus be seen as ambiguous and ambivalent. Both the manner in which we perceive animals and the way in which we treat them show the contradictory nature of this relationship. However, this chapter is not specifically concerned with unravelling the contradictions inherent in the human–animal relationship. Instead, it will explore these disparities in terms of how other animals are treated in everyday life. The ambiguity and ambivalence that typically characterize our relationship with domesticated animals are most clearly reflected in settings in which human–animal interactions play a central role. Although, in recent years, the relationship between humans and other animals has received increasing attention within the social sciences, only a few authors have explicitly drawn attention to the ambiguities that pervade everyday human–animal interactions. For example, Wieder (1980), Lynch (1988) and Arluke (1988) discuss ambivalence towards animals within experimental laboratory settings. Similarly, Sanders (1995a) highlights the ambivalent nature of guide dog trainers' relationships with their canine pupils. However, perhaps the best settings within which ambiguity and ambivalence in the human–animal relationship can routinely be observed are those involving veterinarians. Veterinary settings provide interesting fora within which one can observe both a wide range of human–animal interactions and the professional activities of the veterinarian, who is, in essence, a mediator between the human and animal world (Swabe, 1996).[1]

Veterinarians are engaged in a broad spectrum of activities involving animals, ranging from the treatment of domesticated animals, both large and small, to meat inspection, wild animal medicine, laboratory animal science, veterinary pharmaceutics and public health management. However, when one thinks of veterinarians, one generally conjures up images of Herriotesque animal doctors who rescue and care for sick and injured animals. To this we can perhaps partially owe the plethora of literature, films, documentaries and television series that have been produced highlighting and endearing the role of the veterinarian to wider society. The popular image of the kindly animal doctor has become firmly fixed in our collective imagination and has been repeatedly reinforced by the images that are routinely broadcast into our living rooms: the TV vet, be he real or fictional, is a familiar sight to all. More often than not, the television programme makers have tended to focus upon the work of small animal practitioners, rather than upon the veterinarians who take care of food-producing animals. There are several factors that may explain this apparent preference. First, the sight of small, furry, pet animals is a particularly appealing one to viewers, capable of eliciting the kind of 'cute response' (Serpell, 1996) that can make the programme popular and attain high viewing figures. Second, in our modern, urbanized, industrial society, people seem to be more readily able to identify and empathize with – and prepared to watch – the plights of both pet animals and their owners than those of livestock and farmers. The veterinary treatment of large animals and the environments in which – particularly intensively farmed – animals are kept may often appear distasteful and would confront audiences – often looking for light entertainment, rather than enlightenment – with the origins of their food.

Small animal practitioners are also the topic of this chapter, although, as will become clear, they do not and cannot always live up to either the romantic paragons of fiction or televised heroism. This chapter is based largely upon sociological research that I conducted in a variety of veterinary practices in both urban and rural settings in The Netherlands. By spending several months in the field as a participant observer, I was able to involve myself in the ongoing, daily world of the veterinarians whom I studied. This afforded me with an insight into the everyday work and social significance of the veterinary profession. Moreover, it provided me with the opportunity to converse with a wide range of veterinary professionals and animal owners, and to witness the nature of their interactions with animals and each other. This chapter will examine – from a sociological perspective – some of the routine transactions and procedures that were observed and discussed with veterinarians during my ethnographic study of veterinary practice. Furthermore, it will illustrate how the contradictions that pervade the human–animal relationship are clearly visible in veterinary situations, and will draw attention to some of the dilemmas with which veterinarians are routinely confronted, which may conflict with the interests of their animal patients. It will also suggest that the popular collective image of the small animal practitioner is perhaps unrealistic. However, before discussing the activities and dilemmas of the small animal practitioner any further, it is pertinent to situate the role of the modern animal doctor in its historical context.

A BRIEF HISTORY OF SMALL ANIMAL MEDICINE

As already suggested, the white-coated veterinarian expertly examining a pet cat or dog upon a surgical table is a familiar image today, even to those who have never owned a pet or visited a veterinary surgery. Yet the role of the veterinarian as the defender of canine and feline health is a very recent one. Historically speaking, the diseases and afflictions of dogs and cats have received very limited veterinary attention. Unlike horses and livestock, whose inherent economic and nutritional value has motivated human attempts to preserve their health and cure their disease, these small domesticates and their attendant diseases – with the notable exception of rabies – have posed little threat to the human economy or public health. In centuries past, dogs and cats simply did not warrant the therapeutic attentions of medical science; whilst useful creatures – for protection, pest control and companionship – they were essentially of little economic value and easily replaceable (Swabe, 1999: 167–8). Instead, early medical scientists tended to view small animal species –

particularly those of the canine variety – in quite a different light as the highly suitable subjects for the experimental study of anatomy and physiology; dogs were, after all, cheap, abundant and easy for the experimenters to control. Furthermore, small animals – or, rather, parts or by-products of them – provided useful ingredients for the medical pharmacopoeia of the past (Boor-van der Putten, 1986: 9–10).

Although medical science had a tendency to use small animals either as the objects of anatomical study or as the providers of potentially useful ingredients to cure human ailments, there is evidence that they sometimes received a degree of therapeutic attention and care, generally in accordance with their usefulness. In the agriculturally based society of ancient Rome, for example, shepherd and guard dogs played an important role in protecting humans and their livestock and property. Folk remedies for the prevention of ticks and fleas, and cures for the ulceration and scab that may accompany infestations, can be found in the works of renowned agricultural writers such as Columella (1745). Perhaps the most fertile source of historical information on the early care and diseases of dogs in Europe can be found in the mediaeval literature on hunting and hawking. During the Middle Ages, hunting and falconry enjoyed great popularity amongst the nobility of feudal Europe (see Menache, Chapter 4). The care of hunting hounds and birds became increasingly important and demanded the attentions of specialist animal attendants who could oversee the health and care of these valued creatures (Boor-van der Putten, 1986: 13).

From the sixteenth century onwards, a number of publications appeared that were devoted exclusively to the dog – e.g. Caius' *Of Englishe Dogges* (1570) and Paullinus' *Cynographia Curiosa* (1677) – but these were more concerned with describing the wide variety of breeds and their functions, rather than specifically addressing issues of canine health. It was, in fact, only in 1783 that the first popular work specifically devoted to the diseases of the dog appeared, written by the empiricist John Clater (Boor-van der Putten, 1986: 15). During the early nineteenth century, additional works dedicated exclusively to the veterinary treatment of canine disease were published. However, as Delabere Blaine – the self-styled patriarch of canine medicine – remarked in the preface to his pioneering work *Canine Pathology* (Blaine, 1817), devoting time and energy to this subject inevitably met with considerable social disapprobation. He lamented that 'my attention to the medical treatment of dogs subjected me to an imputation of want of common pride, and an utter disrespect for my former character and habits'. The prevailing view at that time was that animal medicine was not only greatly inferior to human medicine, but also that the horse was the *only* species of animal that was believed to deserve *any* veterinary medical attention whatsoever (Wilkinson, 1992; Swabe, 1999).

Yet, while the fledgling veterinary profession in general failed to acknowledge the importance of employing its skills and directing its attention to the diseases of dogs and other small animals, the increasing popularity of the practice of pet-keeping throughout nineteenth-century western European society revealed that there was evidently a growing market for professional advice on this very subject. Blaine clearly recognized this and carved himself a niche in the market as an animal doctor, author on canine health and purveyor of veterinary medicines, being joined in 1813 by William Youatt, who later became a pivotal figure in the development of British veterinary medicine. Their successful London practice provided the incentive for other veterinary surgeons to follow suit and open up specialist veterinary practices for dogs in other cities. Although the market for canine veterinary services gradually expanded within Britain, the question remained of whether veterinary school-educated veterinarians truly possessed the competence and skills necessary to treat dogs. Veterinary education continued to remain – as it had always done – firmly focused upon the treatment of horses and – to a much lesser extent – livestock species. In the majority of nineteenth-century textbooks that were produced to assist veterinary training, the dog and its complaints generally only received a cursory mention. By the mid to late nineteenth century, the market for specialist canine veterinary services also emerged within mainland Europe and North America. In cities such as Amsterdam and The Hague, for instance, urban veterinary practices were established to deal exclusively with dogs and horses (Boor-van der Putten, 1986: 17–25). Small animal medicine was, however, still very much in its infancy; it was only during the course of the twentieth century that it would come of age.

At the turn of the twentieth century, the attitude of the veterinary profession was still largely ambivalent towards the whole idea of studying and treating pet animals and their diseases. The increasing sentimentality towards animals at this time, particularly amongst the urban middle classes, seems to have been quite alien to most veterinary practitioners, who only saw profit, both in monetary and societal terms, in treating creatures of clear economic value. This attitude was also echoed throughout the veterinary schools of Europe that continued to regard the study and treatment of small animals with considerable disdain. The veterinary profession had after all striven hard throughout the nineteenth century to elevate itself far above the 'vulgar' level of gelders and blacksmiths and wished to be taken seriously as a scientifically enlightened and socially useful profession (Porter, 1993: 28–9). These educated veterinarians did not now wish to lower themselves by tending to, what they essentially regarded as, 'useless' animals (Offringa, 1983).

One of the chief consequences of this attitude was that, at least until the dawn of the twentieth century, there was comparatively little knowledge of, or concern for, the nature and pathology of small animal disease. Up until this time, veterinary attention to this subject seems to have been justified only when the study of small animal disease was seen as instrumental to increasing the understanding of the pathology of horses or food-producing animals (Boor-van der Putten, 1986: 236). More significantly still, most of the advancement that occurred in understanding the diseases and physiology of small animals occurred indirectly through scientific research aimed at understanding and improving human health. For example, experiments conducted on dogs led to the understanding of the process of endocrine secretion in humans, eventually resulting in the isolation of insulin during the 1920s. It was, however, to take many years before such discoveries were actually applied in the treatment of dogs and cats (Dunlop & Williams, 1996: 600–1).

Yet, while pet animals were regarded with a great deal of contempt by the veterinary establishment, they nonetheless seem to have crept insidiously into the veterinary schools as patients. From the late nineteenth century onwards, increasing numbers of small animals and birds were brought by their owners to the veterinary schools for treatment at out-patients' clinics, eventually necessitating the establishment of accommodation for animals requiring hospitalization. In The Netherlands, for example, the *Rijksveeartsenijschool* (State Veterinary School) in Utrecht witnessed a steady increase in the numbers of dogs, cats and birds that could not be ignored. By 1881, dogs made up the greatest part of the patient body upon which the school's veterinarians could practise their clinical skills. In quantitative terms, therefore, dogs had become the most important group of animal patients visiting the school. The clinicians, however, continued to regard these patients with contempt, still preferring to treat the larger and, to them, inherently more interesting patients such as horses and cattle that attended the same out-patients' clinics in lesser numbers (Boor-van der Putten, 1986: 51–65). Elsewhere in Europe, the veterinary interest in small animals also continued to grow, along with the numbers of canine, feline and avian patients demanding veterinary attention. Britain, in particular, continued to produce innovators in this area, leading to medical advancements, such as the development of anaesthesia, that would later become essential to small animal practice.

However, in spite of both the increasing numbers of pet animals requiring treatment, and the emerging medical technologies and innovators able to provide it, the rise of small animal medicine during the twentieth century perhaps owes more to the invention of the internal combustion engine than anything else (Swabe, 1999: 180). The rise of motorized transport led to the inevitable decline of the importance of the horse in European society. This development had its greatest consequences for the urban veterinary practitioners who had, until then, made their living by tending the horses of private citizens, local businesses and the local municipality. However, as the horses that pulled the carriages, wagons, carts, trams and even the fire-brigade were replaced by motorized vehicles, the urban veterinarians were left with little local work other than meat inspection (Offringa, 1981: 41). Thus, it was more by accident than design that the veterinary practitioners of the early twentieth century set aside their contempt for small animals and instead began to earn a living from them.

After the First World War, increasing numbers of practices devoted, often exclusively, to the veterinary treatment of small animals were established in urban areas, generally deriving their income and clientele from the more affluent middle-class members of the community. Some small animal clinics were also established in association with animal protection organizations and the newly emerging animal sanctuaries to provide veterinary care for pet animals (Davids, 1989: 92). This trend accelerated after the end of the Second World War. The increased material affluence of post-war society has not only influenced our tendency to keep pets, but has also provided us with the means to go to considerable lengths to ensure their lives are happy and healthy ones. During the subsequent 50 years or so, an entire industry evolved to provide and service pet animal needs. From the breeders who produce tailor-made animals to the pet food manufacturers who feed them, it was realized that there are considerable profits to be made by exploiting our attachments to small animals and encouraging people to keep them as pets. The veterinary world was also part of this development, and consequently small animal medicine rapidly became the most progressive area within veterinary medicine; it also became a rather profitable one. This was, indeed, a far cry from the previous century when dogs and cats were more or less shunned by the veterinary profession (Swabe, 1999: 180–1).

VETERINARY DILEMMAS

Nowadays, it is quite normal to take one's dog or cat to the veterinarian, whereas as little as 100 years ago doing so was very much more the exception than the rule. The role of the small animal practitioner, as suggested above, is now one that is very familiar to us. Yet, to understand the nature of veterinary work, one must forget the popular image of the animal doctor, battling to save the lives of sick and injured animals. Although activities such as saving accident victims and helping animals in labour are a constituent – and, indeed, as the veterinarians in my study reported, the most fulfilling and challenging – part of their work, the fact remains that veterinarians spend a large proportion of their time performing extremely routine tasks. It certainly surprised me at the outset of my research to observe that in their daily work small animal practitioners seem to be largely preoccupied with performing routine procedures relating to the management, control and prevention of disease and parasitic infection. In the practices studied, the bulk of routine veterinary work tended to be preventive, rather than curative. In fact, animals that had been taken to the veterinarian to be vaccinated against infectious disease or to be treated for parasitic infection accounted for the highest proportion of visitors to regular surgery hours. Such routine consultations – although affording opportunities for the veterinarian to check the animal over more generally, thus allowing for the identification of other health problems – tended to be rather unspectacular. This is in sharp contrast to the kind of veterinary consultations that are filmed and broadcast for popular television viewing. Given the routine nature of most consultations, it is no wonder that the makers of reality television have preferred to focus upon unusual and life-saving operations when portraying veterinarians at work. However, this kind of sensationalism, while making captivating viewing for many, tends to give the public a rather atypical view of veterinary work and all that it entails.[2]

The management of animal reproduction is also a task situated high upon the everyday veterinary agenda. The veterinarian is responsible for helping to control the size of animal populations and overseeing the (re)production of healthy pet animals. This work includes neutering pet animals and providing obstetric and post-natal care. The neutering of pet animals certainly accounted for the largest proportion of surgical procedures observed in each practice through the course of my research. Further to this, the veterinarian is also entrusted the task of, what can best be described as, 'routine animal maintenance'. In other words, he or she is responsible for the repair, rehabilitation or destruction of sick and injured animals. Aside from this, the veterinarian will also treat dermatological conditions, advise on appropriate animal care, nutrition and housing, provide counsel with regard to behavioural problems, attend to dental problems and other matters relating to animal health and welfare.

The procedures that veterinarians are asked to perform and the decisions that they are required to make about their animal patients can potentially present a whole host of practical, moral and ethical problems. In some instances, they may be legally restricted from performing particular surgical interventions, such as declawing cats or debarking dogs, and can avoid having to deal with the ethical implications of such procedures. In many situations, however, the veterinarian's own personal judgement is relied upon and he or she must make weighty decisions as to what course of action should be taken. There are, of course, basic guidelines that the veterinarian is professionally obliged to conform with. For example, article 1 of the Dutch Veterinary Code (Koninklijke Nederlandse Maatschappij voor Diergeneeskunde, 1992: 10) states that a veterinarian should act in accordance with:

(a) the benefit of the health and welfare of the animal and the interests of the owner;
(b) public interest and general veterinary interests;
(c) the benefit of public health and environmental hygiene;
(d) the position and function of veterinary medicine in society.

The first of these basic principles for veterinary conduct is perhaps the one that is the most inherently problematic for veterinarians. The interests of the animal and its owner are not always reconcilable. The course of action that a veterinarian must take is often dependent upon the species or breed with which he or she is dealing and its economic or emotional value, irrespective of the nature of the medical complaint and whether or not it can be successfully treated. Frequently, the veterinarian is well aware that material considerations may overshadow the actual treatability of an animal's condition.

Notwithstanding such guidelines, the veterinarian is still required to perform surgery and make decisions that those more idealistically involved with animal welfare issues may question. Even within small animal medicine, the veterinary treatment that an animal receives may often be guided by practical and financial considerations, rather than sentiments or idealism. The veterinarian must be aware of the client's ability or willingness to pay for services rendered. A consequence of this is that certain decisions and procedures – as distasteful as they may seem to some – are standard to veterinary practice. The following discussion considers some of the procedures and decisions that are routine to small animal practice. The first issue addressed is perhaps the weightiest one: namely, the grounds upon which decisions to continue with veterinary treatment or to euthanize an animal are made. Secondly, standard non-therapeutic surgical interventions such as tail-docking, declawing and debarking are discussed and, finally, the issue of neutering is examined.

TO KILL OR CURE?

As has already been suggested, there is great variety in the value that people place upon small animals in our society. To put it crudely: some pet animals are treated as cherished members of the family for whom no time, energy and expense are to be spared, whereas others may be neglected, abandoned or abused. One cannot, therefore, make generalizations about pet owners' attachments to their animals and the lengths that they may be prepared to go to in order to preserve their pets' health. Indeed, the attitudes of owners with regard to paying for and deciding to proceed with veterinary treatment vary greatly. For some animal owners the decision to go ahead with a life-saving operation or drug therapy is a highly emotive and problematic one; for others, it is very simple and is decided purely upon pragmatic grounds. Furthermore, an animal's condition need not actually be life-threatening for a decision to end its life to be made.

In small animal practice, ending an animal's life is commonly known as 'euthanasia'. This term is seen as particularly problematic when used to refer to healthy animals being killed – from a moral point of view – unnecessarily. Animal rights philosophers, such as Tom Regan (1983), have strongly objected to the use of the term 'euthanasia' as a blanket description for the deliberate killing of pet animals. They argue that the term is entirely inappropriate unless certain conditions are met: first, that the animal is killed by the most painless means possible; second, that the individual who ends its life truly believes that a painless death is in the animal's own interests; and finally, that the individual who euthanizes it is motivated to euthanasia out of a concern for the animal's interests, good or welfare (Regan, 1983: 110–11). In veterinary practice, euthanasia is simply defined and understood as the 'act of inducing a painless death' (Tannenbaum, 1989: 209). This term is, however, only deemed appropriate to the deliberate killing of pet animals and horses: food-producing animals are simply slaughtered, not euthanized. Both of these terms in themselves say much about societal attitudes to, and treatment of, animals.

Unlike the majority of their clients (unless they are involved in farming, hunting or abattoir work), veterinarians are accustomed to death, for they are routinely required to euthanize animals in the course of their work. Dealing with the death of animals is an almost unavoidable part of the job. The task of performing euthanasia is, however, not one which veterinarians relish for it is inherently problematic: first, because performing it can bring serious ethical and personal dilemmas to the fore, and, second, because of the possible emotional responses of the client whose animal is being euthanized. Although a veterinary education provides the practitioner with all the necessary clinical and diagnostic skills, many veterinarians are inadequately trained to deal with client emotion and pet loss. In The Netherlands, for example, scant attention is paid to this subject within the veterinary curriculum. Throughout their lengthy formal training, a mere two afternoon sessions are devoted to teaching students – through role-playing games – how to deal with euthanasia consultations.[3] In Britain, too, a recent survey of small animal practitioners revealed that 96% of respondents had had no formal training in how to explain to clients that a pet is critically or terminally ill. Moreover, 72% of these respondents believed that such training would be potentially useful (Fogle & Abrahamson, 1990: 145).

Observing a wide array of euthanasia consultations during my own research, it became increasingly clear that the veterinarians studied tended to deal with euthanasia consultations in a very *ad hoc* manner. Their sensitivities towards clients' feelings varied greatly, depending to a large extent upon their own ability to empathize, and on their individual social and communicative skills. It is, however, essential that the clients' emotional needs are adequately met and responded to in the event of euthanasia. Indeed, the veterinarian has a vested interest in keeping his or her clients happy, for contented clients are likely to return to the practice with other animals in the future (Swabe, 1994). Today, there is a growing awareness of the necessity for proper training and guidelines for dealing with euthanasia consultations. In recent years, professional journals and organizations have devoted increasing attention to the subject and have attempted to provide practical suggestions about how veterinarians should deal with both the procedure of euthanasia itself and their clients' (and their own) emotional responses to it (Hart, Hart & Mader, 1990). Articles and books designed to explain euthanasia to pet owners and help people to deal with it more adequately have also recently begun to appear on the market (e.g. Quackenbush, 1985; Lommers & van Amsterdam, 1993). Today, access to information on pet loss is also widely available from veterinary and animal welfare agencies via the Internet. Many veterinary colleges operate pet loss support services, such as telephone hotlines (Turner, 1997), for bereaved animal owners. Others, including the Veterinary Faculty at Utrecht in The Netherlands, have a professional psychologist or social worker on-hand for grief counselling.

Irrespective of the ambivalence that is often displayed towards them, pet animals, as this volume attests, have undeniably come to acquire a unique position in human society and are often accorded an almost quasi-human status. Sometimes, it has been argued, they have even been treated as substitutes for people or other human relationships. While this may be so in extreme cases where, for example, acute grief may follow the demise of an animal (e.g. Fogle, 1981; Voith, 1981), it is more plausible that the majority of pet owners' relationships with their pets are supplementary to human contact rather than substitutes for it, animals perhaps offering a kind of relationship that people do not provide and that must, therefore, be sought elsewhere (Endenburg, 1991: 16; see also Chapters 11, 12 and 13). Nevertheless, people can often be rather attached to – or feel a responsibility towards – their pets, and when these animals become ill, veterinary treatment is frequently sought. The nature of the actual veterinary treatment that can be provided is often dependent, however, upon the client's willingness or ability to pay for it and/or the degree to which he or she is prepared or able to care for the animal at home as part of the designated therapy. Conditions such as diabetes, for example, may remain untreated if a client is reluctant to administer regular insulin injections or to maintain a strict dietary regimen for his or her pet. Similarly, if an animal requires a costly operation that the owner cannot afford or is unwilling to pay for, its condition may not be treated. Such action may be detrimental to the animal's welfare, although this will depend upon the exact nature of its predicament. If an animal's well-being is likely to be seriously compromised by a lack of surgical or therapeutic intervention, euthanasia may be the only available and affordable option.

Veterinarians are, however, often reluctant to end an animal's life unnecessarily, particularly when the animal is young and/or stands a good chance of making a full recovery. Often they will try to dissuade the client from choosing euthanasia, and will attempt to make alternative financial arrangements, such as payment by instalments, for any necessary surgery or drug therapy. This reluctance to terminate animal life unnecessarily is also typified by veterinary responses to requests to euthanize animals with behavioural problems, especially when these animals are otherwise physically healthy. When an animal's aggressive behaviour is seen to endanger the health and safety of humans and other pets – and often when all other alternatives such as obedience training and muzzling have been exhausted – euthanasia will generally be seen to be a viable and ethically defensible course of action. However, requests to euthanize animals are commonly made by pet owners when the animal and its behaviour have simply become an inconvenience to them.

A prime example of this relates to feline elimination problems. Inappropriate elimination, most particularly house-soiling by urination, is a behavioural problem that motivates many requests to have cats 'put down'. Months and even years of continually soiled furniture and carpets may drive cat owners to seek a final solution to their problem, but sometimes just one mistake on the part of the cat is enough for it to be condemned to death by its owner. There may well be a physical and treatable cause for such inappropriate feline elimination behaviour, such as diabetes, bladder or urethral inflammation, or bladder damage. Alternatively, such elimination may be related to territorial marking, though it may also occur when the cat is dissatisfied with its litter tray or is unhappy for some other reason. In the event of the latter, the owner's negligence or ignorance of the cat's needs is frequently singled out as the root cause of the problem. Even though the owner may be deemed responsible, the veterinarian is still presented with the dilemma of whether or not he or she should accede to the client's request to have the animal euthanized. The modification of the owner's behaviour towards the animal may provide a solution to these elimination problems (Blackshaw, 1992). However, although attempts may be made to educate the owner to provide adequate toilet facilities for the animal – or to teach him or her how to employ aversion techniques at home to alter the animal's behaviour – it is often the case that once the animal has become a nuisance, the owner will wish to get rid of it by any means possible. When faced with the dilemma of being asked to euthanize a healthy, though behaviourally problematic, cat, veterinarians are often inclined to look for alternative solutions, such as re-housing the animal.

At the other extreme, small animal practitioners may be faced with pet owners who are resolute in their refusal to consider euthanasia as a viable option, in spite of the fact that having the animal euthanized may be both in their interests financially, and in the interests of the animal's welfare. In contrast to the people who are predisposed to having their animals euthanized, some pet owners will be prepared to spend substantial amounts of money (that they cannot necessarily afford) to keep their beloved animals alive and well for as long as possible. However, when an animal's quality of life is significantly reduced and all reasonable therapeutic options have been exhausted, veterinarians will logically conclude that euthanasia is the best and kindest course of action. Getting the owner to consent to it is another matter altogether. Pet owners may insist that the veterinarian attempts a new course of treatment or that the animal undergoes another operation before abandoning hope. Frequently, clients in this situation will refuse to accept the veterinarian's judgement and will go elsewhere to seek a second opinion. Although it may be economically profitable for the veterinarian to do as the client requests, the question remains whether it is morally defensible to do so, especially when such intervention may only extend an animal's life by a matter of weeks, or may cause unnecessary stress or pain.

In this regard, the extension of the use of medical treatments, such as chemotherapy, to small animal medicine from human medicine has raised serious ethical questions about the extent to which one should go to preserve animal life. Few animal tumours, for example, can be cured by chemotherapy. Although animals will generally tolerate chemotherapy fairly well, the majority of patients will only experience a temporary or partial remission from their symptoms; this will, at least in the short term, prolong survival and improve the quality of life. Nevertheless, given that chemotherapy is often not curative – and also because the use of cytotoxic drugs may be potentially harmful – the ethics of employing such a treatment may still be disputed (Dobson & Gorman, 1993: 3). The use of such medical treatments may also be questioned if profit can be made from them. In The Netherlands – at least at the time of writing – chemotherapy for pet animals is only available at the country's sole veterinary school. The veterinarians who employ such veterinary medical therapy are university employees and do not make any personal financial profit from treating their patients. Elsewhere, however, such state-of-the-art veterinary treatment is available in private and profit-making veterinary clinics. There may, therefore, be a potential risk of financial considerations influencing the course of treatment that the veterinarian advises, particularly if the client is willing and able to pay for it.

For all the compassion that they may often have for both their animal patients and human clients, the inescapable fact remains that most veterinarians operate their practices as businesses: essentially, animals are more lucrative to them alive than dead. The decision to kill or cure can thus often be quite arbitrary, irrespective of medical indications. One terminally ill pet may be kept alive by all means possible, whilst another perfectly healthy animal may be euthanized simply because the owner has requested that this service be provided. This state of affairs clearly illustrates not only the ambivalence that exists within the human–animal relationship, but also ethical dilemmas with which the small animal practitioner is routinely faced in weighing up the animal's, client's, and his or her own interests.

ANIMAL MUTILATION[4]

A consistent trait in humankind's relationship with domesticated animals is that, if an animal does not satisfy human requirements, it is simply altered, either through a process of selective breeding or by surgical intervention. Selective breeding enables specific and desirable characteristics, be they physical or behavioural, to be 'artificially' generated. Human interference in the genetic make-up of other species has inevitably led to great problems for animal well-being, particularly when mutant genes, which if left up to nature would probably die out, are deliberately selected to change an animal in order to produce a new breed (Robinson, 1982; Macdonald, 1985; Tabor, 1991). Sometimes, such deliberate selection can seem fairly innocuous, for instance when genes are selected to produce a particular coat colour. But more often than not, genes are selected to produce specific facial features (e.g. ones that are paedomorphic), shape and size that are attractive to humans, yet can seriously impair the animal's well-being (Serpell, 1996). The extent of human intervention through breeding can be illustrated by looking at the great diversity within one particular species: the dog. Today, after centuries of selective breeding, an adult dog can weigh between 2 and 80 kg, in contrast to the wolf, the wild species closest to dogs, whose natural body weight can vary between 20 and 50kg, or heavier still in the case of Canadian timber wolves (Bouw, 1991).

Along with selective breeding, surgical interventions to alter the appearance or functioning of an animal are also sometimes performed. In recent years, procedures such as tail-docking, ear cropping, declawing and debarking have come under increasing attack. The freedom for small animal practitioners to perform such procedures has, in some instances, been greatly restricted, since these surgical interventions have been deemed by professional bodies or legal statutes to be an unacceptable infringement of the animal's interests. It should also be noted that the aforementioned surgical alterations are seldom performed for any therapeutic purpose, but rather they are done entirely for cosmetic purposes or for the convenience of the owner. It is possible that, with advances in genetic engineering, such mutilations may in fact become unnecessary. Molecular geneticists of the future could, for example, attempt to isolate the genetic factors that produce cat claws and produce a race of clawless cats, obviating the need for declawing altogether. Although this would presumably mean that some cats would no longer have to endure declawing, the crucial question remains whether such genetic manipulation would also inhibit their natural and social behaviour.

At present, and for the foreseeable future, however, surgery remains the only option for those wishing to physically alter their pet animals permanently. Tail-docking is probably the oldest form of mutilation. It is thought that it originally developed as a practical measure to allow working dogs to perform their tasks more efficiently. Furthermore, during the Roman era, it was believed that tail-docking would afford a degree of protection against canine rabies (Columella, 1745). As the primary function of such animals as workers has largely been supplanted by their roles as companions, this practice has become more embedded in notions of how each breed should appear physically; the functional aspects of tail-docking have faded or have been lost to history. Today, tail-docking is generally viewed as a purely cosmetic mutilation that has a negative impact on the animals. Dogs use their tails to communicate with others. Opponents to tail-docking have therefore argued that it is cruel since it deprives the docked animal of this function. Similarly, ear cropping is also objected to, not only because it is unnecessary, apart from an aesthetic point of view, but also because it is of no direct benefit to the animal and, again, like tail-docking, it deprives the animal of the ability to communicate fully with others. Although both practices are forbidden in The Netherlands, one can frequently observe dogs on the streets whose ears have clearly been trimmed and whose tails have been docked. These procedures are thus performed illegally, often by non-veterinarians and without adequate anaesthetic or pain-killers (de Waal, 1987: 35–6).

Similarly, the declawing of cats and 'debarking', i.e. the devocalization of dogs, are banned in The Netherlands, although these procedures, particularly the former, are widely and commonly performed in other western countries, particularly in the USA. In The Netherlands, legal and professional regulations stipulate that these surgical procedures can only be performed if they are the only viable alternative to euthanasia. Veterinary ethicists and animal welfare experts consider both procedures to be an infringement of the animal's integrity, and to prevent it from behaving naturally (de Waal, 1987; Tannenbaum, 1989; Rutgers, 1993). Furniture-destroying cats and incessantly barking dogs might be a nuisance to their owners, but operating on them to solve the problem has been deemed unacceptable since such surgery is considered to be against the interests of animal well-being. All of the surgical procedures described deprive pet animals of performing natural behaviours. In recognition of this, they have – at least in The Netherlands – been severely restricted, and veterinarians (should) no longer perform them unless they are of therapeutic value, for instance when a dog's tail is broken. The dilemma of whether or not veterinarians should perform such procedures has been taken out of their hands by the legislators. In this instance, the interests of the animal have been placed above those of the owner. Yet, even where animal mutilations have been severely restricted, veterinarians may still be faced with having to deal with clients who wish such surgical interventions to be performed. Where these mutilations are perfectly legal, the matter is left up to the veterinarian's own conscience and/or economic interests. To some extent, it is also a question of supply and demand. Even though a veterinarian may find such mutilations unnecessary or distasteful, if he or she is not prepared to provide the service, clients may be lost to competitors who will.

The final pet animal mutilation that will be considered here is neutering. This commonplace surgical alteration of animals is deemed unproblematic by veterinarians, although it quite clearly alters the natural function of the animal in a similar fashion to those interventions performed for aesthetic reasons. Within small animal practice, neutering is the most common surgical intervention performed upon pets. Neutering is, however, sometimes viewed by animal owners as a violation of an animal's natural right to procreate and bear young. Consequently, it is often only performed after an animal has had one litter and has briefly experienced its natural 'privilege'. There appears to be little scientific evidence to support this belief that animals should bear at least one litter before being sterilized. It has been widely argued that, since it is unlikely that the animal will have any concept of what it is missing, neutering does it no harm (Council for Science and Society Report, 1988). In view of the surplus of pet animals and the problem of pet abandonment, it is logical that pet procreation and population be controlled. Since pets cannot voluntarily control their own fertility, human owners are required to take responsibility for them by having them neutered. The veterinarian's task is both to educate his or her clients as to the wisdom of taking neutering as a course of action, and to perform the surgery itself. Neutering is also sometimes advised by small animal practitioners as a means of hindering the onset of various disorders of the reproductive organs to which various breeds of animal are often prone. It has also been employed more recently as a means of encouraging dangerous breeds, such as pit-bull terriers, gradually to die out.

The desire to prevent unwanted litters often provides the grounds for the neutering of an individual animal, but neutering is also often performed entirely for the owner's convenience. Uncastrated tomcats, for example, habitually spray pungent urine which is both offensive in smell and often damages furniture and carpets. When castrated, they largely stop this undesirable behaviour and can become far more sedate and submissive. In short, a castrated cat tends to make a more ideal pet than one left fully intact. Neutering is often performed this way as a form of behavioural control since it often makes animals less aggressive and reduces their proclivity to stray or fight with other animals. It could thus also be said that by neutering their animals, humans have ensured that their pets will never, in a sense, really grow up. In other words, they will retain more infantile behavioural characteristics that seem to be particularly attractive to humans, they will not become sexually active, and they will better appeal to human idealizations of what pets should be. Moreover, the control of animal sexuality, in addition to the restriction of pet animals' eating, hunting and toileting behaviour, could be viewed as an attempt to 'civilize' them, to regulate and constrain their natural functioning so that it is no longer offensive to us. The 'natural' behaviour of pet animals has been transformed so that it bears greater semblance to that of humans and, as a consequence, it is not so reminiscent of our own suppressed animality. It can, however, easily be argued that pet-neutering is performed in the interest of the animals themselves, as much as in the interest of humans, given that the excess of pets causes much distress and suffering to many thousands of unwanted animals each year. In this way, the neutering of pets can be justified as a necessary and fairly harmless evil, as long as it is performed using anaesthetic in as sterile conditions as possible. For both veterinarians and their clients alike, neutering is the most defensible mutilation that a pet animal can undergo. Indeed, the fact that it is not commonly perceived as a mutilation speaks volumes about the acceptability of the practice.

CONCLUSION

The contradictions and ambivalence in the relationship between humans and animals are clearly embedded in routine veterinary practice. These ambiguities, as this chapter has sought to illustrate, find expression in the kinds of dilemmas with which small animal practitioners are faced during the course of their everyday work. To some extent, the legal and ethical parameters of their profession help to determine the course of action that they may take in dealing with the requests and demands of their clients. However, it seems that the action taken, particularly when it comes to either preserving or ending the life of an animal, is often highly dependent upon the veterinarian's discretion, and the client's wishes or ability to pay for treatment. As suggested above, a variety of factors may influence the decision-making process; financial gain for the veterinarian sometimes ranks among them. The notion that profit may possibly influence veterinary decision-making undoubtedly conflicts strongly with the collective image of the kindly animal doctor that we hold. Yet, we should remember that providing a veterinary medical service for pet animals is not just about animal doctoring, but is – at least as far as private practice goes – also about running a viable business. This does not, of course, necessarily exclude compassion, humanity or the adherence to ethical standards, but it does mean that the activities and work of veterinarians should not be over-romanticized. The fact remains that veterinary professionals are often faced with difficult dilemmas when dealing with their animal patients and human clients. In deciding the course of action to take, practical and financial considerations may well often outweigh sentiment and idealism.

NOTES

1. Few sociologists have devoted their attention to the veterinary profession and its activities. A notable exception is Clinton R. Sanders (1994a, 1994b, 1995b), who has conducted extensive research into veterinary interaction in a mixed animal practice in New England. Recently, Arnold Arluke issued a plea to encourage more sociologists to look at additional aspects of veterinary practice, such as the training and recruitment of veterinarians (Arluke, 1997).
2. I would even go so far as to suggest that such an atypical perception of veterinary work has potentially significant consequences for the future. With increasing numbers of prospective vets – particularly women – being attracted to study veterinary medicine specifically to go into small animal practice, it is possible that other crucial areas of veterinary medicine, such as large animal practice, meat-inspection and public health management, may suffer. As both Arluke (1997) and Herzog and his colleagues (1989) have suggested, further research into the motivation and attitudes of veterinary students at the outset and during their studies is necessary.
3. The ethical aspects of euthanasia are, however, dealt with elsewhere in the curriculum of the Veterinary Faculty in Utrecht.
4. I have adopted this term from veterinary and animal welfare literature. Whilst the word 'mutilation' invokes quite graphic and shocking imagery, I have used it, in its 'technical' sense, to refer to surgery that is performed for entirely non-therapeutic purposes to deliberately change the appearance or natural function of an animal.

REFERENCES

Arluke, A. (1988). Sacrificial symbolism in animal experimentation: object or pet? *Anthrozoös*, **2**, 98–117.

Arluke, A. (1997). Veterinary education: a plea and plan for sociological study. *Anthrozoös*, **10**, 3–7.

Arluke, A. & Sanders, C.R. (1996). *Regarding Animals.* Philadelphia: Temple University Press.

Blackshaw, J.K. (1992). Feline elimination problems. *Anthrozoös*, **5**, 52–6.
Blaine, D.P. (1817). *Canine Pathology*. London: Boosey.
Boor-van der Putten, I.M.E. (1986). *75 Jaar Geneeskunde van Gezelschapsdieren in Nederland*. Utrecht: Faculteit Diergeneeskunde.
Bouw, J. (1991). Domesticatie. *ARGOS: Bulletin van het Veterinair Historisch Genootschap*. Speciale Uitgave, Zomer 1991, 23–8.
Columella, L.J.M. (1745). *Of Husbandry* (translated into English). London: A. Millar.
Council for Science and Society Report. (1988). *Companion Animals in Society*. Oxford: Oxford University Press.
Davids, C.A. (1989). *Dieren en Nederlanders: Zeven Eeuwen Lief en Leed*. Utrecht: Matrijs.
Dobson, J.M. & Gorman, N.T. (1993). *Cancer Chemotherapy in Small Animal Practice*. Oxford: Blackwell Scientific Publications.
Dunlop, R. & Williams, D. (1996). *Veterinary Medicine: an Illustrated History*. St Louis, MO: Mosby.
Endenburg, N. (1991). Animals as companions: demographic, motivational and ethical aspects of companion animal ownership. PhD thesis, Rijksuniversiteit Utrecht.
Fogle, B. (1981). Attachment–euthanasia–grieving. In *Interrelations between Pets and People*, ed. B. Fogle, pp. 331–44. Springfield, IL: Charles C Thomas.
Fogle, B. & Abrahamson, D. (1990). Pet loss: a survey of the attitudes and feelings of practicing veterinarians. *Anthrozoös*, **3**, 143–50.
Hart, L.A., Hart, B.L. & Mader, B. (1990). Humane euthanasia and companion animal death: caring for the animal, the client, and the veterinarian. *Journal of the American Veterinary Medical Association*, **197**, 1292–9.
Herzog, H. A., Vore, T. L. & New, J. C. (1989). Conversations with veterinary students: attitudes, ethics and animals. *Anthrozoös*, **2**, 181–8.
Koninklijke Nederlandse Maatschappij voor Diergeneeskunde (1992). *Code voor de Dierenarts*.
Lommers, H. & van Amsterdam, H. (1993). *Het is Stil in Huis: Afscheid Nemen van je Huisdier*. Lisse: Etiko.
Lynch, M. (1988). Sacrifice and the transformation of the animal body into a scientific object: laboratory science and ritual practice in the neurosciences. *Social Studies of Science*, **18**, 265–89.
Macdonald, D. (ed.) (1985). *Honden*. (Original title: *The Complete Book of the Dog*, translated from English by H. Cornelder.) Utrecht: Uitgeverij Het Spectrum.
Offringa, C. (1981). *Van Gildenstein naar Uithof: 150 Jaar Diergeneeskundig Onderwijs in Utrecht. Deel 2*. Utrecht: Faculteit der Diergeneeskunde.
Offringa, C. (1983). Ars veterinaria: ambacht, professie, beroep. Sociologische theorie en historische praktijk. *Tijdschrift voor Geschiedenis*, **96**, 407–32.
Porter, R. (1993). Man, animals and medicine at the time of the founding of the royal veterinary college. In *History of the Healing Professions: Parallels between Veterinary and Medical History*, Vol. 3, ed. A.R. Michell, pp. 19–30. London: C.A.B. International.
Quackenbush J. (1985). *When Your Pet Dies: How To Cope With Your Feelings*. New York: Simon & Schuster,.
Regan, T. (1983). *The Case for Animal Rights*. London: Routledge & Kegan Paul.
Robinson, R. (1982). *Genetics for Dog Breeders*. Oxford: Pergamon Press.
Rutgers, L.J.E. (1993). Het Wel en Wee van Dieren: Ethiek en Diergeneeskundig Handelen. PhD thesis, Rijksuniversiteit Utrecht.
Sanders, C.R. (1994a). Biting the hand that heals you: encounters with problematic patients in a general veterinary practice. *Society and Animals*, **2**, 47–66.
Sanders, C.R. (1994b). Annoying owners: routine interactions with problematic clients in a general veterinary practice. *Qualitative Sociology*, **17**, 159–70.

Sanders, C.R. (1995a). Killing with kindness: veterinary euthanasia and the social construction of personhood. *Sociological Forum*, **10**, 195–214.

Sanders, C.R. (1995b). Ambiguity, individuality and trainers' interactions with guide dogs. Paper presented at the ISAZ (International Society for Anthrozoology) Symposium on Cultural & Historical Perspectives on Human–Animal Interactions, Geneva, Switzerland, 7th & 8th September, 1995.

Serpell, J. (1996). *In the Company of Animals: a Study of Human–Animal Relationships*. Cambridge: Cambridge University Press.

Swabe, J. (1994). Preserving the emotional order: the display and management of emotion in veterinary interaction. *Psychologie en Maatschappij*, **68**, 248–60.

Swabe, J.M. (1996). Dieren als een natuurlijke hulpbron: ambivalentie in de relatie tussen mens en dier, binnen en buiten de veterinaire praktijk. In *Milieu als Mensenwerk*, ed. B. van Heerikhuizen, B. Kruithof, K. Schmidt & E. Tellegen, pp. 12–37. Groningen: Wolters-Noordhoff.

Swabe, J. (1999). *Animals, Disease and Human Society: Human–Animal Relations and the Rise of Veterinary Medicine*. London: Routledge.

Tabor, R. (1991). *Cats: the Rise of the Cat*. London: BBC Books.

Tannenbaum, J. (1989). *Veterinary Ethics*. Baltimore: Williams & Wilkins.

Turner, W.G. (1997). Evaluation of a pet loss support hotline. *Anthrozoös*, **10**, 225–30.

Voith, V. (1981). Attachment between people and their pets: behavior problems that arise from the relationship between pets and people. In *Interrelations between Pets and People*, ed. B. Fogle, pp. 271–94. Springfield, IL: Charles C Thomas.

de Waal, D. (1987). *Welzijn van Dieren in Nederland*. 's-Gravenhage: LNO Reeks, Staatsuitgeverij.

Wieder, L.D. (1980). Behaviouristic operationalism and the life-world: chimpanzees and chimpanzee researchers in face-to-face interaction. *Sociological Inquiry*, **50**, 75–103.

Wilkinson, L. (1992). *Animals and Disease: an Introduction to the History of Comparative Medicine*. Cambridge: Cambridge University Press.

PIERS BEIRNE

17

Rethinking bestiality: towards a concept of interspecies sexual assault

INTRODUCTION

In this chapter I seek to contribute to an as-yet-unconstituted sociology of animal abuse (Beirne, 1995), though my specific focus derives less from overtly theoretical labours directed to this end than from the practical needs of pedagogy. In trying to develop an undergraduate course on the sociology of animal abuse, I was immediately confronted with conveying to my students adequate responses to the deceptively simple question 'What is animal abuse?'. Class time devoted to the spectre of such dramatic and well-publicized practices as factory farming, laboratory experimentation, trapping, circuses, etc., would, I believed, tend to stimulate among students a visceral reaction rather than the desired goal of sustained enquiry about the nature of animal abuse.

It happened that, in casting a wide net for some heuristic device that would enable me to examine animal abuse in a pedagogic context, I stumbled upon a would-be erotic video provocatively entitled *Barnyard Love*. This crudely produced, undated German film graphically depicts numerous human and other animals engaged in acts of interspecies sexual relations. Among these are human males, who engage in sexual intercourse with cows and hens and, more often – given that heterosexual males are presumed to be the film's chief audience – human females, who have sexual intercourse with dogs, insert eels into their vaginas and perform fellatio on dogs and horses.

Even from my amateurish perspective and despite the risks of anthropomorphism, I noticed how immensely varied were the filmed reactions of the different non-human animals (henceforth 'animals') to attempted sexual union with, and initiated by, humans. At one extreme, the dogs in *Barnyard Love* who were engaged in sexual activities with women seemed energetically to enjoy such human attention. To me, at least, it did not seem possible that such canine enthusiasm could be feigned by off-camera training designed to suppress more genuine emotions of grief and pain. At the opposite extreme, some animals, such as eels and hens, were obviously unwilling recipients of human sexual advances. None of my students would have much trouble, I thought, in identifying as animal abuse the case of one unfortunate hen who died as a terrifying consequence of enforced sexual intercourse with a human male. Yet, in the case of large quadrupeds, such as the horses and cows depicted in the film, their reaction seemed closer to boredom or perhaps indifference than it did to pain or to bliss – eating, urinating and defecating as they were during intercourse or while their genitalia were being manipulated. Indeed, it was unclear whether these larger animals were even aware of the prolonged sexual relations which humans had foisted on them. What I saw here as animals' indifference might actually have been calculated detachment on their part – a coping strategy for numbing the pain inflicted on them by yet another of the myriad ways in which humans routinely invade, inspect and dispose of their lives.

The events depicted in films like *Barnyard Love* raise interesting questions about bestiality as a social practice. Is bestiality an outrageous and perhaps perverse act or, as the criminal law's tendency to tolerate it suggests, a relatively benign form of social deviance? Why have human–animal sexual relations been so vociferously and ubiquitously condemned yet so little studied?

To these questions let me at once add how remarkable it is, given the intense levels of ideological and physical coercion often applied to bestiality, that the social sciences have almost completely neglected to study an unusual social practice that is traditionally viewed with moral, judicial and aesthetic outrage. When, during his work on the medieval prosecution and capital punishment of animals, E.P. Evans gruffly dismissed bestiality as 'this disgusting crime' (1906/1987: 148), he was probably expressing not an idiosyncratic prejudice but an enduring sentiment that he shared with the great majority of his colleagues. To him, and probably to most of us, bestiality is a disturbing form of sexual practice that invites hurried bewilderment rather than sustained intellectual inquiry. Indeed, in academic discourse the topic of bestiality tends to surface only in lectures on the evolution of criminal law given by professors who, with embarrassed chuckles, refer to the declining volume of bestiality prosecutions since the early nineteenth century in order to instantiate the secularized tolerance and the supposed rationality of western law. While fictional and quasi-autobiographical accounts of bestiality occasionally appear in serious works of literature (e.g. Tester, 1991;

Høeg, 1996), accessible descriptions of it tend to be produced only by libertine presses and cinematographers as erotic commodities for consumption by a popular, albeit limited, audience.

In what follows, I seek to introduce a view of bestiality which differs radically both from the anthropocentrism enshrined in the dogma of Judaeo-Christianity and also from the pseudo-liberal stance of tolerance fashionable today. I suggest, specifically, that bestiality should be understood as 'interspecies sexual assault'. (Since we should not be in the business of policing interspecies sexual relations between non-humans, my argument is limited to humans' sexual abuse of non-human animals.) However, to begin with, I must comment briefly on the evolution of different images of bestiality and the stated justifications for its censure.

'AMONG CHRISTIANS, A CRIME NOT TO BE NAMED'

The cultural universe of bestiality is necessarily an anthropocentric one, though in many societies, past and present, it inhabits an ambiguous ideological terrain. On the one hand, it is exalted in mythic and folkloric traditions. Although they are not my concern here, it is worth noting that these favourable depictions of bestiality are often lodged in the sexual antics, the conquests and the offspring of numerous gods, in the lineage of earthly monarchs and rulers, and in the texts of fairy stories and other morality tales. On the other hand, all known societies have probably applied some form of censure to human–animal sexual relations. Moreover, the judicial accusation of bestiality occasionally blurs into, or is employed in concert with, other charges, such as witchcraft. Thus, some early medieval European accusations of witchcraft involved the claim that the defendant had partaken in a ritual salute of the Devil's backside, the *osculum infame* or obscene kiss (Russell, 1982: 63). In one case of unknown date, a certain Françoise Sécretain was burned alive because she had had carnal knowledge of domestic animals – a dog, a cat and a cock – and because, she admitted, she was a witch and her animals were actually earthly forms of the devil (Dubois-Desaulle, 1933: 58).

What we refer to as 'bestiality' has been denominated variously in different places and times. Besides a hodge-podge of more or less polite colloquialisms, bestiality has also been termed 'zoophilia', 'zooerasty', 'sodomy' and 'buggery'. The seventeenth-century English word 'bestiality' derives from the Latin *bestialitas*, the latter being used in Aquinas' *Summa Theologica* severally to refer to primitive behaviour, to human-animal sexual intercourse, and to how animals copulate.[1] Until the mid-nineteenth century, the term referred broadly to the beast-like, earthy and savage qualities allegedly inherent in non-human animals. Nowadays, bestiality tends exclusively to denote human–animal sexual relations. Usually, in law, it refers to sexual intercourse when a human penis or digit enters the vagina, mouth, anus or cloaca of the animal. However, it often also entails any form of oral–genital contact, including those between women and animals and even, in psychiatry, fantasies about sex with animals.

Bestiality is sometimes classified as a crime against nature; in this it is a bedfellow of other crimes involving 'pollution' such as sodomy, buggery, masturbation and paedophilia. At other times, the terms 'sodomy' and 'buggery' are used interchangeably to describe bestiality, though they have also been employed to denote homosexuality. Each of these terms carries with it pejorative baggage that varies in its moral bases, in its intensity and in the duration of its condemnation. Moreover, in some societies, such as in New England from the Puritan 1600s until the mid-nineteenth century, bestiality has been generally regarded with such trepidation that all mention of it is censured. Accordingly, it is also referred to as 'that unmentionable vice' or 'a sin too fearful to be named' or 'among Christians a crime not to be named'.

ANTHROPOCENTRISM AND THE ABOMINATIONS OF LEVITICUS

From its inception, Christianity applied austere standards and a strict discipline to those of its followers who violated its injunctions against the irremissible major sins of idolatry, the shedding of blood and fornication, including bestiality (McNeill & Gamer, 1938: 4–6). In all cases, the prescribed penalty was death. The earliest and most influential justifications for censures of bestiality are the Mosaic commandments. Deuteronomy declares '[c]ursed be he that lieth with any manner of beast' (27: 21), while Exodus commands that '[w]hosoever lieth with a beast shall surely be put to death' (22: 19) – the 'whosoever' here referring to both men and women (Leviticus, 20: 15–16). Besides mandating death for humans, Leviticus also dictates that the offending animal be put to death, a practice that reached its zenith in certain late-medieval European societies (Beirne, 1994). The precise intentions of those who originally condemned bestiality are probably not open to reclamation, but over the ages three beliefs have persisted about its wrongfulness: it ruptures the natural, God-given order of the universe; it violates the procreative intent required of all sexual relations between Christians; and it produces monstrous offspring that are the work of the Devil.

Let us uncover each of these three beliefs in turn.

A rupture of the 'natural' order of the universe

Prefaced by the general command 'Ye shall be holy: for I the Lord your God am holy', Leviticus declared 'Neither shalt thou lie with any beast to defile thyself therewith; neither shall any woman stand before a beast to lie down thereto: it *is* confusion' (18: 23). This theme continues:

> Ye shall keep my statutes. Thou shalt not let thy cattle gender with a diverse kind; thou shalt not sow thy field with mingled seed; neither shall a garment mingled of linen and woollen come upon thee. (Leviticus, 19: 19)

The rules that cattle should not 'gender with a diverse kind' and that a field should not be sown 'with mingled seed' lie at the heart of the Mosaic injunctions about bestiality. On this very basis the early Christian Church regarded copulation with a Jew as a form of bestiality and applied the penalty of death to it. So, too, from the time of Leviticus to that of seventeenth-century English moralists and beyond, bestiality has been regarded as sinful or criminal because it represents a rupture of the natural order of the universe, whose categories it is immoral to mix. Similarly, in his history of Plymouth Plantation, Governor William Bradford (1650/1970: 404–12) recorded the opinions of three ministers about the acts of 'unnatural vice' to be punished with death, among which were to be women who commit bestiality. Seeking affirmation in Leviticus, the ministers condemned bestiality, whether penetration had occurred or not, because it is 'against the order of nature', 'unnatural' and a 'confusion'. Again, Richard Capel, a seventeenth-century Stuart moralist, argued that bestiality is the worst of sexual crimes because 'it turns man into a very beast, makes a man a member of a brute creature' (quoted in Thomas, 1983: 39).

Violation of procreative intent

In matters of sexual relations, 'Be thee holy' means more than 'Be thee separate', for Christian morality has long required that sexual intercourse flow not from pleasure or play but exclusively from a procreative intent. Bestiality has thus also been condemned because it is held to be a violation of the Christian rule that procreation is the sole purpose of sexual intercourse. Crimes against nature have therefore been proclaimed to be those in which the emission of seed is not accompanied by a procreative intent, as in masturbation, anal and oral sex, incest, adultery, rape and bestiality.

Monstrous offspring

Bestiality has also been condemned because of the offspring a sexual union between human and beast is thought to produce or because of the evil that such offspring are held to signify or portend (Davidson, 1991: 41–3). This particular condemnation has itself been part of a complex cultural framework that includes animism, paganism and a fascination with monsters. Classical antiquity, for example, provides numerous seemingly non-judgmental references to interspecies sexual intercourse, including stories in which animals were thought to be in love with humans. Such cases are very prominent in *De natura animalium*, for example, the Roman historian and sophist Aelian's (1958) miscellany of facts about animals and humans, genuine or supposed, which he gleaned from Greek writers, including Aristotle. Drawing on material from and about Rome, Greece, India, Libya and Egypt, Aelian documented how widespread was the belief in the actual offspring of animal–human unions ('creatures of composite nature'). Although Aelian provided his readers with no clues as to how such offspring were regarded, they cannot always have been viewed with disfavour given his ubiquitous and often reverential references to creatures such as satyrs, centaurs and minotaurs.

How easily the rigid boundaries between animals and humans can become blurred is recorded in a history of Ireland by the twelfth-century chronicler Giraldus of Wales. Without further comment, he related how in the Glendalough mountains a cow gave birth to a man–calf, the fruit of a union between a man and a cow, the local folk 'being especially addicted to such abominations' (Cambrensis, 1863: 85). He reported elsewhere how Irish men and women had sexual intercourse with cows, goats and lions and how the populace believed that such unions were occasionally fertile. Similar superstitions appear in seventeenth-century New England, in the New Haven court records and in the poetry of John Donne and the speeches and sermons of John Winthrop, Cotton Mather and his brother John, Samuel Danforth and William Bradford, which are infected with the fear that colonial agricultural society was a frontier existence beset not only with the internal dangers of alcohol, idleness and lust but also surrounded by forests, wild animals and savages (Thomas, 1983: 38–41; Canup, 1988). Superstition combined with religious doctrine to assail bestiality and to portray its progeny as monsters resulting from the decay of civilization and the encroachment of the wilderness. Monstrous progeny were a visible reminder of how evil it was to transgress the God-given boundaries separating man from beast.

The social control of the object of such fears has been subject to great cultural variation in both style and volume. In some societies, the censure of bestiality has been accompanied by surprisingly few prosecutions. For example, despite the horror with which bestiality was viewed by puritan zealots and some jurists in England (Sharpe, 1983: 65–6) and in colonial America (Chapin, 1983: 127–9), it was rarely indicted and was unlikely to result in a conviction. In other societies, the number of convicts executed is staggering. For example, in Sweden from 1635 to 1778 there were as many as 700 executions for bestiality and an even greater number of males were sentenced to flogging, church penalties and public forced labour in chains (Liliequist, 1991). Upon conviction, both human and animal were usually put to death, often by burning at the stake but occasionally by beheading, hanging or from blows to the head. The bodies of the condemned, both human and animal, were finally burned or butchered and buried together.

If the penalties for bestiality and the entire range of unnatural acts had been strictly enforced, as Goodich (1979: 66–7) has noted, then Europe and colonial America would have become vast penal institutions inhabited by populations restricted in diet and dress, excluded from church services, and condemned to a joyless life of fasts, prayers and flagellation. While the relative frequency with which bestiality was condemned in early modern societies partly reflects the greater contacts between humans and animals in rural societies, such public displays of atonement have largely been dispensed with in modern urban societies. It is far more efficient for the state to deal with bestiality behind closed doors, or even to ignore it, and for the local folk community either to ridicule those who engage in it or to ostracize them.

Indeed, since the mid-nineteenth century, many 'unnatural offences', including bestiality, have effectively been decriminalized. In the USA there is no federal bestiality statute and only 27 states now have such a statute (though Utah has recently recriminalized bestiality). Nowadays, a defendant will probably be charged with a misdemeanor such as public indecency, a breach of the peace or cruelty to animals. Recently, the social control of bestiality has formally passed from religion and criminal law to a psychiatric discourse, at whose centre lie diseased individuals who are often depicted as simpletons or imbeciles with psychopathic personalities, and who allegedly sometimes also have aggressive and sadistic tendencies. However, at once subverting this psychiatrization and also echoing certain aspects of the spirit of decriminalization, there has gradually emerged a pseudo-liberal tolerance of bestiality. This tendency implies that because bestiality is an interesting and vital part of almost every known culture, it should not only be tolerated but, within certain limits, celebrated (e.g. Shell, 1993: 148–75; Dekkers, 1994).

NAMING INTERSPECIES SEXUAL ASSAULT

Are the decriminalization and the psychiatrization of bestiality and the drift to toleration of it signs of increasing civility and social progress? A superficial answer to this question is 'yes', if by it one means that censured humans are no longer brutalized by execution or by solitary confinement with hard labour. However, that would be to look at bestiality solely from an anthropocentric position, which is what the juridico-religious dogma surveyed here does exclusively. Seldom, either in times past or now, do popular images of social control include recognition of the terror and the pain that judicial examination and execution inflict on animals convicted of sexual relations with humans. Neither in the Mosaic commandments nor in the records of past or present court proceedings, neither in the rantings of puritan zealots nor in psychiatric testimony, is bestiality censured because of the harm that it inflicts on animals. But, especially in the case of smaller creatures like rabbits and hens, animals often suffer great pain and even death from human–animal sexual relations. While researchers have examined the physiological consequences of bestiality for humans, they pay no such attention to the internal bleeding, the ruptured anal passages, the bruised vaginas and the battered cloaca of animals, let alone to animals' psychological and emotional trauma. Such neglect of animal suffering mirrors the broader problem that, even when commentators admit the discursive relevance of animal abuse to the understanding of human societies, they do not perceive it, either theoretically or practically, as an object of study in its own right.

In principle, the attempt to understand bestiality as a form of animal abuse might profitably draw on the perspectives and insight of the three major tendencies that lie at the philosophical and theoretical heart of the animal protection community, namely, utilitarianism, liberal rights theory and feminism. We might insist, following liberal rights theory, for example, that if bestiality is engaged in with a mammal, then it is a harm inflicted on a 'moral patient' – to use Regan's (1983) term – entitled to the fundamental right of respectful treatment. But discursive support for this specific task is very difficult to find, either in the writings of the animal protection community or in its everyday activities. Moreover, though in the last decade some of the most important contributions to the understanding of animal abuse have been made by feminism, except for brief statements by Carol Adams (1995a, 1995b: 65–9) and Barbara Noske (1993), feminists have completely ignored the harmful effects of bestiality on animals. Departing from this curious silence, Adams (1995a) insists that we should understand bestiality as forced sex with animals because sexual relationships of unequal power cannot be consensual. In making this argument, and in asserting that all forms of masculinist oppression are linked, Adams thereby begins to claim the perspective of animals as a central concern of feminism.

I agree with Adams that, in seeking to replace anthropocentrism with an acknowledgement of the sentience of animals, we must start with the fact that in almost every situation humans and animals exist in a relation of potential or actual coercion. Whether as domestic pets or as livestock, when they are thoroughly dependent on humans for food, shelter and affection, or as feral creatures, when humans have the capacity to ensnare them and subject them to their will, animals' interaction with humans is always infused with the possibility of coercion. So it is with sex. In the same way that sexual assault against women differs from normal sex because the former is sex obtained by physical, economic, psychological or emotional coercion – any of which implies the impossibility of genuine consent – so, too, Adams' assertion that bestiality is always sexual coercion ('forced sex') is surely a correct description of most, if not all, human–animal sexual relations.

However, I am not convinced that bestiality must entail sexual coercion simply because human–animal sexual relations always occur in a context of 'unequal power' (however theorized). If unequal power is the definitive criterion, then sexual coercion would be an essential characteristic not only of intercourse between human adults and infants or children but of most adult heterosexual and even homosexual intercourse as well. Sexual coercion is not sex that occurs always and only in a context of unequal power, though on occasion, of course, situations of inequality imply coercion because for a variety of reasons the party with less power cannot freely dissent from participation. Ultimately, sexual coercion occurs whenever one party does not genuinely consent to sexual relations or does not have the ability to communicate consent to the other. Sometimes, one participant in a sexual encounter may appear to be consenting because she does not overtly resist, but that does not, of course, mean that genuine consent is present. For genuine consent to sexual relations to be present – somewhat to modify Box's (1983: 124) formulation – both participants must be conscious, fully informed and positive in their desires.

If genuine consent – defined in this way – is a necessary condition of sex between one human and another, then there is no good reason to suppose that it may be dispensed with in the case of sex between humans and other sentient animals. Bestiality involves sexual coercion because animals are incapable of genuinely saying 'yes' or 'no' to humans in ways that we can readily understand. A different way of putting this is to suggest that if it is true that we can never know what it is like to be a non-human animal (Nagel, 1974), then presumably we will never know if animals are able to assent – in their terms – to human suggestions for sexual intimacy. Indeed, if we cannot know whether animals consent to our sexual overtures, then we are as much at fault when we tolerate human–animal sexual relations as when we fail to condemn adults who have sexual relations with infants or with children or with other 'moral patients' who, for whatever reason, are unable to refuse participation. If it is right to regard unwanted sexual advances to women, to infants and to children as sexual assault, then I suggest sexual advances to animals should be viewed similarly.

Moreover, like infants, young children and other 'moral patients', animals are beings without an effective voice. Some animals, such as the cows and other farmyard animals – including those viewed in *Barnyard Love* – are not equipped to resist human sexual advances in any meaningful way because of their docile and often human-bred natures. Other animals, in trying to resist human sexual advances, can certainly scratch, bite, growl, howl, hiss and otherwise communicate protest about unwanted advances. However, in most one-on-one situations, an animal is incapable of enforcing her will to resist sexual assault, especially when a human is determined to effect his purpose. Moreover, animals are disadvantaged in yet another way, for when they are subjected to sexual coercion and to sexual assault, it is impossible for them to communicate the facts of their abuse to those who might give them aid.

In short, because bestiality is in certain key respects so similar to the sexual assault of women, children and infants, I suggest that it should be named *interspecies sexual assault.*

For many of the same reasons that, as it applies to humans, the concept of sexual assault is more widely applicable than that of rape, so too interspecies sexual assault comprises a wider range of actions than those found in dictionary definitions of bestiality or in notions embedded in popular culture, both of which tend to focus narrowly on penetration of the vagina, anus or cloaca of an animal by a human penis. However, if the concept of interspecies sexual assault is not exhausted by penile or digital insertion, then how wide should its scope be? Should it include touching, kissing and fondling? If it is extended to fondling, for example, then to the fondling of what, with what and by whom? Given animals' inability to communicate consent to human sexual overtures, I would like to establish – or at least to aim for – the general principle that *interspecies sexual assault comprises all sexual advances by humans to animals.* Admittedly, such a principle clearly has inherent problems which I cannot pretend to know how to solve. For example, how do we establish a general rule for identifying actions that are physically identical to those defined as interspecies sexual assault but which have a different intent? Consider the following tale related to me by a colleague.

> When I was a little girl I didn't take my dog to bed – she was too big for that – but instead lay regularly in her basket. I even sucked her nipples since I had seen her pups do that. She allowed it and didn't prevent it, even though she wasn't suckling at the time. My mother, a doctor herself, was thank goodness not too narrow-minded and left us alone in our tactile relationship. (Personal communication, September 20, 1996)

This innocent and affectionate suckling was probably not sexual in nature, it certainly was not assaultive and it doubtless caused the dog no harm.

Many actions like this can, of course, be either sexual or affective in nature, or both, depending on their social contexts or on the physiological responses of the actors (for both human and non-human animals, innocent, non-sexual physical touching and stroking slow the pulse and respiration and lower the blood pressure, but quite the opposite responses are produced by sexual arousal). However, where, precisely, should a sociological line of demarcation be drawn? It is clear, to me at least, that the milking of a cow, for example, has nothing to do with sexual assault. However, how about electrically induced ejaculation for insemination? Is this interspecies sexual assault? Simple assault? Neither?

In arguing that interspecies sexual assault comprises all sexual advances by humans to animals, I do not mean to dilute the severity of the condemnation of the sexual assault of one human by another. However, I suspect that, for different reasons, some feminists and most conservative opponents of the animal protection community will wish to accuse me of just that. Such a response assumes, wrongly I believe, not only that there is some anthropocentric chain of moral claims and priorities wherein those of humans are necessarily far above those of animals, but also that the interests of humans and animals are incompatible. On the contrary, sexism and speciesism operate not in opposition to each other but in tandem. Interspecies sexual assault is typically the product of a masculinity that sees women, animals and nature as objects that can be controlled, manipulated and exploited. Listen only to some of the sexist language that prepares the way for bodily sexual assault (and see Dunayer, 1995). Much of this is voiced in speciesist terms. When a man describes women as 'cows', 'bitches', '(dumb) bunnies', 'birds', 'chicks', 'foxes', 'fresh meat', and their genitalia as 'beavers' or 'pussies', he uses derogatory language to distance himself emotionally from, and to elevate himself above, his prey by relegating them to a male-constructed category of 'less than human' or, more importantly, 'less than me'. Reduced to this inferior status, both women and non-human animals are thereby denied subjectivity by male predators, who can then proceed to exploit and abuse them without guilt. Unchallenged, sexist and speciesist terms operate in concert to legitimize sexual assaults on women and animals.

TOWARDS A TYPOLOGY OF INTERSPECIES SEXUAL ASSAULT

Thus far, in outlining and opposing conventional notions of bestiality, I have suggested their replacement with a concept of interspecies sexual assault. Sexological surveys and historical studies of court records of bestiality prosecutions have usefully revealed glimpses of the number and variety of species thus abused, among them mules, cows, sows, dogs, mares, ducks, sheep, goats, rabbits and hens. These diverse creatures include companion animals, farmyard animals, livestock and animal labourers. Although interspecies sexual assault often results from the same malicious masculinity and comprises the same harmful actions as those that constitute the sexual assault of one human by another, it is clearly not a unitary social practice but one with differing social forms.

In what follows, I try to identify some key categories of a typology of interspecies sexual assault, including: (i) sexual fixation; (ii) commodification; (iii) adolescent sexual experimentation; and (iv) aggravated cruelty. These four categories are structured in terms of both differing human–animal social relationships and also the degree of harm that is suffered by abused animals.

Sexual fixation (or 'zoophilia')

This is the form of interspecies sexual assault that occurs when animals are the preferred sexual partners of humans. It is hard to believe that this was not the case when, for example, in colonial New England in 1642, Thomas Granger was indicted for buggery with 'a mare, a cow, two goats, five sheep, two calves and a turkey' (Bradford, 1650/1970: 320). Rare descriptions of sexual fixation with animals are provided by von Krafft-Ebing (1886/1978: 376–7), who designates it as 'impulsive sodomy'.

Sexual fixation with animals is probably the least common form of interspecies sexual assault, especially if the fixation with animals is of an exclusive nature; one observer (Dekkers, 1994: 149) estimates that the percentage of humans who have sex *exclusively* with animals is far below 1%, though this figure lacks suitable evidence. The psychological literature contains no adequate accounts of it, yet Adams (1995a: 30) asserts that there is a similarity in the respective worldviews of the zoophiliac, the rapist and the child sexual abuser. 'They all view the sex they have with their victims as consensual,' she claims, 'and they believe it benefits their sexual "partners" as well as themselves' (*ibid.*). Perhaps Adams' claim is correct, but it will remain unsupported until a significant number of methodologically sensitive life histories have been completed on zoophiliacs. It is just as likely that 'fixated humans' assault animals sexually not because they believe it benefits their sexual 'partners', but because they enjoy inflicting pain on other creatures who, in this particular case, happen to be animals because animals are more available to them than humans.

Commodification

This is the predominant element in interspecies sexual assaults that are packaged as commodities for sale in a market. It often involves a twofold assault – one by a man on a woman, who is assaulted and humiliated by being forced to have sex with an animal, the other on the animal who is coerced, without the possibility of giving genuine consent, into having sex with a human. Examples include live shows of women copulating with animals in bars and sex clubs or depictions of interspecies sexual assaults in films such as *Barnyard Love* and *Deep Throat*. In the latter, for example, Linda Marchiano is filmed having intercourse with a large dog resembling a German shepherd. During this act and long afterwards, Marchiano 'felt nothing but acute revulsion' (Lovelace, 1980: 107–14; and see Hollander, 1972: 35), and she agreed to be filmed in this two-hour episode only because her boyfriend and batterer threatened to kill her.

Consider also the more problematic case of Deena the stripping chimpanzee (Adams, 1990). For $100, Deena and her trainer would appear at a social gathering, during which Deena would perform a striptease act for the partygoers. Is this interspecies sexual assault? Clearly, this case is one that combines commodification with aspects of sexual objectification. The chimp had been trained to perform *like* a human female stripper – a marketable action that it could not possibly have freely chosen to do, and whose social context it could not have fully understood. Though it is true that sexual abuse does not necessarily involve actual physical contact, perhaps this particular act should be understood less as sexual assault than – like Adams (*ibid.*) suggests – as a violation of an animal's right to dignity.

Adolescent sexual experimentation

This seems to be typically practised in rural areas by young males with easy access to animals. It is probably the most common form of interspecies sexual assault, as shown by quite disparate studies of seventeenth-century and eighteenth-century Sweden (Liliequist, 1991) and of mid-twentieth-century rural America. With regard to the latter, for example, it has been documented that about 8% of the male population has some sexual experience with animals, and that 'a minimum of 40–50% of all American farm boys experience some form of sexual contact with animals' (Kinsey, Pomeroy & Martin, 1948: 671), as do 5.1% of American females (Kinsey *et al.*, 1953: 505). However, these findings are highly suspect, both because Kinsey's methodology lacked probability sampling and because his aggressive personal interviewing techniques ensured elevated levels of reporting. Moreover, in most western societies – where pet ownership has dramatically increased and where, with the rise of 'factory farming', there has been a steady decline in the percentage of the human population living in agricultural areas or residing with farm animals inside their houses – it cannot be certain that it is animals used on farms who are nowadays the most common objects of interspecies sexual assault by humans.

Precisely what the practice of adolescent sexual experimentation with animals represents symbolically and culturally, and how it contributes to gender socialization, vary from one social context to another. It can be performed either alone or with other adolescents, who either watch or participate. In a group context, some boys, of necessity, teach how it is done while others learn. It can be performed for a variety of reasons, including mere curiosity, cruelty, showing off for other boys, and acquiring the techniques of intercourse for later use on girls. An anonymous colleague has told me, for example, that when she was doing anthropological fieldwork in rural Algeria, she and a co-worker witnessed a very nervous young male (on the night before his wedding) 'practising' sexual intercourse with a donkey for the explicit purpose of not appearing hopelessly unskilled with his wife the following night. Presumably, too, there is some point towards the end of their adolescence when young males desist from experimental sexual activities with animals because such practices are regarded as unmanly or, perhaps, as perverse.

Aggravated cruelty

It is reasonable to suppose, given their great predominance in sexual experimentation with animals, that young males also disproportionately engage in aggravated cruelty during acts of interspecies sexual assault (i.e. a level of cruelty over and above that already presented in most such acts). It is true that no specific pattern of aggravated cruelty has yet been uncovered among young males who engage in interspecies sexual assault, but this is perhaps only because this category has not yet been properly researched. Psychologists have shown that children and adolescents who assault animals appear to be overwhelmingly young males of normal intelligence (Felthous, 1981), who are often sexually abused at home and whose family situations also often contain spousal abuse (Friedrich, Urquiza & Beilke, 1986; Hunter, 1990: 214–6); in one case a mother was reported to have forced her 16-year-old son to have sex with a German shepherd (Money, Annecillo & Lobato, 1990: 127).

Quite apart from the occurrence of cruelty during adolescent sexual experimentation, aggravated cruelty can be a major element in interspecies sexual assault in other ways. In mid-nineteenth-century England, for example, one case was reported in which two-feet-long knotted sticks were thrust into mares' wombs, which were then vigorously rented, and another in which the penises of cart horses and donkeys were cut off (Archer, 1985: 152). Multiple cases of such atrocities were confirmed in several English counties in 1993 (*The Times*, 1993, March 2, May 8, June 4). Similarly, in 1991 at a zoo in New Bedford, Massachusetts, a deer was found with fatal wounds that included a fractured jaw and extensive bleeding from the rectum and vagina (*Standard Times*, 1991, July 26). Sometimes, aggravated cruelty against animals takes place in conjunction with the humiliation of women. This has been documented in Nazi concentration camps (Fleismann, 1968: 50–71), in the photographic representation of women in pornographic magazines (see Russell, 1993) and in the course of partner abuse (Adams, 1995b: 65–9). In the last case, it can take the form of battering, which involves the use of animals for humiliation and sexual exploitation by batterers and/or marital rapists. Moreover, if one allows that, like humans, animals are capable of experiencing non-physical pain, then aggravated cruelty also occurs whenever interspecies sexual assault produces emotional or psychological pain and suffering (Ascione, 1993; Masson & McCarthy, 1995).

CONCLUSION

This chapter has tried to replace anthropocentric censures of bestiality with a concept that I term 'interspecies sexual assault'. My argument about the meaning and causes of interspecies sexual assault has derived largely from how the situation of animals as abused victims parallels that of women and, to some extent, that of infants and children. Specifically, bestiality should be understood as interspecies sexual assault because: (1) human–animal sexual relations almost always involve coercion; (2) such practices often cause animals pain and even death; and (3) animals are unable either to communicate consent to us in a form that we can readily understand or to speak out about their abuse. Though space does not permit it here, this concept of interspecies sexual assault can doubtless be strengthened with the discursive support of utilitarianism and of liberal rights theory.

As I have proposed it, the concept of interspecies sexual assault clearly needs further elaboration. Key problems remain. For example, given the lack of studies of interspecies sexual assault, my fourfold typology is quite provisional. Between the categories of aggravated cruelty and adolescent sexual experimentation, especially, there is obvious overlap. One must be able to distinguish, too, not only between the malicious masculinity that typically lies behind aggravated cruelty and other situations of adolescent sexual experimentation and exploration, but also between the latter and innocent and affective fondling. Some difficulties seem to resist a clear answer – for example, is electrically induced ejaculation for insemination a form of interspecies sexual assault and, if so, is it an instance of commodification or of aggravated cruelty, or both?

In advancing the concept of interspecies sexual assault, I must stress that I do not wish to add to either the psychiatrization or the criminalization of a practice which nowadays occupies a place at the outer margins of public and legal concern. But this leaves me in an uncomfortable position. If a sexual assault on an animal by a human is a harm that is objectionable for the same reasons as is an assault on one human by another – because it involves coercion, because it produces pain and suffering and because it violates the rights of another being – then it would seem to constitute a sufficient condition for the censure of the human perpetrator. Clearly, we need to confront the nature of the censure that inevitably accompanies the relocation of bestiality as interspecies sexual assault. Should the censure involve criminalization? If so, of what severity? Should culpability be strict, or should the scales of justice depend on such factors as the moral significance of what was done, the degree of harm and the species of animal assaulted?

Even if a cultural consensus could be established about the harmfulness of interspecies sexual assault – or any other form of animal abuse – for animals that are kept in confinement by humans, its effectiveness as a right would nevertheless be undermined by the rival cultural powers associated with private property and privacy. It is, of course, precisely invocations of these rival rights that men (and sometimes women, too) use when they sexually abuse women and children. The right to privacy would undermine the detection and prosecution of interspecies sexual assault; the right to private property would be invoked to defend it. As Ted Benton (1994: 147–8) has argued about the latter, those who wish to ascribe rights to animals, including the right to respectful treatment, would eventually be forced to challenge the very existence of animals as private property.

ACKNOWLEDGEMENTS

I am indebted to Frank Ascione and to an anonymous reviewer for their generous comments on this chapter, which is a shorter version of an original essay in *Theoretical Criminology*, 1997, 1(3), 317–40, reprinted with permission.

NOTE

1. See further Boswell (1980: 323, n.69). Alternatively, 'bestiality' might have derived from the Latin *animal*, which was originally translated into old English as 'beste' or 'beast' from the French *bête* which, in turn, probably derived from the Sanskrit 'that which is to be feared' (Collard, 1989: 24).

REFERENCES

Adams, C.J. (1990). Deena – the world's only stripping chimp. *Animals' Voice Magazine*, **3**, 72.

Adams, C.J. (1995a). Bestiality: the unmentioned abuse. *The Animals' Agenda*, **15**, 29–31.

Adams, C.J. (1995b). Woman-battering and harm to animals. In *Animals and Women: Feminist Theoretical Explorations*, ed. J. Donovan & C.J. Adams, pp. 55–84. Durham, NC: Duke University Press.

Aelian (1958). *On the Characteristics of Animals*, translated by A.F. Scholfield. Cambridge, MA: Harvard University Press.

Archer, J.E. (1985). A fiendish outrage? A study of animal maiming: 1830–1870. *Agricultural History Review*, **33**, 147–57.

Ascione, F.R. (1993). Children who are cruel to animals: a review of research and implications for developmental psychopathology. *Anthrozoös*, **6**, 226–46.

Beirne, P. (1994). The law is an ass: 'Reading' E.P. Evans' The Medieval Prosecution and Capital Punishment of Animals'. *Society and Animals*, **2**, 27–46.

Beirne, P. (1995). The use and abuse of animals in criminology: a brief history and current review. *Social Justice*, **22**, 5–31.

Benton, T. (1994). *Natural Relations: Ecology, Animal Rights and Social Justice*. London: Verso.

Boswell, J. (1980). *Christianity, Social Tolerance, and Homosexuality*. Chicago: University of Chicago Press.

Box, S. (1983). *Power, Crime, and Mystification*. London: Tavistock.

Bradford, W. (1650/1970). *Of Plymouth Plantation, 1620–1647*, ed. S.E. Morison. New York: Alfred A. Knopf.

Cambrensis, G. (1863). *Historical Works*, ed. T. Wright. London: H.G. Bohn.

Canup, J. (1988). The cry of Sodom enquired into: bestiality and the wilderness of human nature in seventeenth-century New England. *American Antiquarian Society*, **98**, 113–31.

Chapin, B. (1983). *Criminal Justice in Colonial America, 1606–1660*. Athens, GA: University of Georgia Press.

Collard, A. (1989). *Rape of the Wild*. Bloomington: Indiana University Press.

Davidson, A.J. (1991). The horror of monsters. In *The Boundaries of Humanity: Humans, Animals, Machines*, ed. J.J. Sheehan & M. Sosnapp, pp. 36–67. Berkeley: University of California Press.

Dekkers, M. (1994). *Dearest Pet: On Bestiality*, translated by Paul Vincent. London: Verso.

Dubois-Desaulle, G. (1933). *Bestiality: an Historical, Medical, Legal and Literary Study*, translated by 'A.F.N'. New York: Panurge Press.

Dunayer, J. (1995). Sexist words, speciesist roots. In *Animals and Women: Feminist Theoretical Explorations*, ed. J. Donovan & C.J. Adams, pp. 11–31. Durham, NC: Duke University Press.

Evans, E.P. (1906/1987). *The Criminal Prosecution and Capital Punishment of Animals*. London: Faber and Faber.

Felthous, A.R. (1981). Childhood cruelty to cats, dogs and other animals. *Bulletin of the American Academy of Psychiatry and Law*, **9**, 48–53.

Fleismann, S. (1968). *Bestiality: Sexual Intercourse Between Men and Women and Animals*, translated by Robert Harris. Baltimore: Medical Knowledge Press.

Friedrich, W.N., Urquiza, A.J. & Beilke, R.L. (1986). Behavior problems in sexually abused young children. *Journal of Pediatric Psychology*, **11**, 47–57.

Goodich, M. (1979). *The Unmentionable Vice: Homosexuality in the Later Medieval Period*. Santa Barbara, CA: Clio.

Høeg, P. (1996). *The Woman and the Ape*. New York: Farrar, Straus and Giroux.

Hollander, X. (1972). *The Happy Hooker*. New York: Dell Publishing.

Hunter, M. (1990). *Abused Boys: the Neglected Victims of Sexual Abuse*. New York: Lexington.

Kinsey, A.C., Pomeroy, W.B. & Martin, C.E. (1948). *Sexual Behavior in the Human Male*. Philadelphia: W.B. Saunders.

Kinsey, A.C., Pomeroy, W.B., Martin, C.E. & Gebhard, P.H. (1953). *Sexual Behavior in the Human Female*. Philadelphia: W. B. Saunders.

Liliequist, J. (1991). Peasants against nature: crossing the boundaries between man and animal in seventeenth- and eighteenth-century Sweden. *Journal of the History of Sexuality*, **1**, 393–423.

Lovelace, L. (1980). *Ordeal*. New York: Bell Publishing.

Masson, J.M. & McCarthy, S. (1995). *When Elephants Weep: the Emotional Lives of Animals*. New York: Delacorte.

McNeill, J.T. & Gamer, H.M. (1938). *Medieval Handbooks of Penance*. New York: Columbia University Press.

Money, J., Annecillo, C. & Lobato, C. (1990). Paraphilic and other sexological anomalies as a sequel to the syndrome of child-abuse (psycho-social) dwarfism. *Journal of Psychology and Human Sexuality*, **3**, 117–50.

Nagel, T. (1974). What is it like to be a bat? *The Philosophical Review*, **83**, 435–50.

Noske, B. (1993). Hoe heet is een ezelin? *Opzij, Feministisch Maandblad*, **21**, 26.

Regan, T. (1983). *The Case for Animal Rights*. Berkeley: University of California Press.

Russell, D.E.H. (1993). *Against Pornography*. Berkeley: Russell Publications.

Russell, J.B. (1982). *A History of Witchcraft*. New York: Thames and Hudson.

Sharpe, J.A. (1983). *Crime in Seventeenth-Century England*. Cambridge: Cambridge University Press.

Shell, M. (1993). *Children of the Earth: Literature, Politics and Nationhood*. New York: Oxford University Press.

Tester, W. (1991). *Darling*. New York: Alfred A Knopf.

Thomas, K. (1983). *Man and the Natural World*. New York: Pantheon.

von Krafft-Ebing, R. (1886/1978). *Psychopathia Sexualis*, translated by Franklin S. Klaf. New York: Stein and Day.

Index